AF395175

Wireless Communications and Machine Learning

This focused textbook demonstrates cutting-edge concepts at the intersection of machine learning (ML) and wireless communications, providing students with a deep and insightful understanding of this emerging field. It introduces students to a broad array of ML tools for effective wireless system design, and supports them in exploring ways in which future wireless networks can be designed to enable more effective deployment of federated and distributed learning techniques to enable AI systems. Requiring no previous knowledge of ML, this accessible introduction includes over 20 worked examples demonstrating the use of theoretical principles to address real-world challenges, and over 100 end-of-chapter exercises to cement student understanding, including hands-on computational exercises using Python. Accompanied by code supplements and solutions for instructors, this is the ideal textbook for a single-semester senior undergraduate or graduate course for students in electrical engineering, and an invaluable reference for academic researchers and professional engineers in wireless communications.

Le Liang is a professor of information science and engineering at Southeast University, Nanjing. He was a former AI/ML research scientist at Intel Labs and was the Founding Technical Program Co-chair of the IEEE International Conference on Machine Learning for Communication and Networking.

Shi Jin is a professor of information science and engineering at Southeast University, Nanjing, and a fellow of the IEEE for contributions to MIMO and reconfigurable intelligent surface-assisted communications.

Hao Ye is an assistant professor of electrical and computer engineering at the University of California, Santa Cruz, and a former researcher in machine learning at Qualcomm AI Research.

Geoffrey Ye Li is a professor of electrical engineering at Imperial College London. He is a fellow of the IEEE, the IET, and the Royal Academy of Engineering for his contributions to machine learning and signal processing for wireless communications.

"A timely book by pioneering researchers, which captures the latest innovations that will generate consequential impact for our industry."

Wen Tong, *CTO Huawei Technologies, Canada*

"A must-read for anyone working on the machine learning aspects of wireless communications. Authored by some of the world's leading researchers in the field, it is brilliantly structured and articulates all the key concepts in an illuminating manner."

Michalis Matthaiou, *Queen's University Belfast*

Wireless Communications and Machine Learning

Le Liang
Southeast University, Nanjing

Shi Jin
Southeast University, Nanjing

Hao Ye
University of California, Santa Cruz

Geoffrey Ye Li
Imperial College of Science, Technology and Medicine, London

CAMBRIDGE
UNIVERSITY PRESS

Shaftesbury Road, Cambridge CB2 8EA, United Kingdom

One Liberty Plaza, 20th Floor, New York, NY 10006, USA

477 Williamstown Road, Port Melbourne, VIC 3207, Australia

314–321, 3rd Floor, Plot 3, Splendor Forum, Jasola District Centre, New Delhi – 110025, India

103 Penang Road, #05–06/07, Visioncrest Commercial, Singapore 238467

Cambridge University Press is part of Cambridge University Press & Assessment, a department of the University of Cambridge.

We share the University's mission to contribute to society through the pursuit of education, learning and research at the highest international levels of excellence.

www.cambridge.org
Information on this title: www.cambridge.org/highereducation/isbn/9781009232203

DOI: 10.1017/9781009232210

© Cambridge University Press & Assessment 2026

This publication is in copyright. Subject to statutory exception and to the provisions of relevant collective licensing agreements, no reproduction of any part may take place without the written permission of Cambridge University Press & Assessment.

When citing this work, please include a reference to the DOI 10.1017/9781009232210

First published 2026

A catalogue record for this publication is available from the British Library

A Cataloging-in-Publication data record for this book is available from the Library of Congress

ISBN 978-1-009-23220-3 Hardback

Additional resources for this publication at www.cambridge.org/wirelesscomms.

Cambridge University Press & Assessment has no responsibility for the persistence or accuracy of URLs for external or third-party internet websites referred to in this publication and does not guarantee that any content on such websites is, or will remain, accurate or appropriate.

For EU product safety concerns, contact us at Calle de José Abascal, 56, 1°, 28003 Madrid, Spain, or email eugpsr@cambridge.org

Contents

Preface

The potential of machine learning in addressing challenging wireless communication problems has been increasingly acknowledged by experts in the community. A significant amount of research has been dedicated to this emerging area, whose reach and impact keep growing. The writing of this book was directly prompted by the vast research accomplishments at the intersection of machine learning and wireless communications. We believe this is a good time to publish an informative yet accessible book that systematically documents these exciting scientific advances. This book provides a coherent and unified treatment of machine learning in wireless system design and the role of wireless networks to enable more efficient machine learning. It has a concise and focused structure that strategically positions major concepts and methods to form a landscape that makes it much easier for readers to gain a deep and insightful understanding of the subject. The book is intended to serve as a textbook for advanced undergraduate or graduate students in electrical engineering and a useful reference for academic researchers and practicing engineers working in the wireless communication field or computer scientists with research interests in wireless systems.

Organization of the Book

Chapter 1 introduces the background of the book and briefly discusses the emergence of machine learning-enabled wireless communication system design. A concise introduction to basic machine learning concepts is included for the ease of understanding the subsequent chapters.

Chapter 2 is dedicated to the investigation of wireless channels with the new tool of machine learning. In particular, this chapter elaborates on the use of machine learning techniques to capture complex wireless channel effects with generative models, e.g., generative adversarial networks; develop novel learning-based channel estimation methods; and learn efficient channel state information compression and reconstruction mechanisms.

Chapter 3 treats the design of learning-based methods to perform key communication processing functionalities at the receiver, including signal detection and channel decoding. For the two themes, our presentation is consistent in that data-driven learning methods are first introduced to recover data (transmitted symbols or codewords), and then model-driven learning methods are discussed that leverage both communication expertise and the power of learning from data to address receiver design challenges.

Chapter 4 introduces end-to-end learning-enabled wireless communication system design, seen as a fundamental change from the traditional wireless design philosophy. Semantic communication aiming for efficient transmission of semantic information rather than raw bits is also explained on the basis of end-to-end learning-based communication.

In Chapter 5 the focus shifts to radio resource allocation in communication networks. We introduce deep learning-assisted optimization for resource allocation, where the different roles and uses of deep learning in the context of optimization are discussed in detail. Reinforcement learning is then introduced as a paradigm shift that learns to make sequential resource allocation decisions based on interactions with an unknown environment, without explicitly formulating an optimization problem.

Chapter 6 presents a study of distributed and federated learning in wireless networks. In contrast to the preceding chapters where machine learning is used as a set of tools to solve communication problems – seen as AI for wireless – this chapter views wireless networks as a computing platform to enable more efficient machine learning training and inference – i.e., wireless for AI. We present the architecture, training procedures, and major challenges of federated and distributed learning in general and gradient compression and joint learning and communication design in particular, which are typical of federated and distributed learning in wireless networks.

How to Use the Book

This book can be used to teach advanced undergraduate or graduate students about the principles and methods of machine learning in wireless communication systems. Some example course titles include "Wireless Communications and Machine Learning" or "Intelligent Communication Systems Design." The expected prerequisite knowledge for this book is a basic understanding of probability, signals and systems, and wireless communications. Knowledge of machine learning is helpful though not required as the book provides a self-contained introduction to basic machine learning concepts.

Although the book is fairly expansive in coverage, it is short in length, which makes it ideal for a one-semester course. The six chapters are intended to be consistent in extent such that similar numbers of lectures are expected for each chapter. The introductory Chapter 1 may or may not need the same number of lectures, depending on the background of the students. For a short course, the instructors can selectively cover parts of the book, which is organized around three themes: Chapters 2–4 center around physical communication layer design, while Chapter 5 is focused on resource allocation or the medium access control layer design. The last chapter is more on the other side of the story, i.e., wireless for AI.

We have designed around 100 exercises to help readers comprehend the concepts discussed in the main text. In particular, we include quite a few "get-your-hands-dirty" programming exercises in Python, based in part on several open-source code repositories. The code to reproduce most of the figures in the book and for completing the exercises is stored in the GitHub repository https://github.com/le-liang/wcmlbook.

Acknowledgments

We would first like to thank the students in our research groups for their tireless contributions to many parts of the book. They helped revise the texts, draw the figures, collate the codes, etc. The efforts of Kai Huang, Xingyu Zhou, Jiajia Guo, Weijie Jin, Yucheng Sheng, Yandi Liu, Jiaming Yu, Kexin Wang, Jiacheng Wang, Xinya Peng, Hao Fang, Yifan Fan, Guowei Liu, Keying Zhu, Xiaotian Fan, Tianhao Mao, Zijian Cao, Jipeng Gan, Xinjie Li, Fangyu Liu, Yan Lv, and Chengyong Jiang are greatly appreciated.

We are grateful to Prof. Chongtao Guo, Prof. Hua Zhang, Prof. Zhijin Qin, Prof. Xinping Yi, Dr. Jing Zhang, and Jiacheng Yao for their very helpful comments on various chapters of the book.

The anonymous reviewers provided considerable useful feedback, and thanks are due to them as well.

Finally, we would like to express our sincere thanks to our Development Editor Helen Shannon and Commissioning Editor Elizabeth Horne. They have provided invaluable guidance on the book writing and accommodated our delays with grace throughout the journey.

Notation

$\triangleq$	Defined as equal to ($a \triangleq b$: a is defined as b)
$\approx$	Approximately equal to
$\propto$	Proportional to
$\odot$	Hadamard product (entry-wise multiplication)
$\otimes$	Kronecker product operator
$\circledast$	Circular convolution operator
$f: \mathcal{X} \mapsto \mathcal{Y}$	Function f from elements of set $\mathcal{X}$ to elements of set $\mathcal{Y}$
$\leftarrow$	Assignment
$\mathbb{R}$	Field of all real numbers
$\mathbb{R}^+$	Field of all positive real numbers
$\mathbb{C}$	Field of all complex numbers
$\mathbb{M}$	Set of all candidate transmission symbols
$\mathbf{1}\{predict\}$	Indicator function ($\mathbf{1}\{predict\} = 1$ if the $predict$ is true, else 0)
$\mathbf{A}^H$	Hermitian (conjugate transpose) of matrix $\mathbf{A}$
$\mathbf{A}^T$	Transpose of matrix $\mathbf{A}$
$\mathbf{A}^{-1}$	Inverse of matrix $\mathbf{A}$
$\|\mathbf{A}\|_F$	Frobenius norm of matrix $\mathbf{A}$
$\det(\mathbf{A})$	Determinant of matrix $\mathbf{A}$
$\mathrm{diag}([x_1, \ldots, x_N])$	The $N \times N$ diagonal matrix with diagonal elements $x_1, \ldots, x_N$
$\mathrm{vec}(\mathbf{A})$	Linear transformation that converts $\mathbf{A}$ into a vector
$\mathbf{A} \succeq \mathbf{0}$	The matrix $\mathbf{A}$ is positive semidefinite
$\mathrm{tr}(\mathbf{A})$	Trace of matrix $\mathbf{A}$
$\|\mathbf{x}\|$	Norm of vector $\mathbf{x}$
$\exp(x)$	e^x
$\ln(x)$	The natural log of x
$\log_x(y)$	The log, base x, of y
$\mathbf{x}^*$	Complex conjugate of vector $\mathbf{x}$
$\mathbf{x}^H$	Hermitian (conjugate transpose) of vector $\mathbf{x}$
$\mathbf{x}^T$	Transpose of vector $\mathbf{x}$
$\mathbf{0}$	The zero vector
$\mathbf{I}$	The identity matrix
$\mathbb{E}[X]$	Expectation of a random variable X
$\mathrm{Cov}[X, Y]$	Covariance of random variables X and Y
$\Pr\{X = x\}$	Probability that a random variable X takes on the value x
$X \sim p_X(x)$	The random variable X has distribution $p_X(x)$
$\mathcal{N}(\mu, \sigma^2)$	Real Gaussian (normal) distribution with mean μ and variance σ^2

$\mathcal{CN}(\mu,\sigma^2)$	Circularly symmetric complex Gaussian (normal) distribution with mean μ and variance σ^2
$\mathrm{Re}\{x\}$	Real part of x
$\mathrm{Im}\{x\}$	Imaginary part of x
$\lvert x \rvert$	Absolute value (amplitude) of x
$\lvert \mathcal{X} \rvert$	Size of set $\mathcal{X}$
$\frac{\partial y}{\partial x}$	Partial derivative of y with respect to x
$\lfloor x \rfloor$	The largest integer not exceeding x
$\langle \mathbf{x}, \mathbf{y} \rangle$	The inner product of vectors $\mathbf{x}$ and $\mathbf{y}$

1   Introduction

The quest for faster and more reliable wireless communications has never stopped. This trend is further reinforced by the recent emergence of several much-touted new applications, including the internet of things, connected autonomous driving, augmented and virtual reality, etc. To support these services, wireless networks need to target greater numbers of simultaneous connections, higher throughput, lower latency, and higher reliability. Such diverse service requirements create daunting challenges for system design and have fueled a flurry of interest in rethinking the fundamental principles of wireless communications. Conventional communication system design relies on simplified mathematical models to develop neat solutions with analytic insights. However, in real-world systems, it is difficult to capture the variety of hardware-induced impairments and the growing heterogeneity and complexity of wireless networks with tractable mathematical models. In addition, discrete design of each communication block, e.g., channel coding/decoding, modulation, channel estimation, and signal detection, leads to inevitable loss of optimality from an end-to-end perspective. As a result, it is crucial to have a more flexible framework that accounts for various distortions and takes a holistic approach to the communication problem.

Machine learning, particularly deep learning, has drawn considerable attention in recent years due to its exceptional performance in areas such as computer vision, speech recognition, and natural language processing. It helps build intelligent systems to operate in complicated environments and provides powerful data-driven solutions to many problems that are deemed difficult due to, e.g., lack of accurate models or prohibitive computation complexity. These problems have motivated researchers and practitioners to exploit machine learning techniques to address the aforementioned challenges in wireless communications. A synergistic combination of machine learning tools and traditional wireless signal processing algorithms promises to bring transformative changes to the wireless system landscape. Better still, we can leverage the rich expert knowledge in the communication domain to complement the data-driven learning methods to improve data efficiency and push the performance boundary further. Adopting machine learning in virtually every aspect of wireless system design has thus become a prevalent idea, and concerted efforts from academia and industry alike have been expended to study both the theory and the practice of learning-enabled wireless systems.

In this introductory chapter, we will briefly review the history of using machine learning in wireless communications and then provide a concise introduction to the key machine learning concepts and methods that form the foundation for much of the subsequent discussion in the book.

1.1 History of Machine Learning in Wireless Communications

Pioneering machine learning research started in the 1940s [1], with early works providing the theoretical foundations for neural networks and pattern recognition. However, these works did not see significant improvements until the emergence of massively parallel processing architectures with distributed memory in the 2010s [2, 3]. These architectures have provided the required computational capabilities to manage large-scale machine learning models and extensive datasets, thereby promoting rapid advancements in the field. Recent years have witnessed significant breakthroughs in machine learning across various domains, such as computer vision and natural language processing, alongside the ongoing emergence of disruptive technologies like large language models, inspiring researchers to explore their application in wireless communication systems.

The interplay between machine learning and wireless communications has a long history. The use of machine learning techniques in solving communication problems dates back to the 1990s: the use of neural networks for decoding error correction codes was explored in 1989 [4], while their application for modulation recognition was investigated as early as 1985 [5]. Moreover, neuro-evolution was pioneered in 1994 [6], laying the groundwork for applying neural networks to challenging regression and classification tasks in wireless communication systems. These foundational studies and pioneering papers were instrumental in introducing neural networks into digital communications to achieve larger performance gains and more manageable implementation complexity in physical layer design, such as channel identification and equalization, coding and decoding, and signal detection and estimation [7]. Subsequently, significant research has been conducted in the fields of cognitive radio, resource allocation, and positioning, each following distinct design principles [8, 9]. However, the idea of applying machine learning to system design was not a mainstream research direction in wireless communications at that time because of the limited computational capability available then and the lack of adequate high-quality datasets for training.

Fortunately, visionary researchers have diligently advanced the field of machine learning and neural networks, continuously enhancing and optimizing the tools needed to scale network capacities to levels previously deemed unattainable. These works laid the foundation for subsequent technological breakthroughs. The emergence of deep learning in 2012, coupled with compelling results showcasing its remarkable performance and scalability, signaled the potential of modern machine learning techniques to revolutionize wireless communications and signal processing. Therefore, a resurgence of interest in this field started in 2017 following the publication of several prominent papers. Among them, the seminal work in [10] stood out as the single most important piece of research, which introduced the concept of end-to-end learning to communication system design. Benefiting from the capability of deep learning models to analyze extensive data and autonomously design the most suitable policy, this end-to-end optimization strategy brings significant break-throughs to wireless communications. The transmitter and receiver can be jointly optimized using real-world data, thereby enabling efficient management of non-ideal distortions without the need to construct a mathematical model to establish robust wireless communication systems [11]. As a result, substantial performance gains and reduced implementation complexity are anticipated as compared to conventional block-based systems [12].

The emergence of deep learning has significantly transformed problem-solving approaches in physical layer communications. In [13], traditional channel estimation has been revolutionized by deploying a deep-learning-based approach for beamspace millimeter wave massive multiple-input multiple-output (MIMO) systems, particularly excelling in scenarios with a constrained number of radio frequency chains. Further extending deep learning's reach, a mechanism that effectively addresses the challenge of channel state information (CSI) feedback in massive MIMO systems was introduced in [14], significantly enhancing reconstruction quality under extreme compression and thereby achieving notable beamforming gains. In [15], a robust receiver design was developed for orthogonal frequency division multiplexing (OFDM) systems where deep learning models were leveraged to implicitly estimate channel states and directly recover transmitted symbols. This scheme showed robustness against channel distortions and offered a compelling alternative to traditional receiver processing methods. A large number of follow-up works in a similar spirit have appeared since then, inspired by this pioneering research. Significant strides have also been achieved in the realm of channel decoding [9, 16]. The power of deep learning models to decode a variety of linear codes, including high- and low-density parity-check (LDPC) codes, has been demonstrated through either improving the belief propagation algorithm or learning decoding algorithms without explicit CSI. In [17], the implementation of a machine learning-enhanced air interface for enabling optimized communication schemes across various hardware, radio environments, and applications demonstrated the practical value of machine learning techniques in wireless communications.

The applications of machine learning are also prevalent in the medium access control (MAC) layer, where many researchers have designed intelligent wireless resource allocation methods based on supervised, unsupervised, and reinforcement learning according to diverse requirements for quality of services (QoS) and spectrum efficiency in different scenarios. These methods can be primarily classified into two categories: deep learning-assisted optimization and reinforcement learning-based resource allocation. On the one hand, numerous works employ deep learning to assist in optimizing wireless resource allocation, given the expressive power of neural networks and their efficient computational structures [18, 19, 20, 21]. In these approaches, deep learning methods are leveraged to learn the relationship between the parameters and solutions of an optimization problem through supervised learning or to directly learn to improve communication metrics of interest through unsupervised learning. On the other hand, reinforcement learning has demonstrated its power in solving sequential decision problems in recent years. Consequently, many researchers have begun to model wireless resource allocation as a sequential decision-making process and have pioneered the utilization of reinforcement learning for resource allocation to cope with the highly dynamic nature of wireless environments [22, 23, 24, 25].

In the recommended framework for the vision of IMT-2030 (more frequently referred to as 6G), completed by the International Telecommunication Union (ITU) in June 2023, integrated artificial intelligence (AI) and communications has become a crucial component. Reinforced by this trend, machine learning is recognized as a key enabling technology for future mobile communication systems. Both academia and industry are working toward making machine learning a fundamental functionality embedded within future communication networks, providing AI as a service anywhere and anytime. It is widely believed that disruptive AI and machine learning technologies, such as large language or foundation models, will introduce revolutionary changes into wireless communications in the near future.

1.2 Machine Learning

In this section, we introduce some of the most basic yet important concepts in machine learning. The main ideas are covered concisely. For more detailed coverage, see references [26, 27, 28, 29].

1.2.1 Basic Concept of Machine Learning: Supervised and Unsupervised Learning

Machine learning is a branch of AI that allows computer systems to learn from examples, data, and experience. The goal of machine learning is to automatically figure out a good strategy from the training data so that these systems can perform well under novel situations without human intervention. Depending on the type of training dataset, machine learning can be roughly categorized into supervised learning, unsupervised learning, and reinforcement learning. We will concisely discuss supervised and unsupervised learning in this section and examine reinforcement learning in Section 1.2.3.

Supervised Learning

If the dataset consists of pairs of inputs and targets, *supervised learning* can be applied to learn the relationship between them. The target, also called a label, in the dataset can be considered an instructor or a teacher, which provides the machine learning algorithm with the desired output given the current input. The relationship to be learned is usually characterized by a parameterized function whose parameters will be continuously updated in the training stage to best fit the training data. Popular function forms include polynomials (but linear in adjustable parameters), neural networks. Once the parameters in the function are determined, the model is said to be trained and can be used to predict the target value given a new input. To illustrate these concepts, we show an instance of supervised learning problems in Example 1.1 that uses linear regression to predict house prices, where the living area is the input feature and the house price is the target.

Supervised learning problems can be further categorized into classification and regression tasks, depending on whether the output value is continuous or discrete. Image classification is a prototypical example of the classification task, where the input image is assigned one label from a finite number of discrete categories. This is one of the core problems in computer vision and has a large variety of practical applications, such as handwritten digit recognition. Widely-used classification algorithms include Bayesian classifiers, k-nearest neighbors, decision trees, support vector machines, and neural networks. These algorithms have been adopted in the domain of wireless communications; for instance, support vector machines are used in [30] to identify whether a channel is busy or idle. We will cover neural networks in Section 1.2.2; these underpin many important learning-based methods for communication design and refer readers to standard texts [26, 27] for detailed coverage of the other algorithms. Instead of discrete predictions, regression tasks assign a continuous value to each input, such as estimating the house price given its living area. Linear regression, although conceptually simple, is an important class of regression problems and is illustrated in Example 1.1 for house price prediction.

Example 1.1   In this example, we investigate linear regression in addressing house price prediction and illustrate the main components of supervised learning along the way. Assume there is a dataset consisting of the living areas and prices of 47 houses from Portland, Oregon, as given in Table 1.1. Our task is then to predict the price of another house in Portland given its living area.

We use x^i to denote the input feature (i.e., living area) and y^i to denote the target value (i.e., price). A pair, (x^i, y^i), is a training sample. A list of n training samples compose a training dataset, $\{(x^i, y^i)\}_{i=1}^{n}$. We use $\mathcal{X}$ to represent the space of input

Table 1.1 Portland house price prediction.

Living areas (feet^2)	Price ($000)
2104	400
1600	330
2400	369
1416	232
3000	540
$\vdots$	$\vdots$

features and $\mathcal{Y}$ to represent the space of output values. Here, $\mathcal{X} = \mathcal{Y} = \mathbb{R}$. Our goal is to learn a function $h\colon \mathcal{X} \mapsto \mathcal{Y}$ with the given dataset such that $h(x)$ is a good predictor for the corresponding value of y.

In this house price prediction problem, we decide to approximate the price y as a linear function of the living area in square feet x: $h_\theta(x) = \theta_0 + \theta_1 x$, where θ_0 and θ_1 are the weights parameterizing the space of linear functions mapping from $\mathcal{X}$ to $\mathcal{Y}$. We update these weights by minimizing the empirical risk

$$\mathcal{L}(\theta) = \frac{1}{n} \sum_{i=1}^{n} (h_\theta\left(x^i\right) - y^i)^2 . \tag{1.1}$$

The optimized linear function is shown in Figure 1.1. Suppose now we have a new house whose living area is x_{new}. By using the trained linear regression model, the house price is readily available as $y_{\text{new}} = h_\theta(x^i)$ with θ assuming the optimized value.

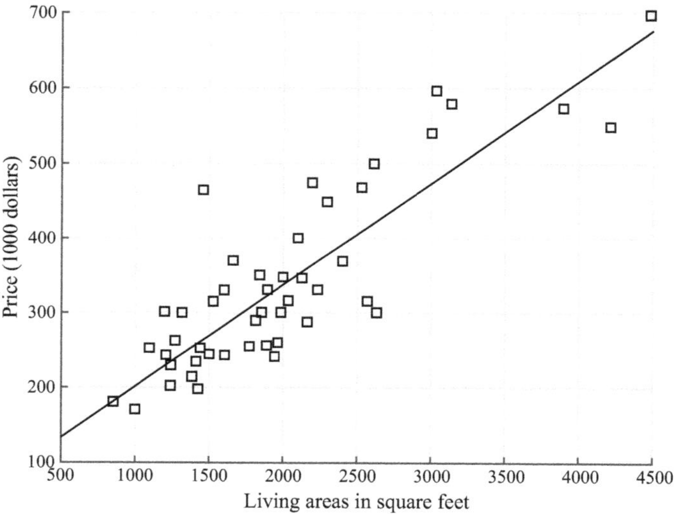

Figure 1.1 An example of supervised learning: House price prediction.

Unsupervised Learning

Although the usage of labels can ease the learning process, the collection of labeled data is often cumbersome and infeasible in many cases. To solve this issue, *unsupervised learning* is investigated to learn from unlabeled datasets. This paradigm of learning aims to find efficient representations of the data samples, which might be explained by hidden structures or hidden variables. Clustering, whose goal is to group data points into different clusters based on similarities among the points, is a representative task of unsupervised learning. The input could be either the absolute description of each data point or the pairwise similarities between two data points. Popular clustering algorithms include k-means clustering, hierarchical clustering, and spectrum clustering [26].

Dimension reduction, in which the inputs are transformed from a high dimensional space onto a low dimensional space while preserving the semantic or structural information, is another important area of unsupervised learning. In many scenarios, the raw data comes with a high dimension, and we may want to reduce the input dimension since with a low-dimension representation, the complexity of learning, saving, and transmission of the data can be significantly reduced. Data visualization is possible if the dimension is reduced to two or three. Conventional dimension reduction approaches mainly rely on linear projections, such as principal component analysis [26, 27], while with deep learning, non-linear transformations can be achieved, such as the deep autoencoder, which will be discussed in Section 1.2.2.

1.2.2 Deep Learning

During the last decade, machine learning has brought a number of unprecedented breakthroughs in such diverse fields as computer vision, natural language processing, and speech recognition. One of the primary driving forces behind these advancements is the evolution of artificial neural networks toward increasingly deep neural networks (DNNs), collectively summarized under the umbrella term of deep learning. In this section, we introduce basic concepts and important examples in deep learning, beginning with an explanation of the nuts and bolts of neural networks, then moving on to introduce popular neural networks for images and sequences, and finally concluding with a presentation of the deep autoencoder and prominent generative models.

Multilayer Perceptron

A multilayer perceptron (MLP) is a quintessential example of the deep learning architecture. It is composed of several layers of hidden neurons. Each neuron is a parameterized function $f(\mathbf{x}; \mathbf{w}, b)$ with input $\mathbf{x}$ and trainable parameters $\{\mathbf{w}, b\}$, where $\mathbf{w}$ denotes the linear coefficient and b denotes the bias. The function $f(\mathbf{x}; \mathbf{w}, b)$ is composed of a linear function of the input $\mathbf{x}$ and a non-linear activation function $\phi(\cdot)$, namely,

$$f(\mathbf{x}; \mathbf{w}, b) = \phi\left(\sum_j x_j w_j + b\right). \tag{1.2}$$

There are several commonly-used activation functions, including

- Sigmoid function:

$$\phi(z) = \frac{1}{1 + \exp(-z)}. \tag{1.3}$$

- Hyperbolic tangent function:

$$\phi(z) = \tanh(z). \tag{1.4}$$

- Rectified linear unit (ReLU) function:

$$\phi(z) = \max(0, z). \tag{1.5}$$

Although the function of a neuron is very simple and stereotyped, when a large number of neurons are aggregated, they can represent a wide range of complicated functions. The neurons in an MLP are arranged into layers, and each neuron in one layer is linked to all the neurons of the next layer but has no link with the neurons of the same layer. The first layer is the input layer, and its neurons take in the values of the input features. The last layer is the output layer for making predictions. All the other layers in between are called hidden layers. Mathematically, the weights associated with the links that connect all neurons in the lth layer to all neurons in the previous layer can be represented by a weight matrix $\mathbf{W}^{(l)}$ and bias vector $\mathbf{b}^{(l)}$. In this way, the output of the lth layer can be represented by

$$\mathbf{h}^l = f_l(\mathbf{h}^{l-1}) = \phi\left(\mathbf{W}^{(l)}\mathbf{h}^{l-1} + \mathbf{b}^{(l)}\right), \tag{1.6}$$

where $\mathbf{h}^{l-1}$ denotes input of the lth layer.

Before deployment, the parameters, i.e., the weights $\mathbf{W}^{(l)}$ and bias $\mathbf{b}^{(l)}$, need to be optimized using training data. In supervised learning, the training set consists of M labeled samples $\{\mathbf{x}_i, y_i\}_{i=1}^{M}$, and the training loss $\mathcal{L}$ is defined to measure the distance between the output of the neural network $\hat{y}_i$ and the label y_i, i.e.,

$$\mathcal{L} = \sum_{i}^{M} \ell(\hat{y}_i, y_i), \tag{1.7}$$

where $\ell(\hat{y}_i, y_i)$ represents the loss function. For regression problems, the commonly used loss functions includes the quadratic loss, or ℓ_2 loss, given by

$$\ell(\hat{y}_i, y_i) = \|\hat{y}_i - y_i\|^2, \tag{1.8}$$

while for binary classification problems, cross-entropy loss is the primary loss function, given by

$$\ell(\hat{y}_i, y_i) = -\left(y \log(\hat{y}_i) + (1 - y) \log(1 - \hat{y}_i)\right). \tag{1.9}$$

Optimization of the model weights usually requires the gradients of the training loss calculated on the entire training set, which are difficult to obtain when the set is large in scale. To solve this problem, optimization of the weights often uses the stochastic gradient descent (SGD) algorithm, which randomly picks one sample from the dataset at each iteration to reduce the computation required to determine gradients. The detailed algorithm is illustrated in Algorithm 1.1. The η denotes the learning rate of SGD, which determines the step size for each parameter update. An excessively large learning rate may cause the model parameters to oscillate or even diverge, preventing convergence. Conversely, a learning rate that is too small can significantly slow down the training process as the model may require many iterations to reach the optimal solution. Therefore, during training, techniques such as learning rate decay or adaptive learning rate methods (e.g., Adam, RMSprop) are sometimes employed

---

Algorithm 1.1 SGD for DNN Training

1:   Initialize the weights $\theta = \{\mathbf{W}^{(l)}, \mathbf{b}^{(l)}\}$ and learning rate η.

2:   **while** not convergence **do**

3:       Randomly shuffle samples in the training set.

4:       **for** $i = 1, 2, \cdots, M$ **do**

5:           Calculate the gradient g_i on the ith sample, $g_i = \frac{\partial l(\hat{y}_i, y_i)}{\partial \theta}$.

6:           $\theta = \theta - \eta \cdot g_i$.

---

to dynamically adjust the learning rate based on the progress of training. These techniques allow the model to learn quickly in the early stages and fine-tune parameter adjustments as it approaches a local minimum. In practice, the gradients are often computed using a mini-batch of samples, instead of only one sample, and this training procedure is universal for all kinds of DNNs beyond MLPs, including convolutional neural networks (CNNs) specialized for processing image data, recurrent neural networks (RNNs) specialized for processing sequence data, and so on. Mini-batch training balances the computational efficiency of full-batch gradient descent with the noise reduction benefits of stochastic gradient descent by processing a small subset of data at each iteration. This method reduces computational overhead, accelerates convergence, and improves generalization by averaging gradients over multiple samples. The choice of mini-batch size is crucial: Smaller batches introduce beneficial stochasticity but may lead to noisy updates, while larger batches provide more stable updates at the expense of increased computational complexities for each update. Therefore, selecting an appropriate mini-batch size is essential for optimizing training performance and achieving a balance between convergence speed and model accuracy.

Convolutional Neural Networks

The CNNs are one important class of neural networks that are powerful in dealing with data with a known grid-like topology, such as image data, which has a two-dimensional spatial structure. It is characterized by using convolution, a specialized linear operation, in place of general matrix multiplication in (1.6) and has been tremendously successful in practical applications such as image recognition and object detection. A CNN typically consists of four kinds of layers, i.e., the convolutional layer, non-linearity layer, pooling layer, and fully-connected layer.

In a convolutional layer, filters or kernels are used to extract features from the data, using the convolution operation. Given an image data $\mathbf{X}$ of size $N \times N$, a filter $\mathbf{W}$ is a trainable vector of size $F \times F$, and the output of the convolutional layer is

$$[\mathbf{W} \circledast \mathbf{X}](i,j) = \sum_{u=0}^{F-1} \sum_{v=0}^{F-1} w_{u,v} x_{i+u, j+v}. \tag{1.10}$$

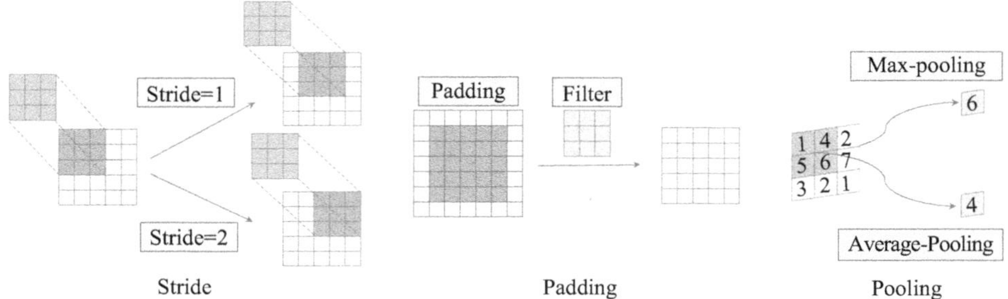

Figure 1.2 Illustration of stride, padding, and pooling operations in CNNs.

This means that the neurons of the output of a convolutional layer are obtained by element-wise multiplying the corresponding image patch, called the receptive field, and the filter. The convolution process is similar to sliding a window of $F \times F$ over the input neurons and mapping the generated output to the corresponding place.

To prevent the size of each layer from shrinking and avoid losing information on image borders since we do not allow the filter to slide off the boundary, zero-padding is used to manage the output size, which means we add a border of 0s to the image, as illustrated in Figure 1.2.

Since each neuron of the CNN layer consists of a weighted combination of pixels in the receptive field of the previous layer, neighbouring outputs can be very similar since their receptive fields overlap. To reduce such overlaps and speed up computation, strided convolution is introduced in CNNs. As shown in Figure 1.2, the stride denotes the number of nodes the filter skips each time. An increase in the stride will reduce both the overlap and the output size.

Finally, the output size of the convolutional layer, O, is

$$O = \frac{N - F + 2P}{S} + 1, \tag{1.11}$$

where S and P denote the stride and padding sizes respectively.

Generally, the input to a CNN is three-dimensional, such as an image (e.g., color images with a width and height of 32×32 pixels and a depth of three to denote RGB values) or a video (e.g., grayscale videos whose height and width are the horizontal and vertical resolution, and the depth is the frame number). The depth of the data is also called number of channels. The definition of convolution in this case can be extended by defining a kernel for each input channel. The size of input data $\mathbf{X}$ becomes $N \times N \times C$, and the size of filter $\mathbf{W}$ changes to $F \times F \times C$, where C denotes the number of input data channels. Then, (1.10) becomes

$$[\mathbf{W} \circledast \mathbf{X}](i,j) = \sum_{u=0}^{F-1} \sum_{v=0}^{F-1} \sum_{c=0}^{C-1} w_{u,v,c} x_{Si+u, Sj+v, c} + b, \tag{1.12}$$

where S is the stride and b is the bias.

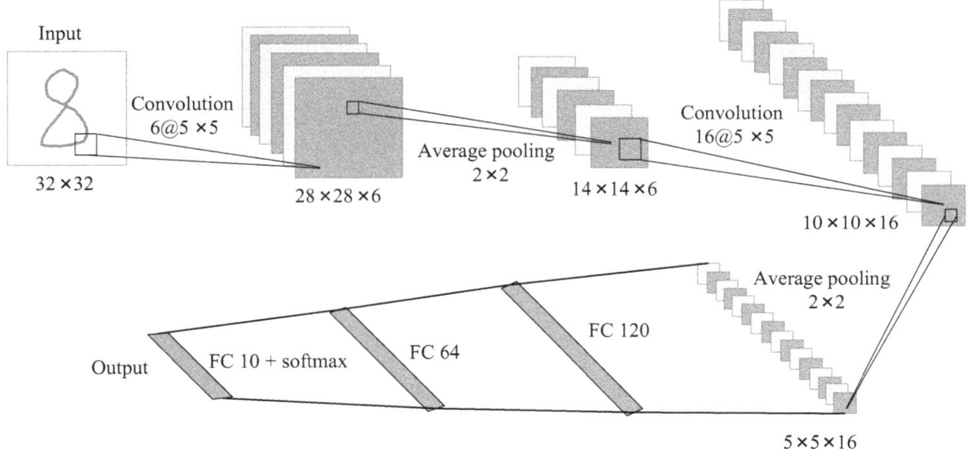

Figure 1.3 Architecture of LeNet-5 for handwritten digit recognition.

In practice, the pooling layer is commonly used in CNNs to make the representation approximately invariant to input translations by summarizing neighboring outputs into a single value. Max-pooling and average-pooling are two popular pooling methods. These methods both partition the input data to sub-region rectangles, and the max-pooling method only returns the maximum value of that sub-region, while the average-pooling returns the average value, as shown in Figure 1.2.

An illustrative CNN architecture, LeNet-5, is proposed in [31] for handwritten digit recognition. The detailed architecture is shown in Figure 1.3.

The input is a 32×32 pixel image, which contains a handwritten digit. The first convolutional layer has six filters of size 5×5, and the size of its output is $28 \times 28 \times 6$ when the stride is set as $S = 1$. The output will be processed by an average-pooling layer, and the size of data is subsampled to $14 \times 14 \times 6$. The second convolutional layer has 16 filters of size 5×5, and the output's size shrinks to $10 \times 10 \times 16$. Afterwards, another average-pooling layer subsamples this output's size to $5 \times 5 \times 16$. This $5 \times 5 \times 16$ feature map is then vectorized as a 400×1 vector and processed by three fully connected layers, whose sizes are 120, 64, and 10, respectively. The last fully connected layer is followed by a softmax layer to produce the probabilities of the digit being zero to nine. Here, the softmax function converts a vector $\mathbf{x} = [\mathbf{x}_1, \mathbf{x}_2, \ldots, \mathbf{x}_I]$ into probabilities that sum to one, and is defined as:

$$
\text{softmax}(\mathbf{x}) = \frac{1}{\sum_i \exp(\mathbf{x}_i)} \begin{pmatrix} \exp(\mathbf{x}_1) \\ \exp(\mathbf{x}_2) \\ \vdots \\ \exp(\mathbf{x}_I) \end{pmatrix}. \tag{1.13}
$$

According to this probability distribution, LeNet-5 can choose the most probable digit, thus completing the digit recognition task.

Example 1.2   Given a single-channel image matrix $\mathbf{X}$ and a 3×3 convolution kernel $\mathbf{W}$, calculate the output matrix $\mathbf{O}$ after the convolution operation. Assume that there is no padding and the stride is set as 1.

$$\mathbf{X} = \begin{bmatrix} 1 & 2 & 3 & 0 & 1 \\ 0 & 1 & 2 & 3 & 0 \\ 2 & 3 & 0 & 1 & 2 \\ 3 & 0 & 1 & 2 & 3 \\ 1 & 2 & 3 & 0 & 1 \end{bmatrix}, \quad \mathbf{W} = \begin{bmatrix} 0 & 1 & 0 \\ 1 & -4 & 1 \\ 0 & 1 & 0 \end{bmatrix}.$$

Solution

The size of input image is 5×5 and the size of kernel is 3×3. As there is no padding and the stride is 1, the size of the output can be obtained by (1.11):

$$\mathbf{O}_{\text{rows}} = \mathbf{O}_{\text{cols}} = \frac{5-3}{1} + 1 = 3.$$

Therefore, the size of the output matrix is 3×3. Then, we perform the convolution operation to calculate each element of $\mathbf{O}$. Using (1.12), we have

$$\mathbf{O}[0,0] = \sum_{i=0}^{2} \sum_{j=0}^{2} \mathbf{X}[i,j] \cdot \mathbf{W}[i,j] = \begin{bmatrix} 1 & 2 & 3 \\ 0 & 1 & 2 \\ 2 & 3 & 0 \end{bmatrix} \cdot \begin{bmatrix} 0 & 1 & 0 \\ 1 & -4 & 1 \\ 0 & 1 & 0 \end{bmatrix} = 3,$$

$$\mathbf{O}[0,1] = \sum_{i=0}^{2} \sum_{j=0}^{2} \mathbf{X}[i,j+1] \cdot \mathbf{W}[i,j] = \begin{bmatrix} 2 & 3 & 0 \\ 1 & 2 & 3 \\ 3 & 0 & 1 \end{bmatrix} \cdot \begin{bmatrix} 0 & 1 & 0 \\ 1 & -4 & 1 \\ 0 & 1 & 0 \end{bmatrix} = -1,$$

$$\cdots$$

$$\mathbf{O}[1,0] = \sum_{i=0}^{2} \sum_{j=0}^{2} \mathbf{X}[i+1,j] \cdot \mathbf{W}[i,j] = \begin{bmatrix} 0 & 1 & 2 \\ 2 & 3 & 0 \\ 3 & 0 & 1 \end{bmatrix} \cdot \begin{bmatrix} 0 & 1 & 0 \\ 1 & -4 & 1 \\ 0 & 1 & 0 \end{bmatrix} = -9,$$

$$\cdots$$

Finally, we can obtain the output matrix:

$$\mathbf{O} = \begin{bmatrix} 3 & -1 & -9 \\ -9 & 7 & 3 \\ 9 & 1 & -3 \end{bmatrix}.$$

Sequence Models: Recurrent Neural Networks

In this and the following two sections, we discuss various kinds of neural networks for sequences, where the inputs or outputs (or both) are sequence data. They are widely used in natural language processing, speech recognition, image captioning, and similar areas. Among these sequence models, RNNs, long short-term memory

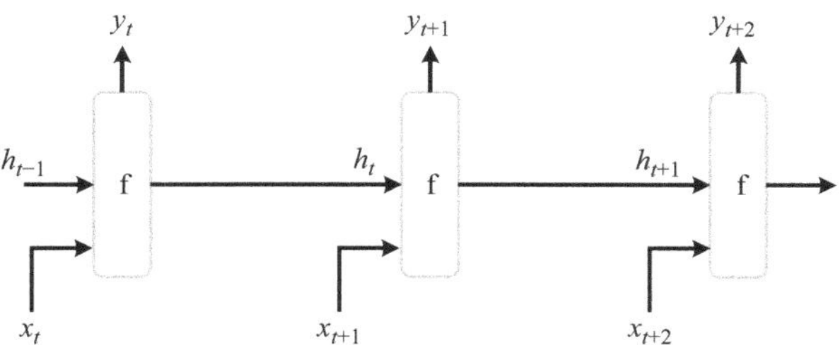

Figure 1.4 Architecture of the RNN.

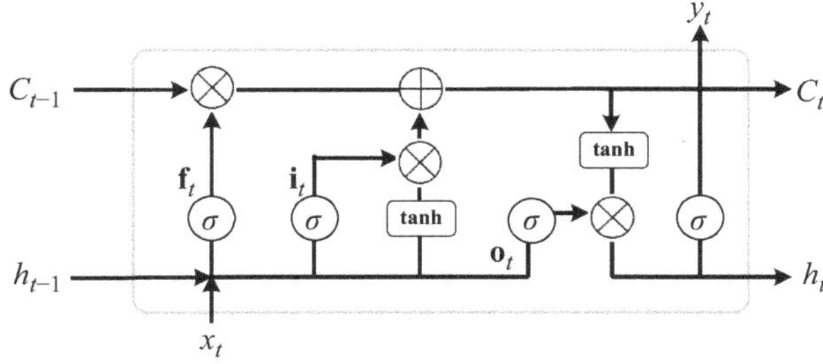

Figure 1.5 Architecture of the LSTM.

(LSTM), attention mechanisms, and transformers have been particularly successful and will be examined in detail.

The architecture of RNNs is illustrated in Figure 1.4, where at each time step t the RNN receives an input vector $\mathbf{x}_t$, updates its hidden state $\mathbf{h}_{t-1}$, and makes a prediction $\mathbf{y}_t$. The high-dimensional hidden state can integrate information over many time steps and use it to make accurate predictions.

The hidden state is updated as follows:

$$\mathbf{h}_t = \sigma(\mathbf{W}^h \mathbf{h}_{t-1} + \mathbf{W}^i \mathbf{x}_t + \mathbf{b}^h), \tag{1.14}$$

where $\mathbf{W}^h$ is the recurrent weight matrix, $\mathbf{W}^i$ is the input-to-hidden-state weight matrix, and $\mathbf{b}^h$ is the bias. σ denotes the non-linear activation, such as the sigmoid function in (1.3) or tanh function in (1.4). The output $\mathbf{y}_{t+1}$ can be related to the hidden state as

$$\mathbf{y}_t = \varsigma(\mathbf{W}^o \mathbf{h}_t + \mathbf{b}^o), \tag{1.15}$$

where $\mathbf{W}^o$ is the hidden-state-to-output weight matrix and $\mathbf{b}^o$ is the bias. ς is often chosen as the softmax non-linearity function.

Based on RNNs, the LSTM model [32] is proposed to address the issue of vanishing gradients when dealing with sequences of extensive length. The LSTM model includes specially-designed memory units to store information over long time periods; its architecture is shown in Figure 1.5. Unlike vanilla RNNs, the LSTM model has two kinds of states. One is called the cell state, which is denoted by $\mathbf{C}_t$. The cell state is mainly used to store the long-term memory of sequence data and can be considered the long-term memory unit of the network. It is capable of retaining important information within the sequence, making it less likely to be forgotten even if the sequence is very long. The other state is the hidden state, $\mathbf{h}_t$, which changes quickly from step to step. In each step, there are three kinds of gates inside the LSTM as (1.16)–(1.18) show, i.e., the forget gate, information gate, and output gate, to control how much information should be let through:

$$\text{Forget gate: } \mathbf{f}_t = \sigma(\mathbf{W}^f \cdot [\mathbf{h}_{t-1}, \mathbf{x}_t] + \mathbf{b}^f), \tag{1.16}$$

$$\text{Information gate: } \mathbf{i}_t = \sigma(\mathbf{W}^i \cdot [\mathbf{h}_{t-1}, \mathbf{x}_t] + \mathbf{b}^i), \tag{1.17}$$

$$\text{Output gate: } \mathbf{o}_t = \sigma(\mathbf{W}^o \cdot [\mathbf{h}_{t-1}, \mathbf{x}_t] + \mathbf{b}^o), \tag{1.18}$$

where $\mathbf{W}^f$, $\mathbf{W}^i$, and $\mathbf{W}^o$ are the weight matrices of these gates, and $[\mathbf{h}_{t-1}, \mathbf{x}_t]$ is the concatenation of the hidden state and the input data. $\mathbf{b}^f$, $\mathbf{b}^i$, and $\mathbf{b}^o$ are the biases, and σ is the sigmoid non-linearity. The cell state is updated as follows:

$$\mathbf{C}_t = \mathbf{f}_t \odot \mathbf{C}_{t-1} + \mathbf{i}_t \odot \tilde{\mathbf{C}}_t, \tag{1.19}$$

where $\odot$ is the Hadamard product, i.e., entry-wise multiplication. $\tilde{\mathbf{C}}_t$ is a vector of new candidate values that could be added to the cell state, given by

$$\tilde{\mathbf{C}}_t = \tanh(\mathbf{W}^c \cdot [\mathbf{h}_{t-1}, \mathbf{x}_t] + \mathbf{b}^c), \tag{1.20}$$

where $\mathbf{W}^c$ and $\mathbf{b}^c$ represent the weight matrix and bias. Through (1.19), the previous cell state is filtered by the forget gate to discard outdated information, and it incorporates current inputs via the input gate, allowing for a precise evaluation of the current state. The hidden state and the output can be obtained by

$$\mathbf{h}_t = \mathbf{o}_t \odot \tanh(\mathbf{C}_t) \tag{1.21}$$

and

$$\mathbf{y}_t = \sigma(\mathbf{W}' \cdot \mathbf{h}_t + \mathbf{b}'), \tag{1.22}$$

respectively, where $\mathbf{W}'$ and $\mathbf{b}'$ represent the weight matrix and bias. The softmax function is usually chosen as the non-linearity for the output $\mathbf{y}_t$.

Sequence Models: Attention

Despite improvements over traditional RNNs, LSTM can still struggle with the vanishing gradient problem, leading to the loss of important information from distant time steps. Their sequential processing nature also limits efficiency, as information from each time step is processed one after another. These issues motivate the use of attention mechanisms, which address these limitations by allowing the model to dynamically focus on various parts of the input sequence. Unlike LSTM, attention mechanisms directly access and weigh all input elements simultaneously, regardless of their distance in the sequence. This approach facilitates better handling of long-range dependencies and contextual information. Essentially, attention mechanisms work by computing attention scores for each part of the input, which are then used to create weighted combinations, allowing the model to focus more effectively

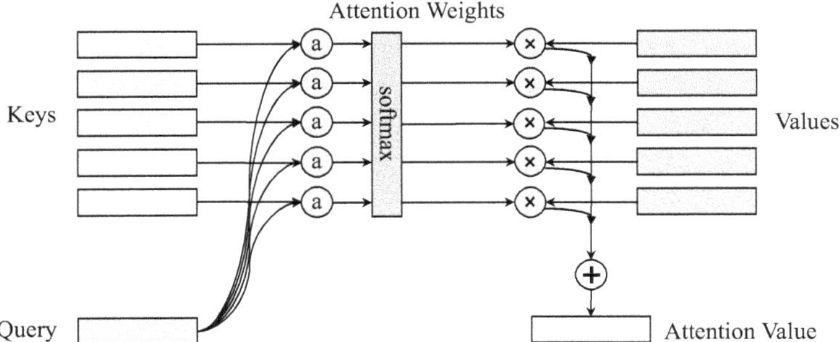

Figure 1.6 Architecture of an attention model.

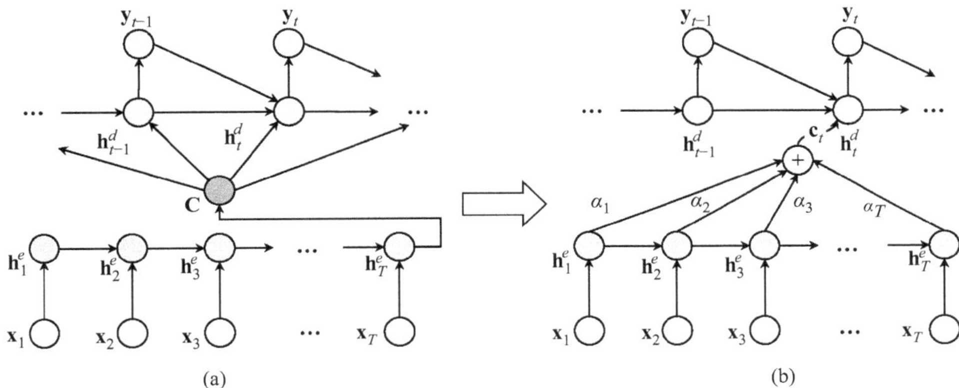

(a)   (b)

Figure 1.7 Illustration of a Seq2Seq model (a) based on the encoder-decoder architecture; (b) based on the attention mechanism.

on the most relevant information. This attention or weight allocation mechanism is analogous to human reading processes. For instance, when reading a complex sentence, a pronoun sometimes refers to a noun that is positioned far from itself. In such cases, humans naturally focus on the distant noun, assigning it a higher weight score, rather than solely concentrating on words immediately adjacent to the pronoun. Specifically, suppose there is a set of I value vectors $\mathbf{V} \in \mathbb{R}^{I \times n}$, and the model can choose one to use based on the similarity between the query vector $\mathbf{q} \in \mathbb{R}^n$ and a set of I keys $\mathbf{K} \in \mathbb{R}^{I \times n}$. The value $\mathbf{v}_i$ will be used when the $\mathbf{q}$ is most similar to key $\mathbf{k}_i$.

To make this operation differentiable, a combination of the values is computed as follows:

$$\text{Attention}(\mathbf{q}, \mathbf{K}, \mathbf{V}) = \text{softmax}(e(\mathbf{q}, \mathbf{K}))\mathbf{V} \in \mathbb{R}^n, \tag{1.23}$$

where $e(\cdot)$ generally denotes the attention score, which computes the similarity of query $\mathbf{q}$ to $\mathbf{k}_i$. Figure 1.6 gives an illustration of this procedure.

As shown in Figure 1.7a, consider a sequence-to-sequence (Seq2Seq) model as an example, where both the input and output are sequences. In this model, a set of RNN-based encoder and decoders transforms the sequence $\mathbf{x} = (\mathbf{x}_1, \ldots, \mathbf{x}_T)$

into $\mathbf{y} = (\mathbf{y}_1, \ldots, \mathbf{y}_T)$, which is a commonly used approach for translation and text generation tasks. In a traditional RNN decoder without an attention mechanism, the decoder generates each element of the sequence step by step based on a fixed-length context vector $\mathbf{c}$ and the previous output. Specifically, the process of generating the sequence $\mathbf{y} = (\mathbf{y}_1, \ldots, \mathbf{y}_T)$ can be described as

$$p(\mathbf{y}|\mathbf{x}) = \prod_{t=1}^{T'} p(\mathbf{y}_{t'}|\mathbf{y}_1, \ldots, \mathbf{y}_{t'-1}, \mathbf{c}), \tag{1.24}$$

where $\mathbf{c}$ is a fixed context vector produced by the encoder, which is denoted by

$$\mathbf{c} = q(\mathbf{h}_1^e, \ldots, \mathbf{h}_T^e), \tag{1.25}$$

where q is a non-linear function, which could be part of typical network units such as the gated recurrent unit (GRU) or LSTM. The hidden state $\mathbf{h}_t^e$ at time step t is based on the input sequence $\mathbf{x} = (\mathbf{x}_1, \ldots, \mathbf{x}_T)$ and follows (1.14). However, this approach has a limitation. When the input sequence $\mathbf{x}$ is long, using a fixed context vector to represent the entire input sequence may result in the loss of important information.

In contrast, the attention mechanism allows the decoder to dynamically select the most relevant parts of the input sequence at each time step t', rather than relying on a fixed context vector. Thus, at each time step, the decoder calculates a context vector $\mathbf{c}_{t'}$, which is a weighted sum of all positions in the input sequence. The attention mechanism is illustrated in Figure 1.7b. Specifically, for the decoder's current hidden state $\mathbf{h}_{t'}^d$ and the encoder's hidden state at time step t, $\mathbf{h}_t^e$, an attention score $e_{t',t}$ is first computed, indicating the relevance of the input sequence at time step t to the current decoder state. The attention score $e_{t',t}$ is denoted by

$$e_{t',t} = \text{score}(\mathbf{h}_{t'}^d, \mathbf{h}_t^e). \tag{1.26}$$

Then, these scores are normalized to obtain the attention weights $\alpha_{t',t}$, indicating the importance of the input at time step t for generating the current output, which can be described as

$$\alpha_{t',t} = \frac{\exp(e_{t',t})}{\sum_{k=1}^{T} \exp(e_{t',k})}. \tag{1.27}$$

Finally, the attention weights $\alpha_{t',t}$ are used to compute a weighted sum of the encoder's hidden states $\mathbf{h}_t^e$, obtaining the context vector $\mathbf{c}_{t'}$ for the current time step, which can be described as

$$\mathbf{c}_{t'} = \sum_{t=1}^{T} \alpha_{t',t} \mathbf{h}_t^e. \tag{1.28}$$

With the attention mechanism, the probability of generating the output $\mathbf{y}_{t'}$ at each time step t' by the decoder can be expressed as

$$p(\mathbf{y}_{t'} \mid \mathbf{y}_1, \ldots, \mathbf{y}_{t'-1}, \mathbf{x}) = f(\mathbf{y}_{t'-1}, \mathbf{h}_{t'}^d, \mathbf{c}_{t'}), \tag{1.29}$$

where f is a non-linear function, which could be part of typical network units such as GRU or LSTM. $\mathbf{h}_{t'}^d$ is the current hidden state of the decoder, determined by the hidden state of the previous time step $\mathbf{h}^d_{t'-1}$, the previous output $\mathbf{y}_{t'-1}$, and the context vector $\mathbf{c}_{t'}$ at the current time step. The generation of the entire sequence $\mathbf{y} = (\mathbf{y}_1, \ldots, \mathbf{y}_T)$ is obtained by multiplying together the conditional probability at each time step:

$$p(\mathbf{y} \mid \mathbf{x}) = \prod_{t'=1}^{T'} p(\mathbf{y}_{t'} \mid \mathbf{y}_1, \ldots, \mathbf{y}_{t'-1}, \mathbf{x}). \tag{1.30}$$

Example 1.3 Consider an input sequence $\mathbf{x} = (\mathbf{x}_1, \mathbf{x}_2, \mathbf{x}_3)$, with corresponding encoder hidden states $\mathbf{h}^e_1$, $\mathbf{h}^e_2$, and $\mathbf{h}^e_3$. The decoder's current hidden state is $\mathbf{h}^d_{t'}$. The values of the hidden states are given as

$$\mathbf{h}^e_1 = \begin{pmatrix} 1 \\ 0 \end{pmatrix}, \quad \mathbf{h}^e_2 = \begin{pmatrix} 0 \\ 1 \end{pmatrix}, \quad \mathbf{h}^e_3 = \begin{pmatrix} 1 \\ 1 \end{pmatrix},$$

$$\mathbf{h}^d_{t'} = \begin{pmatrix} 1 \\ 1 \end{pmatrix}.$$

Suppose that a simple dot product is used as the scoring function to calculate the attention scores. Calculate the context vector $\mathbf{c}_{t'}$.

Solution
First, the attention scores are calculated as

$$e_{t',1} = \mathbf{h}^{d\top}_{t'} \mathbf{h}^e_1 = \begin{pmatrix} 1 & 1 \end{pmatrix} \begin{pmatrix} 1 \\ 0 \end{pmatrix} = 1,$$

$$e_{t',2} = \mathbf{h}^{d\top}_{t'} \mathbf{h}^e_2 = \begin{pmatrix} 1 & 1 \end{pmatrix} \begin{pmatrix} 0 \\ 1 \end{pmatrix} = 1,$$

$$e_{t',3} = \mathbf{h}^{d\top}_{t'} \mathbf{h}^e_3 = \begin{pmatrix} 1 & 1 \end{pmatrix} \begin{pmatrix} 1 \\ 1 \end{pmatrix} = 2.$$

Then, the normalized attention weights can be computed as

$$\alpha_{t',1} = \frac{\exp(e_{t',1})}{\sum_{k=1}^{3} \exp(e_{t',k})} = \frac{2.718}{12.825} \approx 0.212,$$

$$\alpha_{t',2} = \frac{\exp(e_{t',2})}{\sum_{k=1}^{3} \exp(e_{t',k})} = \frac{2.718}{12.825} \approx 0.212,$$

$$\alpha_{t',3} = \frac{\exp(e_{t',3})}{\sum_{k=1}^{3} \exp(e_{t',k})} = \frac{7.389}{12.825} \approx 0.576.$$

This indicates that at the current time step t', the decoder focuses most on the third position in the input sequence, as it has the highest weight (approximately 0.576). The encoder's hidden state at this position contributes the most to generating the current output. Finally, the context vector $\mathbf{c}_{t'}$ can be calculated as a weighted sum of the encoder's hidden states, given as

$$\mathbf{c}_{t'} = \alpha_{t',1}\mathbf{h}_1^e + \alpha_{t',2}\mathbf{h}_2^e + \alpha_{t',3}\mathbf{h}_3^e,$$

$$\mathbf{c}_{t'} = 0.212\begin{pmatrix}1\\0\end{pmatrix} + 0.212\begin{pmatrix}0\\1\end{pmatrix} + 0.576\begin{pmatrix}1\\1\end{pmatrix} = \begin{pmatrix}0.788\\0.788\end{pmatrix}.$$

Sequence Models: Transformers

As shown in Figure 1.4, RNNs structure their computations sequentially along the positions of symbols in the input and output sequences. At each step, the model generates a hidden state, h_t, based on the previous hidden state, h_{t-1}, and the input at the current position, x_t. This sequential structure of computation prevents parallelization within training samples, which becomes a significant limitation as sequence lengths increase. To address this limitation, the transformer model is proposed in [33], relying entirely on an attention mechanism that allows for significantly greater parallelization.

As Figure 1.8 illustrates, before being fed into the encoder or decoder, a sequence of tokens is transformed into a sequence of embeddings with dimension d_{model} through a learnable embedding layer. Subsequently, the embeddings are integrated with positional encodings to add information regarding the relative or absolute positions of tokens within the sequence. The positional encodings are designed to have the same dimension as the embeddings, ensuring that they can be seamlessly added together. Specifically, the positional encodings are generated using sine and cosine functions at varying frequencies, represented by

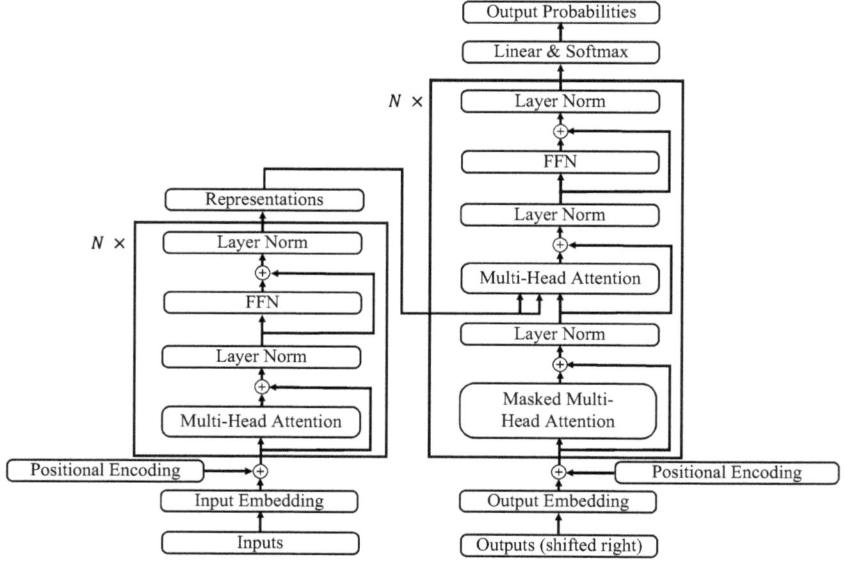

Figure 1.8 Structure of the transformer. In the figure, FFN stands for the feed-forward network.

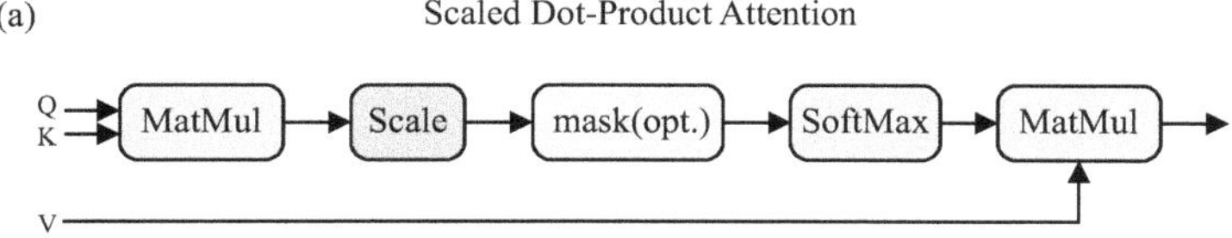

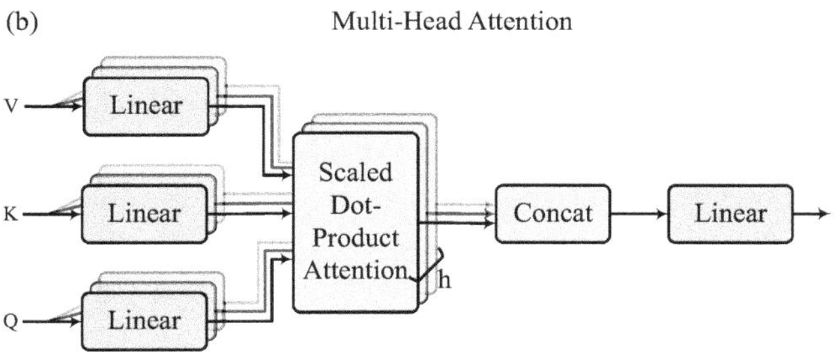

Figure 1.9 Scaled dot-product attention (a). Multi-head attention consists of h attention heads in parallel (b).

$$
PE_{(pos,\,i)} = \begin{cases} \sin\left(\dfrac{pos}{10000^{i/d_{\mathrm{model}}}}\right), & \text{if } i \text{ is odd}, \\[2mm] \cos\left(\dfrac{pos}{10000^{i/d_{\mathrm{model}}}}\right), & \text{if } i \text{ is even}, \end{cases} \tag{1.31}
$$

where pos denotes the position of a token in the embedded sequence, and $i \in [0, d_{\mathrm{model}} - 1]$ is the index of the dimension.

After positional encoding, the embeddings are transformed into a sequence of continuous representations by the encoder, which consists of a stack of N identical layers. Each layer is composed of two sub-layers: the first is a multi-head attention block, and the second is a simple position-wise fully connected feed-forward network. As shown in Figure 1.9, the multi-head attention consists of h scaled dot-product attention heads. The input of the scaled dot-product attention is composed of a query matrix $Q \in \mathbb{R}^{L \times d_k}$, a key matrix $K \in \mathbb{R}^{L \times d_k}$, and a value matrix $V \in \mathbb{R}^{L \times d_v}$, where L is the length of the input sequence. Given Q, K, and V, the output matrix can be computed as

$$
\mathrm{Attention}(Q, K, V) = \mathrm{softmax}\left(\frac{QK^T}{\sqrt{d_k}}\right) V, \tag{1.32}
$$

where the dot product of Q with K is scaled by $\sqrt{d_k}$ and then passed through a softmax function to obtain the weights for V. Instead of applying a single scaled dot-product attention function to the d_{model}-dimensional Q, K, and V, it is more effective to apply a multi-head attention mechanism, which linearly projects the queries, keys, and values h times using different learned linear transformations to dimensions d_k, d_k, and d_v, respectively. The multi-head attention mechanism allows the model to focus on information from different representation subspaces at various positions simultaneously. This can be formulated as

$$\text{MultiHead}(Q, K, V) = \text{Concat}(\text{head}_1, \ldots, \text{head}_h) W^O, \tag{1.33}$$

and

$$\text{head}_i = \text{Attention}(QW_i^Q, KW_i^K, VW_i^V), \tag{1.34}$$

where the projections are defined by the parameter matrices $W_i^Q \in \mathbb{R}^{d_{\text{model}} \times d_k}$, $W_i^K \in \mathbb{R}^{d_{\text{model}} \times d_k}$, $W_i^V \in \mathbb{R}^{d_{\text{model}} \times d_v}$, and $W^O \in \mathbb{R}^{hd_v \times d_{\text{model}}}$. Typically, for each attention head, $d_k = d_v = d_{\text{model}}/h$.

In addition to the multi-head attention sub-layers, each layer in the encoder incorporates a fully connected feed-forward network. This network consists of two linear transformations separated by a ReLU activation function, where both the input and output dimensions are d_{model}, and the hidden layer dimension is d_{ff}. Furthermore, each of these sub-layers is wrapped with a residual connection, followed by layer normalization.

Given the continuous representations generated by the encoder, the decoder sequentially generates an output sequence of tokens. Similar to the encoder, the decoder is composed of a stack of N identical layers. However, unlike the encoder layers, which consist of two sub-layers, each decoder layer includes an additional third sub-layer that performs multi-head attention over the encoder's output. Specifically, in this sub-layer, the encoder's output serves as both the queries and the keys for the multi-head attention. Furthermore, as illustrated in Figure 1.9, the attention mechanism in the first sub-layer of the decoder layers is masked to prevent each position from attending to subsequent positions. Similar to the encoder, each sub-layer in the decoder is wrapped with a residual connection, followed by layer normalization. Finally, after the last layer in the decoder, a learnable linear transformation is applied, followed by a softmax function, to map the decoder's output to predicted next-token probabilities.

Deep Autoencoder

The autoencoder is a neural network model that learns to copy its input to the output and is widely used for dimensionality reduction and feature learning. As illustrated in Figure 1.10, the autoencoder has two main components: an encoder and a decoder. The encoder learns a latent feature representation, $\mathbf{h}$, of the input, $\mathbf{x}$, and the decoder learns to recover the input from the latent feature representation. Formally, an encoder $f_e(\mathbf{x}; \theta_e)$ with parameters θ_e embeds the input to the latent representation $\mathbf{h}$, given by

$$\mathbf{h} = f_e(\mathbf{x}; \theta_e), \tag{1.35}$$

while a decoder $f_d(\mathbf{x}; \theta_d)$ with parameters θ_d reconstructs the input based on the representation $\mathbf{h}$, given by

$$\hat{\mathbf{x}} = f_d(\mathbf{h}; \theta_d). \tag{1.36}$$

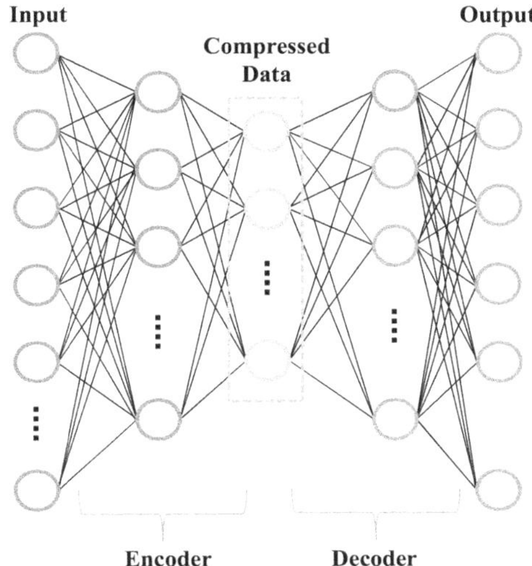

Figure 1.10 Architecture of an autoencoder.

There is much flexibility in choosing the specific DNN architectures for the encoder and decoder. For instance, the MLPs, CNNs, and RNNs that we discussed so far can all be exploited in an autoencoder.

The dimension of the latent feature representation is typically much smaller than the dimension of the input, so the encoder network does natural (by design) dimensionality reduction for input data. In addition, training of the autoencoder network does not require a labeled dataset. The training loss for the autoencoder measures differences between the input $\mathbf{x}$ and output $\hat{\mathbf{x}}$. Therefore the input itself serves as the label, which is often called self-supervision.

The autoencoder architecture has many connections with wireless communications. The encoder network can be adopted at the transmitter to learn a feature representation of the transmitted signal, which is then sent through the wireless channels. The decoder network is used at the receiver to recover the transmitted signals with the received signal. This provides a learned novel end-to-end communication system design. When trained on a massive dataset with a wide range of channel realizations, various channel distortions can be addressed in a data-driven manner. We will elaborate more on this topic in Chapter 4.

Generative Adversarial Networks

The generative adversarial network (GAN) is an emerging class of generative methods for implicitly modeling sophisticated data distributions through an adversarial training process. The objective of GANs is to learn a generative model, usually in the form of DNNs, that can capture the data distribution, p_{data}, and generate new samples approximately according to this distribution. The GAN has been widely applied in image and video synthesis, image super-resolution, and image-to-image

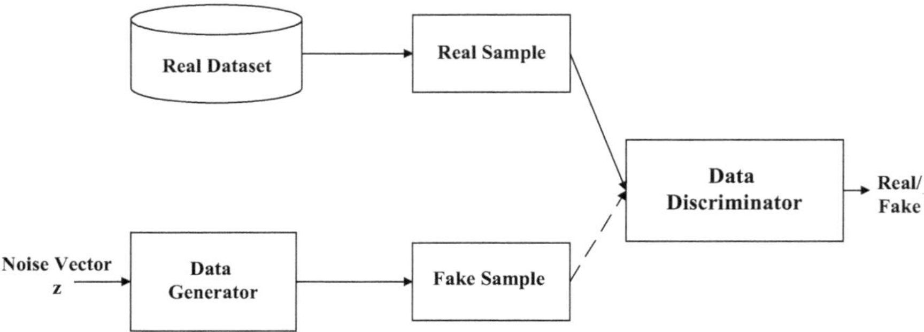

Figure 1.11 Architecture of a generative adversarial network.

translation. In Chapter 2, GANs will be employed for wireless channel modeling, and this GAN-based channel representation will be further investigated in Chapter 4 for end-to-end learning-based communication system design.

As shown in Figure 1.11, the GAN can be characterized by training two models that compete with each other, i.e., the generator, G, and the discriminator, D. The generator is used for generating samples. In particular, the input to the generator, G, is a noise vector $\mathbf{z}$ sampled from a prior distribution p_z, e.g., a uniform distribution. The input $\mathbf{z}$ is transformed by the generator into a generated sample, $G(\mathbf{z})$. The discriminator, D, is a classifier, whose input is either a real sample from the target distribution, p_{data}, or a generated sample, $G(\mathbf{z})$. The discriminator is trained to distinguish between real and generated samples.

The two models, i.e., the generator and discriminator, are usually represented by DNNs and trained together with the mini-batch SGD algorithm. Parameters of the generator and discriminator network are denoted by θ_G and θ_D, respectively. During training, the discriminator network is trained to distinguish real data from fake data generated by the generator. Meanwhile, the generator network is trained to create data to fool the discriminator into making mistakes. The loss functions for training the generator network and the discriminator network can be expressed as

$$\mathcal{L}_G = \min_{\theta_G} \mathbb{E}_{\mathbf{z}\sim p_z}[\log(1 - D(G(\mathbf{z})))], \tag{1.37}$$

$$\mathcal{L}_D = \max_{\theta_D} \mathbb{E}_{\mathbf{x}\sim p_{data}}[\log(D(\mathbf{x}))] + \mathbb{E}_{\mathbf{z}\sim p_z}[\log(1 - D(G(\mathbf{z})))], \tag{1.38}$$

respectively. The generator's loss function $\mathcal{L}_G$ aims to minimize the generator's loss by minimizing $\log(1 - D(G(\mathbf{z})))$, with the goal of deceiving the discriminator. This makes it difficult for the discriminator to distinguish between the fake samples generated by G and the real data, pushing the discriminator's output $D(G(\mathbf{z}))$ closer to one. On the other hand, the discriminator's loss function $\mathcal{L}_D$ aims to maximize the discriminator's ability to differentiate between real and fake data. It seeks to maximize the likelihood that real data is correctly classified as real, i.e., maximizing $\log(D(x))$, while also accurately identifying fake samples, i.e., maximizing $\log(1 - D(G(\mathbf{z})))$. These two objective functions interact through an adversarial process, gradually improving both the generator's ability to produce realistic samples and the discriminator's ability to distinguish between real and fake data. For each training step, two mini-batches of data are sampled, including a mini-batch of real data and a mini-batch of generated data from the generator. Two gradient steps are taken on parameters of the generator network and discriminator networks in order to improve the generator and discriminator loss, respectively. The training procedure will terminate when the two models reach an equilibrium where the discriminator network can do no better than random guessing to distinguish between the real samples and the generated fake samples.

Score-Based Generative Models

Score-based generative models [34], also known as generative diffusion models [35, 36], represent a cutting-edge technique in generative learning. This category of models utilizes the gradients of the data distribution, known as "scores," to synthesize new data samples. By leveraging the concept of score matching and stochastic differential equations, score-based models have achieved state-of-the-art performance in generating high-quality data samples. Moreover, they have shown potential in solving inverse problems such as inpainting, deblurring, and super-resolution, making them highly relevant for wireless receiver design. The application of score-based models in wireless channel estimation will be explored in Chapter 2.

The core idea behind score-based generative models is to estimate the score function, $\nabla_{\mathbf{x}} \log p_{\text{data}}(\mathbf{x})$, which represents the gradient of the log-probability of the data distribution $p_{\text{data}}(\mathbf{x})$ [37]. This score function provides valuable information about the structure of the data, enabling the model to generate new samples that resemble the original data distribution. Score-based generative modeling typically involves two key steps: score matching and sample generation. For the first step, a neural network $s_\theta(\mathbf{x})$ is trained to approximate the score function. Naive score matching utilizes the following loss function to train the model s_θ:

$$\mathcal{L}_{p_{\text{data}}}(\boldsymbol{\theta}) = \mathbb{E}_{p_{\text{data}}(\mathbf{x})}\left[\|s_\theta(\mathbf{x}) - \nabla_{\mathbf{x}} \log p_{\text{data}}(\mathbf{x})\|_2^2 \right]. \tag{1.39}$$

However, the ground truth data score is generally intractable, making the direct computation of (1.39) infeasible. To bypass this challenge, an effective solution is to use noise-perturbed score matching, also known as denoising score matching [37], which perturbs the data $\mathbf{x} \sim p_{\text{data}}(\mathbf{x})$ with noise and trains the model using noisy data points $\tilde{\mathbf{x}} = \mathbf{x} + \sigma\mathbf{w}$, where σ denotes the noise scale, and $\mathbf{w} \sim \mathcal{N}(\mathbf{0}, \mathbf{I})$. Specifically, the objective is transformed into

$$\tilde{\mathcal{L}}(\boldsymbol{\theta}) = \mathbb{E}_{p_{\text{data}}(\mathbf{x})}\mathbb{E}_{p_\sigma(\tilde{\mathbf{x}}|\mathbf{x})}\left[\|s_\theta(\tilde{\mathbf{x}}) - \nabla_{\tilde{\mathbf{x}}} \log p_\sigma(\tilde{\mathbf{x}}|\mathbf{x})\|_2^2 \right], \tag{1.40}$$

where $p_\sigma(\tilde{\mathbf{x}}|\mathbf{x})$ is the conditional distribution, whose score is given by

$$\nabla_{\tilde{\mathbf{x}}} \log p_\sigma(\tilde{\mathbf{x}}|\mathbf{x}) = -\frac{\tilde{\mathbf{x}} - \mathbf{x}}{\sigma^2}. \tag{1.41}$$

The loss function in (1.40) has been proved equivalent (up to a constant) to the explicit score matching objective in (1.39) for the perturbed data $\tilde{\mathbf{x}} \sim p_\sigma(\tilde{\mathbf{x}}) = \int p_\sigma(\tilde{\mathbf{x}}|\mathbf{x})p_{\text{data}}(\mathbf{x})d\mathbf{x}$. Hence, the score of the perturbed distribution $p_\sigma(\tilde{\mathbf{x}})$ can be

---

Algorithm 1.2 Annealed Langevin Dynamics [38]

Require: $\{\sigma_i\}_{i=1}^{L}, \epsilon_0, T.$

    Initialize $\mathbf{x}_1$.

    for $i = 1$ to L **do**

        Set annealed step size $\epsilon_i = \epsilon_0 \cdot \sigma_i^2 / \sigma_L^2$.

        for $t = 1$ to T **do**

            Draw $\mathbf{w}_t \sim \mathcal{N}(\mathbf{0}, \mathbf{I})$.

            $\mathbf{x}_{t+1} = \mathbf{x}_t + \epsilon_i s_\theta(\mathbf{x}_t, \sigma_i) + \sqrt{2\epsilon_i}\mathbf{w}_t.$

        $\mathbf{x}_1 = \mathbf{x}_{T+1}.$

    return $\mathbf{x}_{T+1}$

---

learned using (1.40). Extending this proposition to multiple noise scales $\sigma_1 > \sigma_2 > \cdots > \sigma_L$, we derive the loss function for training a noise conditional score network $s_\theta(\tilde{\mathbf{x}}, \sigma_i)$ [38]:

$$\tilde{\mathcal{L}}_{\text{NCSN}}(\boldsymbol{\theta}) = \frac{1}{L} \sum_{i=1}^{L} \mathbb{E}_{p_{\text{data}}(\mathbf{x})} \mathbb{E}_{p_{\sigma_i}(\tilde{\mathbf{x}}|\mathbf{x})} \left[\| s_\theta(\tilde{\mathbf{x}}, \sigma_i) - \nabla_{\tilde{\mathbf{x}}} \log p_{\sigma_i}(\tilde{\mathbf{x}}|\mathbf{x}) \|_2^2 \right]. \tag{1.42}$$

By minimizing this loss function, the score network can jointly learn the score of perturbed distributions, $\nabla_{\tilde{\mathbf{x}}} \log p_{\sigma_i}(\tilde{\mathbf{x}})$, within different noise scales ($i = 1, \ldots, L$). Large noise scales can cover low-data-density regions to enhance score estimation accuracy, whereas small noise scales lead to distributions closer to the original data distribution.

Once the score function has been estimated, the model can generate new data samples by solving a differential equation that leverages the estimated scores. One popular approach is to use the (discretized) Langevin dynamics,[1] which iteratively updates an initial random noise sample $\mathbf{x}_1$ following

$$\mathbf{x}_{t+1} = \mathbf{x}_t + \epsilon \nabla_{\mathbf{x}} \log p_{\text{data}}(\mathbf{x}_t) + \sqrt{2\epsilon}\mathbf{w}_t, \ t = 1, \cdots, T, \tag{1.43}$$

where ϵ is a step size, and $\mathbf{w}_t \sim \mathcal{N}(\mathbf{0}, \mathbf{I})$ is Gaussian noise. This process results in samples that asymptotically converge to the data distribution $p_{\text{data}}(\cdot)$ as $\epsilon \to 0$ and $T \to \infty$. To further enhance the convergence rate and align with the multi-scale noise-perturbed score matching, annealed Langevin dynamics has been proposed in [38]. This approach, inspired by simulated annealing, sequentially samples from noise-perturbed data distributions with the gradually decreased noise scales $\sigma_1 > \sigma_2 > \cdots > \sigma_L$ involved in score matching. As shown in Algorithm 1.2, at the noise scale σ_i, the algorithm starts from the final sample of the previous noise scale σ_{i-1} and runs Langevin dynamics to sample from $p_{\sigma_i}(\cdot)$ by using the estimate scores from the pre-trained score network $s_\theta(\cdot, \sigma_i)$. Note that the step size of the Langevin dynamics is also annealed according to the noise scale as $\epsilon_i = \epsilon_0 \cdot \sigma_i^2 / \sigma_L^2$, where ϵ_0 is a hyperparameter to be tuned. Continuing in this fashion, the sampling

---

[1] Discretized Langevin dynamics, also known as the unadjusted Langevin algorithm, is derived by performing Euler–Maruyama discretization to the Langevin diffusion process [39].

distribution $p_{\sigma_i}(\cdot)$ gradually approaches the true data distribution $p_{\text{data}}(\cdot)$ as the noise scale decreases, particularly when the final noise scale $\sigma_L \approx 0$.

1.2.3 Reinforcement Learning

Reinforcement learning is a field of machine learning that learns to map situations to actions so as to maximize certain numeric reward signals based on experiences gained through interaction with an unknown environment. Mathematically, the reinforcement learning problem can be framed as a Markov decision process (MDP) that formalizes sequential decision-making, as illustrated in Figure 1.12. At each time step t, the agent observes some representation of the environment state S_t from the state space S and then selects an action A_t from the action space $\mathcal{A}$. Following the action, the agent receives a numerical reward, R_{t+1}, and the environment transitions to a new state, S_{t+1}, with probability

$$p(s',r|s,a) = \Pr\left\{S_{t+1} = s', R_{t+1} = r | S_t = s, A_t = a\right\}. \tag{1.44}$$

In reinforcement learning, the decision-making of the agent is formalized in a policy $\pi(a|s)$, which is a mapping from states in S to probabilities of selecting each action in $\mathcal{A}$. The goal of learning is to find the optimal policy, π_*, that maximizes the expected return of a complete episode from any initial state. The return at time t, denoted G_t, is defined as the cumulative rewards with a discount rate $\gamma \in (0,1]$, given by

$$G_t = \sum_{k=0}^{\infty} \gamma^k R_{t+k+1}. \tag{1.45}$$

The infinite sum has a finite value as long as the reward sequence R_k is bounded. As γ approaches one, the agent places greater emphasis on future rewards, thus avoiding being shortsighted.

To be precise, we define the value function of state s under policy π to be the expected return when starting in s and thereafter following π, written as

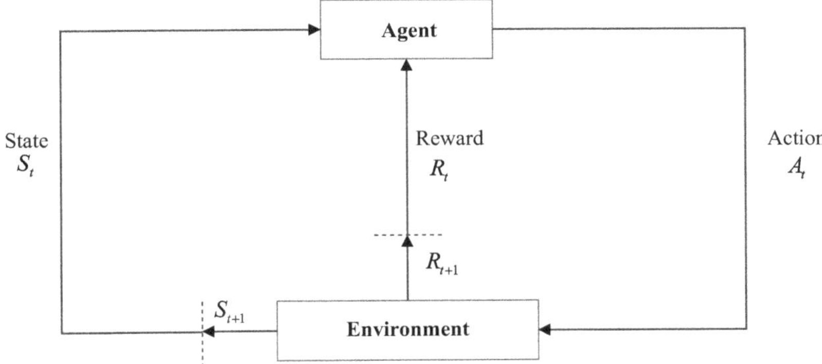

Figure 1.12 The agent–environment interaction in a Markov decision process.

$$v_\pi(s) = \mathbb{E}_\pi[G_t|S_t = s], \text{for all } s \in \mathcal{S}. \tag{1.46}$$

This formulation evaluates the expected performance of an agent in a given state, conditioned on the long-term rewards that can be obtained by following policy π. A fundamental property of value functions used throughout reinforcement learning is that they satisfy recursive relationships as follows:

$$
\begin{aligned}
v_\pi(s) &= \mathbb{E}_\pi[G_t \mid S_t = s] \\
&= \mathbb{E}_\pi[R_{t+1} + \gamma G_{t+1} \mid S_t = s] \\
&= \sum_a \pi(a|s) \sum_{s'} \sum_r p(s',r|s,a)\Big[r + \gamma \mathbb{E}_\pi[G_{t+1}|S_{t+1} = s']\Big] \\
&= \sum_a \pi(a|s) \sum_{s',r} p(s',r|s,a)\Big[r + \gamma v_\pi(s')\Big], \quad \text{for all } s \in \mathcal{S}.
\end{aligned}
\tag{1.47}
$$

This equation is the Bellman equation for v_π. The value function gives rise to the definition of partial ordering over policies: $\pi \geq \pi'$, if and only if $v_\pi(s) \geq v_{\pi'}(s)$, for all $s \in \mathcal{S}$. Then the optimal policy, π_*, is defined to be the set of policies that are better than or equal to any other policy. By definition, optimal policies achieve the optimal state-value function, defined as

$$v_*(s) = \max_\pi v_\pi(s), \text{for all } s \in \mathcal{S}. \tag{1.48}$$

This implies that if the expected return of an agent in any state is maximized by following a particular policy, then that policy is considered to be optimal.

In theory, if complete knowledge about the environment, i.e., the dynamics characterized by $p(s',r|s,a)$, is available, one can solve the MDP and find the optimal policy using dynamic programming. However, this is rarely the case in practice due to the difficulty of acquiring such complete knowledge and the constraints of computation and memory. In fact, one of the most distinctive features of reinforcement learning is its capability to approximately solve the MDP without knowing a complete and accurate model of the environment's dynamics. It is based on interactions with the unknown environment, and the resulting algorithms are model-free in nature. These algorithms need to balance exploration and exploitation, i.e., to exploit the "best" moves based on the current value function estimation while still allowing exploratory moves that continually explore in order to make better actions in the future.

A wide variety of model-free algorithms, including the value-based methods, such as Q-learning and SARSA; policy-based methods, such as policy gradient; and model-based algorithms that integrate planning and learning, such as Dyna, have been developed to solve reinforcement learning problems [29]. They can further be combined with deep learning to bring the performance to a new level, with deep Q-learning as a great example that has achieved remarkable performance in human-level video game play [40].

Here we first discuss two basic reinforcement learning algorithms, i.e., (deep) Q-learning and REINFORCE, as representatives of the value-based and policy-based learning methods, respectively. We also provide a concise introduction to the actor–critic algorithm in its basic form and the proximal policy optimization algorithm, which blend the two categories of methods. For other algorithms, such as trust region policy optimization [41], async advantage actor–critic [42] and many others, please refer to the given references and the tutorial on deep reinforcement learning in [43].

Example 1.4 Figure 1.13(a) gives a simple finite MDP. The grids of this figure represent the states of the environment. At each state, four actions are possible: left, right, up, and down, which deterministically cause the agent to move in the respective direction on the grid. Actions that would take the agent off the grid leave its location unchanged but also result in a reward of -1. Other actions result in a reward of 0, except those that move the agent out of the special state A. If the current state is A, taking any action will move the agent to state A' and yield a reward of $+10$. Suppose the agent selects these four actions with equal probability in all states. Given the discount factor $\gamma = 0.9$, calculate the value function of each state under this equiprobable random policy.

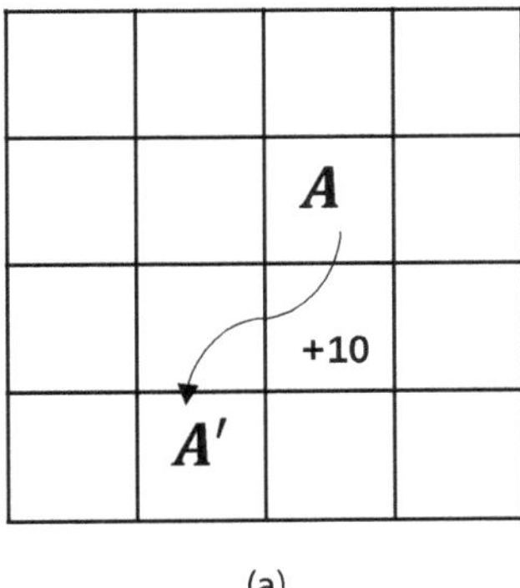

0.12	1.82	3.97	2.34
0.69	3.29	10.6	3.97
0.07	1.51	3.29	1.82
-0.8	0.07	0.69	0.12

(a)   (b)

Figure 1.13 An example of a finite MDP: (a) illustration of the gridworld; (b) the state-value function for the equiprobable random policy.

Solution
Suppose that state A is at $(1, 2)$ and state A' is at $(3, 1)$. For each state $s = (i, j)$, we should consider all possible actions $a \in \{\text{left}, \text{right}, \text{up}, \text{down}\}$. The value functions of all the states could be calculated following (1.47):

$$v^\pi(s_{(0,0)}) = \frac{1}{4}[(-1 + \gamma v^\pi(s_{(0,0)})) + (\gamma v^\pi(s_{(0,1)})) + (\gamma v^\pi(s_{(1,0)}))$$
$$+ (-1 + \gamma v^\pi(s_{(0,0)}))],$$

$$v^\pi(s_{(0,1)}) = \frac{1}{4}[(-1 + \gamma v^\pi(s_{(0,1)})) + (\gamma v^\pi(s_{(0,0)})) + (\gamma v^\pi(s_{(0,2)}))$$
$$+ (\gamma v^\pi(s_{(1,1)}))],$$

$\cdots$

$$v^{\pi}(s_{(1,1)}) = \frac{1}{4}\left[(\gamma v^{\pi}(s_{(0,1)})) + (\gamma v^{\pi}(s_{(1,0)})) + (\gamma v^{\pi}(s_{(1,2)})) + (\gamma v^{\pi}(s_{(2,1)}))\right],$$

$$\cdots$$

$$v^{\pi}(A) = 10 + \gamma v^{\pi}(A'),$$

$$\cdots$$

The solution of the above equations is shown in Figure 1.13b.

Q-Learning

Q-Learning is a popular model-free method of solving reinforcement learning problems. It is based on the concept of the action-value function, $q_{\pi}(s,a)$, which is similar to the value function and is defined as the expected accumulative rewards starting from state s, taking action a, and thereafter following policy π, formally expressed as

$$q_{\pi}(s,a) = \mathbb{E}_{\pi}\left[G_t | S_t = s, A_t = a\right]. \tag{1.49}$$

The action-value function of the optimal policy, $q_*(s,a)$, satisfies a recursive relation, known as the Bellman optimality equation,

$$q_*(s,a) = \sum_{s',r} p(s',r|s,a)\left[r + \gamma \max_{a'} q_*(s',a')\right], \tag{1.50}$$

for any state s, action a, successor state s', and action a'. In principle, one can solve this system of non-linear equations for $q_*(s,a)$ if the underlying dynamics $p(s',a'|s,a)$ are perfectly known. Once q_* is obtained, it is easy to determine the optimal policy $\pi_*(a|s)$ following a greedy procedure,

$$\pi_*(a|s) = \begin{cases} 1, & \text{if } a = \arg\max_{a\in\mathcal{A}} q_*(s,a), \\ 0, & \text{otherwise.} \end{cases} \tag{1.51}$$

Q-learning avoids acquiring complete environment dynamics and approaching the system of non-linear equations directly by taking an iterative update procedure

$$Q(S_t,A_t) \leftarrow Q(S_t,A_t) + \alpha\left[R_{t+1} + \gamma \max_{a'} Q(S_{t+1},a') - Q(S_t,A_t)\right], \tag{1.52}$$

where α is the step-size parameter, and the choice of A_t in state S_t follows some soft policies, e.g., the ϵ-greedy policy, meaning that the action with maximal estimated value is chosen with probability $1 - \epsilon$, while a random action is selected with probability ϵ. With a variant of the stochastic approximation conditions on α and under the assumption that all state–action pairs continue to be updated, Q converges with probability of one to the optimal action-value function q_*.

It is worth noting that in the Q-learning algorithm, the policy used to generate the experience data (i.e., the ϵ-greedy policy), referred to as the behavior policy, is different from the one being updated (i.e., the greedy policy with respect to the current action-value function estimation $Q(S_t,A_t)$), commonly known as the target policy.

These reinforcement learning algorithms are called the off-policy algorithms. This is in contrast to the on-policy algorithm, where the behaviour policy is the same as the target policy, as is the case in SARSA [29].

Deep Q-Network with Experience Replay

In many problems of practical interest, such as the Go game and robotic control, the state or action space can be too large to store all action-value functions in tabular form. As a result, it is common to use function approximation to estimate these value functions. An extra advantage of doing so is the ability to generalize from limited seen state–action pairs and produce approximations in a much larger space.

In deep Q-learning [40], a DNN parameterized by θ, called deep Q-network (DQN), is used to approximate the action-value function. The state–action space is explored with some soft policies, e.g., ϵ-greedy, and at each time step the transition tuple $(S_t, A_t, R_{t+1}, S_{t+1})$ is stored in a replay memory. At each step, a mini-batch of transition tuples is uniformly sampled from the replay memory, defined as experience $\mathcal{D}$. The DQN updates θ with variants of SGD methods on experience $\mathcal{D}$, a process known as experience replay, to minimize the sum-squared error

$$\sum_{\mathcal{D}} \left[R_{t+1} + \gamma \max_{a'} Q(S_{t+1}, a'; \theta^-) - Q(S_t, A_t; \theta) \right]^2, \tag{1.53}$$

where θ^- is the parameter set of a target Q-network, which is duplicated from the training Q-network parameter set θ periodically and fixed for a couple of updates. Experience replay improves sample efficiency through repeatedly sampling stored experiences and breaks correlation in successive updates, thus also stabilizing learning.

Policy Gradient and Actor–Critic

In contrast to value-based reinforcement learning methods that estimate a value function and then use it to compute a deterministic policy as shown in (1.51), policy gradient methods directly search the policy space for an optimal one. The policy is usually represented by a function approximator, such as a DNN, parameterized by θ, i.e., $\pi_\theta(a|s)$. The parameter set θ is updated in the direction of improvement of a certain performance measure, $J(\theta)$, by gradient descent,

$$\theta \leftarrow \theta + \alpha \nabla_\theta J(\theta), \tag{1.54}$$

with an appropriate step size α. In reinforcement learning, the episodic learning tasks refer to tasks that are naturally divided into episodes. Each episode consists of a sequence of states, actions, and rewards that starts from an initial state and ends in a terminal state. Once a terminal state is reached, the environment resets and a new episode begins. Episodic tasks are common in environments where there is a clear start and end, such as games, navigation tasks, or any scenario where there is a goal to be achieved or a failure condition. For the episodic learning tasks, we define $J(\theta)$ as the expected return from the start state,

$$J(\theta) = \mathbb{E}\left\{\sum_{t=0}^{T} \gamma^t R_{t+1}\right\}, \tag{1.55}$$

where T is the final time step. From the policy gradient theorem [29], the gradient $\nabla J(\theta)$ follows

$$\nabla_\theta J(\theta) = \mathbb{E}_\pi \left\{\nabla_\theta \log \pi_\theta(A_t|S_t) q_{\pi_\theta}(S_t, A_t)\right\}. \tag{1.56}$$

In the REINFORCE algorithm, we use the cumulative rewards obtained from time step t onward, i.e., G_t in (1.45), as an unbiased estimate of $q_{\pi_\theta}(S_t, A_t)$ and then run multiple episodes of the task following policy $\pi_\theta(a|s)$ to update the parameter set θ in each time step t during training.

In fact, the REINFORCE algorithm suffers from high variance, and we can alleviate this issue by learning an approximation of the action-value function for variance reduction, given by

$$Q_w(s, a) \approx q_{\pi_\theta}(s, a), \tag{1.57}$$

where w is the set of trainable parameters of the approximation function, usually a neural network. Various policy evaluation methods, such as temporal-difference learning, can be leveraged to address the action-value function approximation. In the actor–critic terminology, the learned action-value function approximation $Q_w(s, a)$ is called the critic, while the approximate policy $\pi_\theta(a|s)$ is called the actor. Then the actor–critic algorithm follows an approximate policy gradient

$$\theta \leftarrow \theta + \alpha \nabla_\theta \log \pi_\theta(A_t|S_t) Q_w(S_t, A_t). \tag{1.58}$$

Finally, it is important to note that compared with value-based reinforcement learning methods, policy gradient algorithms can learn stochastic policies, where the learned mapping from the state to the action is in the form of a probabilistic distribution over the action space, instead of being deterministic. Therefore, with such a policy, the same state can lead to different actions according to certain probabilities, which allows for exploration and better handling of uncertainty. Policy gradient algorithms tend to be more effective in high-dimensional or continuous action space but are more likely to converge to a local rather than global optimum.

Proximal Policy Optimization

Proximal policy optimization (PPO) is a state-of-the-art algorithm proposed in [44]. It is a kind of policy gradient algorithm that has an actor–critic framework. The objective function for training a reinforcement learning model is to maximize the expected return defined in (1.55), written out as

$$J(\theta) = \mathbb{E}_{\tau \sim p_\theta(\tau)}[R(\tau)] = \int_\tau G(\tau) p_\theta(\tau), \tag{1.59}$$

where $\tau = \{S_0, A_0, R_1, S_1, A_1, \ldots, S_T, A_T, R_{T+1}\}$ denotes one trajectory collected from interactions with the environment. $G(\tau) = \sum_{t=0}^{T} \gamma^t R_{t+1}$ is the discounted cumulative reward of the trajectory τ. The probability of the trajectory, $p_\theta(\tau)$, is

$$p_\theta(\tau) = p(S_0) \prod_{t=0}^{T} \pi_\theta(A_t|S_t)p(S_{t+1}|S_t, A_t), \tag{1.60}$$

where $p(S_0)$ is the distribution of the initial state, and $p(S_{t+1}|S_t, A_t)$ is the transition probability. π_θ is the policy parameterized by θ. The gradient of the policy is generally formed as

$$\nabla J(\pi_\theta) = \mathbb{E}_\tau \left[\sum_{t=0}^{T} \nabla \log \pi_\theta(A_t|S_t)A_{\pi_\theta}(S_t, A_t) \right], \tag{1.61}$$

where $A_{\pi_\theta}(S_t, A_t)$ is the advantage function to evaluate whether the action A_t is good or bad compared with other actions relative to the current policy, i.e.,

$$A_{\pi_\theta}(S_t, A_t) = Q_w(S_t, A_t) - V_\phi(S_t), \tag{1.62}$$

where Q_w is the action-value function defined in (1.57), and $V_\phi(s) \approx v_{\pi_\theta}(s)$ is the state-value function with ϕ denoting the trainable parameters. However, this advantage function may introduce high bias in implementation. Therefore, the generalized advantage estimation (GAE) function is utilized to refine it, which is written as

$$\hat{A}_{\pi_\theta}(S_t, A_t) = \delta_t + \gamma\lambda\delta_{t+1} + \cdots + (\gamma\lambda)^{T-t+1}\delta_{T-1}, \tag{1.63}$$

where λ is the GAE parameter. δ_t is the temporal difference (TD) error, which is written as

$$\delta_t = R_{t+1} + \gamma V_\phi(S_{t+1}) - V_\phi(S_t). \tag{1.64}$$

$V_\phi(S_t)$ is the state value of S_t, and it can be approximated by a critic network.

In addition, importance sampling is adopted in the PPO algorithm to improve the sampling efficiency. Let $p(x)$ be a probability density for a random variable X, and suppose we need to compute an expectation, $\mathbb{E}_p[f(x)]$, with only the data sampled from another probability density $q(x)$ available; then we have

$$\mathbb{E}_p[f(x)] = \int f(x)p(x)dx = \int f(x)\frac{p(x)}{q(x)}q(x)dx = \mathbb{E}_q \left[\frac{p(x)}{q(x)}f(x) \right]. \tag{1.65}$$

Along this line of thought, we can use the data collected by a sampling policy, $\pi_{\theta'}$, to update the target policy, π_θ, many times to improve the data efficiency. The gradient in (1.61) is turned into

$$\nabla J(\pi_\theta) = \mathbb{E}_{(S_t, A_t)\sim\pi_{\theta'}} [r_t(\theta)\hat{A}_{\pi_\theta}(S_t, A_t)], \tag{1.66}$$

where $r_t(\theta)$ is the probability ratio of the target policy to the sampling policy, i.e.,

$$r_t(\theta) = \frac{\pi_\theta(A_t|S_t)}{\pi_{\theta'}(A_t|S_t)}. \tag{1.67}$$

However, directly maximizing (1.66) without a constraint will lead to an excessively large policy update. Therefore, PPO clips the objective to limit the magnitude of gradient updates to stabilize the training process. The objective function of training the actor-network is

$$\mathcal{L}^a(\theta) = \mathbb{E}_t[\min(r_t(\theta)\hat{A}_{\pi_\theta}(S_t, A_t), \mathrm{clip}(r_t(\theta), 1 - \epsilon, 1 + \epsilon)\hat{A}_{\pi_\theta}(S_t, A_t))], \qquad (1.68)$$

where $\mathrm{clip}(\cdot, 1 - \epsilon, 1 + \epsilon)$ denotes the clip function, and ϵ is a parameter representing the clip range. The objective function of training the critic network is to minimize

$$\mathcal{L}^c(\phi) = \mathbb{E}_t[(R_{t+1} + \gamma V_\phi(S_{t+1}) - V_\phi(S_t))^2]. \qquad (1.69)$$

1.2.4 Federated Learning

Federated learning is a machine learning paradigm that enables multiple devices to collaboratively train a high-quality global model while retaining their data locally [45, 46]. This is achieved under the coordination of a central parameter server that orchestrates the training process. In each training round, participating devices download a global model from the parameter server and compute an update to the current model using data stored locally. Each device then communicates these updates to the central server, which are aggregated to produce an updated global model.

Federated learning emerges as a result of the increasing abundance of data generated on user devices from, e.g., users' interaction with mobile applications, and the growing computing power of mobile devices. This is further reinforced by the urgent need to protect data privacy, which makes unrestricted access to large collections of datasets ever more difficult. That said, federated learning faces a range of unique challenges, including the huge number of devices; highly unbalanced, non-identical, and independently distributed data available on each device; and the potentially slow and unreliable communication between participating devices and the central server. As a result, numerous research efforts have been dedicated to addressing these issues in recent years, ranging from communication-efficient model update and sharing design to client selection strategy optimization and development of federated learning mechanisms that are robust to stragglers or device absence. More details of federated learning will be provided in Chapter 6.

1.2.5 Other Machine Learning Techniques

Several other important machine learning methods, such as approximate inference and meta-learning, prove useful in wireless communication and network design. Here we provide a brief introduction to them, while a more detailed exposition will be presented in Chapter 3 and Chapter 5 where these methods become relevant.

Approximate inference [26], widely used in Bayesian statistical learning, aims to find a computationally feasible way to approximately evaluate the posterior distribution of some latent variables in a probabilistic graphical model. Approximate inference broadly falls into two classes: stochastic and deterministic approximations. Typical stochastic approximation techniques include Markov chain Monte Carlo (MCMC) sampling, which provides a principled approach to generating samples following the desired posterior distribution. Meanwhile, deterministic approximation methods resort to analytic approximations to the posterior distribution, with variational inference and expectation propagation as typical examples. We will defer until Chapter 3 the discussion of such techniques for signal detection in MIMO communication receivers.

Meta-learning, also known as learning to learn, is an approach to transfer learning that explicitly optimizes transferability and fast learning. It trains a model that quickly adapts to a new task using only a few data points or training iterations. This is accomplished by pre-training the model on a set of tasks, in the so-called meta-training phase, to learn the structure of these tasks such that the learned prior can be combined with small amounts of data from the new task to make generalizable inferences [47]. We will discuss in Chapter 5 how to use meta-learning to develop an intelligent resource allocation scheme that achieves remarkable performance while demonstrating fast adaptation to different environments.

1.3 Exercises

Exercise 1.1 With the development of machine learning technology, its application in the field of wireless communication has gradually attracted attention. What potential applications do you see for machine learning technology in wireless communications? What challenges might machine learning face when applied to communication systems?

Exercise 1.2 Section 1.2.2 introduces several commonly used activation functions. Answer these questions.

(a) For hidden cells, why is the tanh activation function generally more efficient than the sigmoid activation function?
(b) Suppose you are building a binary classifier that identifies cucumbers (labeled as $y = 1$) and watermelons (labeled as $y = 0$). Which activation function would be most appropriate for the output layer, and why?
(c) Compute the derivative of the sigmoid function at $x = 0$, and prove that the derivative of the sigmoid function can be expressed in terms of the sigmoid function itself, i.e., $f'(x) = f(x)(1 - f(x))$.

Exercise 1.3 Answer the following questions regarding the use of CNNs in image processing.

(a) Suppose your input is a 300×300 color (RGB) image, and you are using a plain MLP. If the first hidden layer has 100 neurons, each fully connected to the input layer, how many parameters does this hidden layer have (including bias parameters)?

(b) Suppose your input is a 300×300 color (RGB) image, and you use a convolutional layer and 100 filters, each of which is 5×5 in size. How many parameters (including bias parameters) does this hidden layer have?

(c) Suppose you have an input of $63 \times 63 \times 16$ and convolve it with 32 filters of size 7×7, using a stride of two and no padding. What is the output?

Exercise 1.4 Answer the following questions about CNNs.

(a) Given the input data $\mathbf{X}$ and the filter $\mathbf{W}$ of one convolutional layer, write the outputs y_O when the stride is set as $S = 1$ and $S = 2$.

$$\mathbf{X} = \begin{bmatrix} 1.23 & 0.94 & 0.83 & 0.2 \\ 0.34 & 0.44 & 0.37 & 0.3 \\ 0.34 & 1.14 & 2.34 & 1.24 \\ 0.49 & 0.18 & 2.04 & 1.42 \end{bmatrix}, \quad \mathbf{W} = \begin{bmatrix} 1 & 0.5 \\ 0.3 & 1 \end{bmatrix}.$$

(b) To prevent the change of the data size, we use zero-padding and set $P = 1$. Recalculate the outputs y_O in (a). With the stride is set as $S = 1$, write the outputs when the y_O is processed by max-pooling and average-pooling.

(c) The size of input image is $3 \times 3 \times 3$, which consists of three 3×3 matrices. And there are two filters $\mathbf{W}_1$, $\mathbf{W}_2$ in this convolutional layer. Write the outputs when the stride is set as $S = 1$ and the zero-padding is set as $P = 1$.

$$\mathbf{R} = \begin{bmatrix} 123 & 94 & 83 \\ 34 & 44 & 37 \\ 34 & 114 & 234 \end{bmatrix}, \mathbf{G} = \begin{bmatrix} 123 & 94 & 20 \\ 11 & 30 & 22 \\ 12 & 40 & 23 \end{bmatrix}, \mathbf{B} = \begin{bmatrix} 123 & 94 & 83 \\ 34 & 44 & 37 \\ 34 & 114 & 234 \end{bmatrix},$$

$$\mathbf{W}_1 = \begin{bmatrix} 1 & 0.5 \\ 0.3 & 1 \end{bmatrix}, \mathbf{W}_2 = \begin{bmatrix} 0.5 & 1 \\ 1 & 0.3 \end{bmatrix}.$$

Exercise 1.5 Answer the following questions based on your understanding of RNNs and LSTM.

(a) What are the improvements of LSTM compared to ordinary RNNs?

(b) What is the function of the three gates in LSTM (input gate, forget gate, output gate)?

(c) Why does the gradient vanishing problem occur when training RNNs? How does LSTM solve the gradient vanishing problem? Explain briefly.

Exercise 1.6 When working with sequential data or natural language tasks, would you choose to use an attention mechanism or a CNN? What are their respective strengths and weaknesses?

Exercise 1.7 Answer the following questions based on your understanding of score-based generative models.

(a) What are the similarities and differences between GANs and score-based generative models?

(b) Derive the score shown in (1.41).

(c) Show the equivalence between the loss function in (1.40) and the explicit score matching objective in (1.39) for the perturbed data $\tilde{\mathbf{x}} \sim p_\sigma(\tilde{\mathbf{x}}) = \int p_\sigma(\tilde{\mathbf{x}}|\mathbf{x})p_{\text{data}}(\mathbf{x})d\mathbf{x}$. You can refer to [37] for additional information.

Exercise 1.8   Greedy Play supposes that the reinforcement learning player was greedy, that is, it always played the move that brought it to the position that it rated the best. Might it learn to play better, or worse, than a non-greedy player? What problems might occur?

Exercise 1.9   Answer the following questions:

(a) What is the difference between on-policy algorithms and off-policy algorithms in reinforcement learning?
(b) What is the difference between value-based algorithms and policy-based algorithms in reinforcement learning?
(c) What is the difference between deterministic methods and stochastic methods in reinforcement learning?
(d) Determine the categories of reinforcement learning algorithms below. For example, DQN [40] is a value-based off-policy deterministic reinforcement learning algorithm.

   1) REINFORCE.
   2) Deep deterministic policy gradient (DDPG) [48].
   3) Trust-region policy optimization (TRPO) [41].
   4) PPO [44].
   5) Soft actor-critic (SAC) [49].
   6) Twin-delayed DDPG (TD3) [50].

Exercise 1.10   Consider a continuing task without the terminal state. The agent operates in an environment with a discount factor $\gamma \in (0, 1)$.

(a) Are the signs of these rewards important, or are only the intervals between them essential? Prove that adding a constant b to all the rewards will add a constant v_b to the value functions of all states and thus does not affect the comparisons between any states under any policies. What is v_b in terms of b and γ?
(b) Now consider adding a constant b to all the rewards in an episodic task, such as maze navigation. Would this have any effect on the value function of each state, or would it leave the task unchanged as in the continuing task above? Why or why not? Give an example.

Exercise 1.11   Why is Q-learning considered an off-policy control method? Supposing action selection is greedy; is Q-learning then exactly the same algorithm as SARSA? Will they make exactly the same action selections and action-value function updates?

Exercise 1.12   MountainCar-v0 is a classic reinforcement learning task in the OpenAI Gym library. In this task, an underpowered car must drive up a mountain to reach the goal at the top. The car cannot reach the goal by driving directly up the hill due to insufficient engine power. Instead, it must build momentum by going back and forth between two hills. The state space includes the car's position and velocity, and the action space consists of three actions: push left, push right, or do nothing. The task is considered solved when the agent reaches the flag on top of the hill within a certain number of steps. The code snippet has been provided in https://github.com/le-liang/wcmlbook/blob/main/ch1/Exercise_1.12/PPO_MountainCar-v0.py. Fill in the blanks and run the program to implement the PPO algorithm for solving this task.

2   Channel Modeling, Estimation, and Compression

In this chapter, we will first examine the use of machine learning for channel modeling. In particular, we elaborate on using generative models, such as generative adversarial networks (GANs), to capture the complex wireless channel effects and quickly produce channel parameters that resemble measurement data. The knowledge of channel information is important for receiver design and for transmitters to adapt transmission parameters. We then discuss in detail the use of machine learning methods to estimate the channel in orthogonal frequency division multiplexing (OFDM) systems, after explaining how such channel estimation is performed traditionally. For frequency division duplex systems, timely and accurate channel feedback is critical for link adaptation at the transmitter. It is even more so for multiple-input multiple-output (MIMO) systems, where channel state information (CSI) plays a defining role in designing pre-coding or beamforming matrices to realize the potential gain of multiple antennas. We will study how to leverage machine learning, in particular deep learning, to achieve efficient compression of the CSI to reduce the feedback overhead and meanwhile attain desirable reconstruction.

2.1   Wireless Channel Modeling with Machine Learning

We first introduce the characterization of wireless channels, where important channel modeling concepts are discussed. We then present wireless channel modeling methods using generative models in machine learning.

2.1.1   Wireless Channel Characterization

Wireless signals propagate from the transmitter to the receiver through different paths, each with different time-varying propagation delays, amplitude attenuation, and phase rotations. These multipath components may then add constructively or destructively at the receiver, leading to the channel fading effect. This can be roughly divided into two major types, i.e., large-scale and small-scale fading, depending on their impact distances or periods.

One component of the large-scale fading effect is path loss, which characterizes the average power reduction of a radio signal due to space loss, absorption, diffraction, etc. The other component is shadowing, caused by obstacles between the transmitter and the receiver attenuating signal power through absorption, reflection, and scattering, to name a few. The combined effects of path loss and shadowing can be characterized in the following simplified model [51, 52]:

$$\frac{P_t}{P_r}\,\mathrm{dB} = 10\log_{10} K + 10\gamma \log_{10} \frac{d}{d_0} + \phi_{\mathrm{dB}}, \qquad (2.1)$$

where P_t and P_r denote the transmitted power and the average received power in decibels (dB), respectively, K is the power attenuation at the reference distance d_0, γ is called the path loss exponent, and d is the distance between the transmitter and receiver antennas. ϕ_{dB} in dB denotes the shadowing effects and follows a Gaussian distribution with mean zero and variance $\sigma^2_{\phi_{\mathrm{dB}}}$. Note that this is also commonly known as log-normal shadowing due to the fact that the shadowing is Gaussian distributed after being converted to the dB scale, i.e., taking the logarithm of the original shadowing value.

Small-scale fading takes place over the distance on the order of the carrier wavelength due to the constructive and destructive effects of multiple time-varying delayed replicas of the original signals received from different paths between the transmitter and the receiver. It is characterized by the equivalent lowpass time-variant *channel impulse response*

$$h(\tau, t) = \sum_i \alpha_i e^{j\phi_i(t)} \delta(\tau - \tau_i) \qquad (2.2)$$

and

$$\phi_i(t) = \phi_i - 2\pi f_c \tau_i + 2\pi f_{D,i} t, \qquad (2.3)$$

where $\alpha_i, \tau_i, f_{D,i}$, and ϕ_i are the amplitude, delay, Doppler shift, and random phase offset, respectively, associated with the ith propagation path. For a more concise presentation, we define the time-variant complex amplitude of the ith path (or the sum of multiple non-resolvable multipath components with similar delays associated with the ith cluster) as $\hat{\alpha}_i(t) = \alpha_i e^{j\phi_i(t)}$, which is a wide-sense stationary (WSS) random process with $\mathbb{E}[|\hat{\alpha}_i(t)|^2] = \sigma_i^2$.

The channel power delay profile, specified by the delay, τ_i, and the average power, σ_i^2, of each path, provides an effective way to describe the channel's frequency selective characteristics. The *delay spread, τ_s,*[1] is a compact description of the delay dispersion of a channel and can be derived from the power delay profile by

$$\tau_s = \sqrt{\frac{\sum \sigma_i^2 (\tau_i - \tau_a)^2}{\sum \sigma_i^2}}, \qquad (2.4)$$

where $\tau_a = \sum \sigma_i^2 \tau_i / (\sum \sigma_i^2)$. The delay spread closely relates to the channel frequency selectivity. The time-variant channel frequency response $H(f; t)$ is defined as the Fourier transform of $h(\tau, t)$ with respect to τ. Then the *channel coherence bandwidth,*

---

[1] We consider the root mean square (RMS) delay spread here. For additional definitions of delay spread, interested readers may refer to [51].

B_c, over which $H(f; t)$ does not change appreciably, is inversely proportional to the delay spread, i.e., $B_c \approx c_0/\tau_s$ up to a constant c_0. A linearly modulated signal with bandwidth larger than B_c will experience non-negligible inter-symbol interference when transmitted through the channel, and sophisticated signal processing is then needed at the receiver to mitigate such interference and recover the transmitted signal. In this case, the channel is said to be *frequency-selective*. If the signal bandwidth is much smaller than B_c, the channel is referred to as a *flat fading* channel.

Conversely, the Doppler spectrum is frequently used to characterize the time-varying nature of wireless propagation channels. It is defined as the Fourier transform of $\mathbb{E}[\hat{\alpha}_i(t + \Delta t)\hat{\alpha}_i^*(t)]$, which is the autocorrelation function of the time-variant complex amplitude of the ith path (or the ith cluster of multipath components with similar delays). The Doppler spectrum characterizes the average power at the channel output as a function of the Doppler frequency. The *Doppler spread* is defined as the range of frequency over which the Doppler spectrum function is significant. The *channel coherence time*, T_c, indicates the duration over which the channel impulse response remains roughly constant. Similar to the relationship between the delay spread and coherent bandwidth, the Doppler spread and the channel coherence time are inversely proportional to each other. We usually refer to a channel as *fast fading* if T_c is much shorter than the delay requirement of the wireless application and *slow fading* if T_c is larger [53]. Over fast-fading channels, we can transmit the codewords across multiple realizations of channel fading to average out the fading effects, while over a slow-fading channel, we are not able to do so.

These parameters can be derived from the second-order statistics of the channel, i.e., the channel autocorrelation function, $A_c(\tau, \Delta t)$, the scattering function, $S_c(\tau, \rho)$, and the spaced-time spaced-frequency correlation function, $A_C(\Delta f, \Delta t)$. These three functions are related to each other through Fourier transform pairs, whose details can be found in standard wireless communication texts such as [51, 52]. The spaced-time spaced-frequency correlation function $A_C(\Delta f, \Delta t)$ describes the correlation of the channel in time and frequency, from which the channel coherence bandwidth B_c and coherence time T_c can be accurately defined. The channel autocorrelation function $A_c(\tau, \Delta t)$ provides the same temporal correlation data for the channel as the spaced-time spaced-frequency correlation function $A_C(\Delta f, \Delta t)$ and also provides its power delay profile, given by $A_c(\tau) = A_c(\tau, 0)$. The scattering function $S_c(\tau, \rho)$ gives the power delay profile and Doppler spectrum of the channel. Wireless channels are often assumed to be WSS, which implies that channel statistics do not vary over time. Additionally, channel response at different delays is often assumed to be uncorrelated, leading to uncorrelated scattering (US). Combining these two assumptions yields the widely used WSSUS channels, first introduced in [54].

2.1.2 Channel Modeling with Deep Generative Models

Conventional methods for Channel modeling can be categorized into deterministic and stochastic approaches. Deterministic channel modeling necessitates precise knowledge of communication environments and the governing electromagnetic wave propagation laws dictated by Maxwell's equations. Among these methods, ray tracing stands out as a prominent choice, which approximates the propagation of radio waves through solving simplified wave equations given the geometric and dielectric properties of the surrounding environment. Despite the superior precision offered by these methods, high computational complexity and dependence on the exact description of the environments make them impractical for use as a general modeling tool. Meanwhile, stochastic Channel modeling methods can provide the statistical characteristics, e.g., power delay profile, of different channels over time, frequency, and space. Stochastic methods can be categorized into two groups: geometry-based and non-geometric stochastic models. Geometry-based stochastic approaches generate the geometry (scatters) of the propagation environment according to certain distributions and then use the propagation law or ray tracing techniques to model the channel effects. The scatters can be placed in a one-ring, two-ring, ellipsis, or other shapes, making the approach fairly flexible for modeling different propagation environments. In comparison, non-geometric stochastic models directly adopt a stochastic approach to modeling the channel without assuming the underlying geometry. For example, in the widely used tapped-delay line (TDL) model, the channel is represented by a finite impulse response filter with a few taps, each associated with different delays and different types of Doppler spectra, as well as different amplitude statistics. It is important to note that stochastic methods can become exceedingly complex as the number of statistical variables grows, and they rely on certain mathematical assumptions about the model that may not align with real-world scenarios.

In contrast to complex conventional approaches, machine learning-enabled Channel modeling methods have gained increasing attention due to their generalizability and capacity to learn and extract underlying properties from observed data [55, 56]. Furthermore, machine learning-enabled methods can deal with more complicated communication environments, offering higher channel modeling accuracy. The advantage of machine learning-based channel models compared to conventional approaches is mainly derived from efficiently using the tremendous amount of channel data. Hence, if the training data or the measurement data is insufficient, the accuracy of the trained channel model becomes constrained. Deep generative methods are utilized to generate synthetic data capable of mimicking the training data, addressing the limitations of the measurement dataset. Channel modeling with deep generative models aims to formulate models that capture the statistical distribution of the limited measurement data and generate new samples that are representative of the communication environment. Several examples are provided below to demonstrate how deep generative methods model the channels.

Channel Modeling Based on GANs

Based on the concept of GANs described in Section 1.2.2 of Chapter 1, a learning-based Channel modeling method will be discussed, which constitutes three parts: real channel samples, a channel data generator G, and a channel data discriminator D. Real channel samples are obtained from measurement campaigns, encompassing an array of channel characteristics. The channel data generator tries to generate fake samples by utilizing a noise vector z, and the channel data discriminator tries to distinguish between the real and fake samples. The generator and discriminator are concurrently trained following the loss functions defined in (1.37) and (1.38).

In the first iteration of training, the generated samples are close to noise, and it is easy for the discriminator to distinguish them. As the adversarial learning between these two components progresses, the discriminator becomes smarter and more cautious, while the generator continually improves its ability to produce fake samples that closely resemble the real data. The training stops when the discriminator can no longer distinguish between fake and real samples. At this point, the channel data generator learns the distribution of the real channel samples, which is then extracted as the target channel model requiring no domain-specific knowledge. As the real channel dataset contains the channel coefficients, then after training, the GANs-based model can directly generate coefficients that are typical of the channel used for collecting the training dataset, as discussed in detail in [57] and [58].

In addition, the problem of modeling more complicated propagation channels can also be tackled efficiently via GANs-based methods. In the context of a single-input single-output (SISO) transmission scenario, we consider the channel's two-dimensional time–frequency response. The channel can be represented by a matrix $\mathbf{H} \in \mathbb{C}^{m \times n}$, where m corresponds to the number of subcarriers (or subchannels), and n is the number of time slots. Each value in the channel matrix captures the effect of the channel in a particular frequency and time slot. The real channel samples can be regarded as channel images. A deep convolutional GAN is employed as the channel data generator and discriminator due to its ability to deal with image data. The generator is trained to learn the distribution of channel images and generate similar samples. The discriminator's role is to distinguish between real channel images and the ones generated by the generator. Both of these networks are trained using the loss function (1.37–1.38). This method can also be extended to more complicated scenarios. For example, the effect of the user speed is considered by using a conditional GAN in [59], and the generator generates the channel images conditioned on the velocity of the users.

Recently, GAN-based Channel modeling methods have been adopted to model the MIMO channels. Consider a MIMO system with N_t antennas at the transmitter and N_r antennas at the receiver. The real channel in the time domain can be represented as $\tilde{\mathbf{H}} \in \mathbb{C}^{N_r \times N_t \times N_d}$, where N_d denotes the number of delay paths. To apply GAN-based methods, the real channel samples are reformulated to $\mathbf{H} \in \mathbb{R}^{N_t N_r \times N_d \times 2}$ with normalized amplitude, where the third dimension means the real and imaginary parts of the original complex channel are separated. A small real channel dataset $\mathcal{H} = \{\mathbf{H} | \mathbf{H} \sim p_r\}$ with distribution p_r is used, and the channel data generator aims to generate a fake channel dataset $\hat{\mathcal{H}} = \{\hat{\mathbf{H}} | \hat{\mathbf{H}} \sim p_f\}$ with distribution p_f.

As MIMO channels are more complex than SISO ones, we employ the Wasserstein GAN with gradient penalty (WGAN-GP) [60] to better learn the channel distribution. The Wasserstein GAN is an enhanced version of GANs designed to address the issues of mode collapse and training instability that are often encountered with traditional GANs. It incorporates the Wasserstein distance into the loss function, which effectively measures the distance between two distributions. Simulation results in [60] demonstrate that the Wasserstein GAN allows the generator to learn complex distributions more stably and efficiently. As the MIMO channel matrices tend to be large and contain complex features, the channel data generator and discriminator need to have wide, dense layers and deep convolutional layers to deal with these data. The generator maps a noise vector $\mathbf{z}$ sampled from $p(\mathbf{z})$ to a fake channel $\hat{\mathbf{H}}$, i.e., $\hat{\mathbf{H}} = G(\mathbf{z})$, while the discriminator aims to distinguish the generated fake channels from the real channels. The Wasserstein GAN aims to learn a 1-Lipschitz function with mild gradients so the training stability can be improved [60]. Alongside, a gradient penalty is adopted to enforce the Lipschitz constraint. The minimax objective function of the whole GAN-based framework can be written as

$$\min_{G} \max_{D} \mathbb{E}_{\mathbf{H}\sim p_{\mathrm{r}}}[D(\mathbf{H})] - \mathbb{E}_{\mathbf{z}\sim p(\mathbf{z})}[D(G(\mathbf{z}))] - \lambda\mathbb{E}_{\bar{\mathbf{H}}\sim p_{\mathrm{s}}}[(\||\nabla_{\bar{\mathbf{H}}}D(\bar{\mathbf{H}})\||_2 - 1)^2], \qquad (2.5)$$

where λ is a positive weight. $\bar{\mathbf{H}} = \epsilon\mathbf{H} + (1 - \epsilon)G(\mathbf{z})$ represents the interpolation between the real and generated data samples, with ϵ following the uniform distribution within $(0, 1)$. $\||\nabla_{\bar{\mathbf{H}}}D(\bar{\mathbf{H}})\||_2$ represents the Euclidean norm of the gradient of the discriminator with respect to the interpolated sample. $\mathbb{E}_{\mathbf{H}\sim p_{\mathrm{r}}}[D(\mathbf{H})] - \mathbb{E}_{\mathbf{z}\sim p(\mathbf{z})}[D(G(\mathbf{z}))]$ measures the difference between the generated distribution and the real distribution using the Wasserstein distance. $-\lambda\mathbb{E}_{\bar{\mathbf{H}}\sim p_{\mathrm{s}}}[(\||\nabla_{\bar{\mathbf{H}}}D(\bar{\mathbf{H}})\||_2 - 1)^2]$ is the gradient penalty term and is used to enforce that the gradient norm of the discriminator at the interpolated samples is close to 1. During training, the distance between the distribution of the fake channel, p_f, and the real channel, p_r, will gradually decrease. More details can be found in [61].

The GAN-based approach can also be used to implicitly model a channel by mimicking its input–output relation. For example, in an additive white Gaussian noise (AWGN) channel, the output of the channel $\mathbf{y}$ is the sum of the input signal $\mathbf{x}$ and Gaussian noise $\mathbf{w}$, i.e., $\mathbf{y} = \mathbf{x} + \mathbf{w}$. A conditional GAN can be applied to model this channel output, where the conditioning information includes the transmitted signal $\mathbf{x}$ and the received pilot signal $\mathbf{y}_p$. By doing so, the generated samples can match the distribution of the channel output $\mathbf{y}$ given the input signal $\mathbf{x}$ and received pilot data $\mathbf{y}_p$. We will explore the details of this approach in Chapter 4.

Example 2.1 Here, we choose a flat-fading channel to illustrate the effectiveness of the GAN-based Channel modeling method. A large number of real samples are prepared to train the GAN, and the key parameters of the generator G and discriminator D are shown in Table 2.1. The activation function of these two networks is ReLU, and the size of a minibatch is set as 32.

The envelope of the channel response is modeled as a Gaussian distribution. The GAN framework is trained with the real samples generated from a Gaussian distribution with a mean of 2 and a standard deviation of 0.5, i.e., $\mathcal{N}(2, 0.5)$. We compare the probability density function of the generated samples with that of the true channel. Figure 2.1a shows the distributions of synthetic and real data at the beginning of the training process, revealing notable disparities between them.

Table 2.1 Key parameters of the GAN.

	Generator G	Discriminator D
Number of hidden layers	2	3
Neuron number of each hidden layer	50	100
Learning rate	1e−3	1e−2
Learning rate decay	1e−5	1e−4

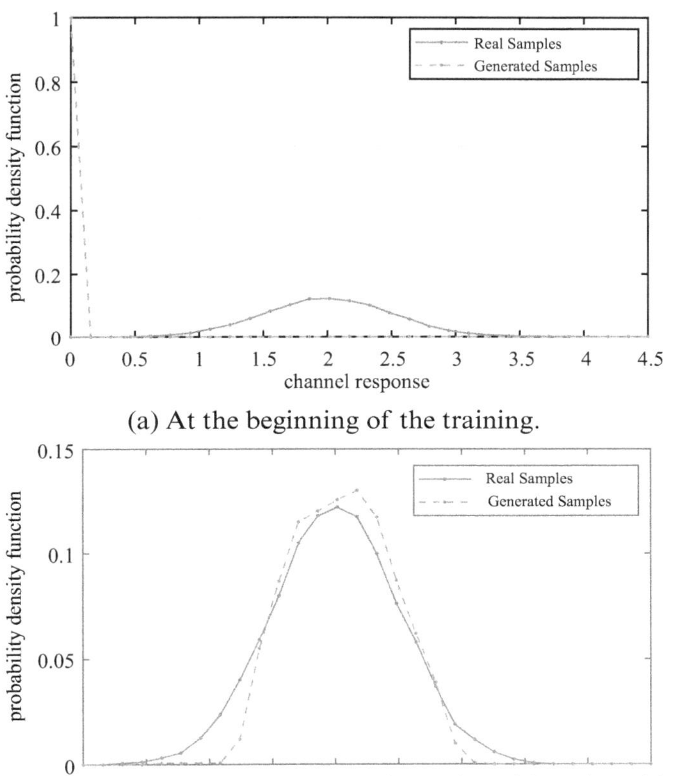

(a) At the beginning of the training.

(b) After training.

Figure 2.1 The probability density functions of the real channel samples and the generated samples.

After training with a sufficient number of iterations, the channel discriminator fails to distinguish the real samples from the generated ones. At this point, we can use the channel data generator to approximate the true channel. The two distributions are presented in Figure 2.1b. As illustrated in the figure, the GAN-based method provides a commendable approximation of the real channel distribution without the need for expert knowledge or accurate information on the wireless communication environment.

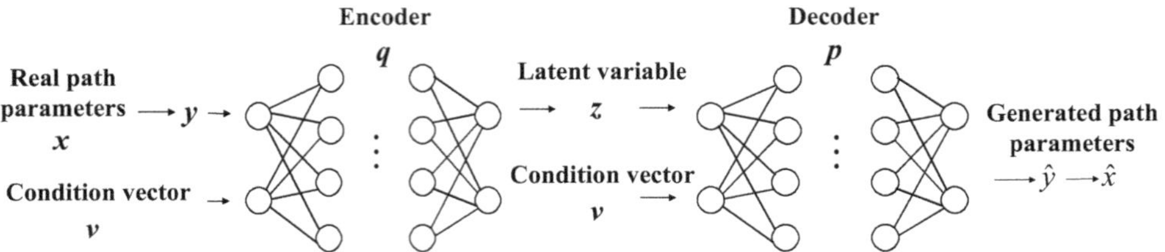

Figure 2.2 Illustration of CVAE-based wireless Channel modeling.

Channel Modeling Based on VAE

In addition to GANs, the variational autoencoder (VAE) is another important generative model that has been utilized for Channel modeling. Consider modeling channels in unmanned aerial vehicle (UAV) communication, where the transmitter is mounted on the UAV, and the base station is the receiver. A collection of parameters is used to describe each communication link denoted by $\mathbf{x} = \left\{(L_k, \tau_k, \phi_k^{\mathrm{rx}}, \theta_k^{\mathrm{rx}}, \phi_k^{\mathrm{tx}}, \theta_k^{\mathrm{tx}})\right\}_{k \in \mathcal{K}}$, where $\mathcal{K} = \{1, \ldots, K\}$ denotes the set of paths, and K is the number of paths. L_k is the loss of the kth path, τ_k is its propagation delay, and $(\phi_k^{\mathrm{rx}}, \theta_k^{\mathrm{rx}})$ and $(\phi_k^{\mathrm{tx}}, \theta_k^{\mathrm{tx}})$ are its azimuth and elevation angles of arrival and departure, respectively.

The framework of a conditional VAE (CVAE)-based Channel modeling method is shown in Figure 2.2. $\mathbf{v}$ is the condition vector, which includes some necessary information for Channel modeling, such as the type of gNB, the link state, and the distance between UAVs and gNBs. The goal is to capture the conditional distribution $p(\mathbf{x}|\mathbf{v})$. Notably, the parameters, $\mathbf{x}$, are heterogeneous, containing different types of parameters, such as path losses, angles, and delays. Therefore, some transformations are required to map the $\mathbf{x}$ onto $\mathbf{y}$, ensuring that all parameters are in a similar range, of which the details can be found in [62].

According to the framework of a standard VAE, the decoder, p, accepts latent vector $\mathbf{z}$ along with the condition vector, $\mathbf{v}$, as input, and it outputs means and variances for the estimated parameters, $\hat{\mathbf{y}}$, which can be denoted by

$$[\boldsymbol{\mu}_{\hat{y}}, \sigma_{\hat{y}}^2] = p(\mathbf{v}, \mathbf{z}). \tag{2.6}$$

The estimated parameters, $\hat{\mathbf{y}}$, are sampled from the means and variances,

$$\hat{\mathbf{y}} = \boldsymbol{\mu}_{\hat{y}} + \boldsymbol{\sigma}_{\hat{y}} \odot \boldsymbol{\eta}, \tag{2.7}$$

where η is a zero-mean, unit-variance, independent and identically distributed (i.i.d.) Gaussian vector, and $\odot$ indicates entry-wise multiplication. The number of entries in η equals the dimension of $\hat{\mathbf{y}}$. Once $\hat{\mathbf{y}}$ has been obtained, the transformation should be undone to recover the parameters, $\hat{\mathbf{x}}$.

The VAE also requires training an encoder, q, that maps real parameters to the latent vector, $\mathbf{z}$. The encoder aims to approximate the posterior probability of $\mathbf{z}$ given $\mathbf{y}$ and $\mathbf{v}$, which can be represented as

$$[\boldsymbol{\mu}_z, \sigma_z^2] = q(\mathbf{v}, \mathbf{y}), \tag{2.8}$$

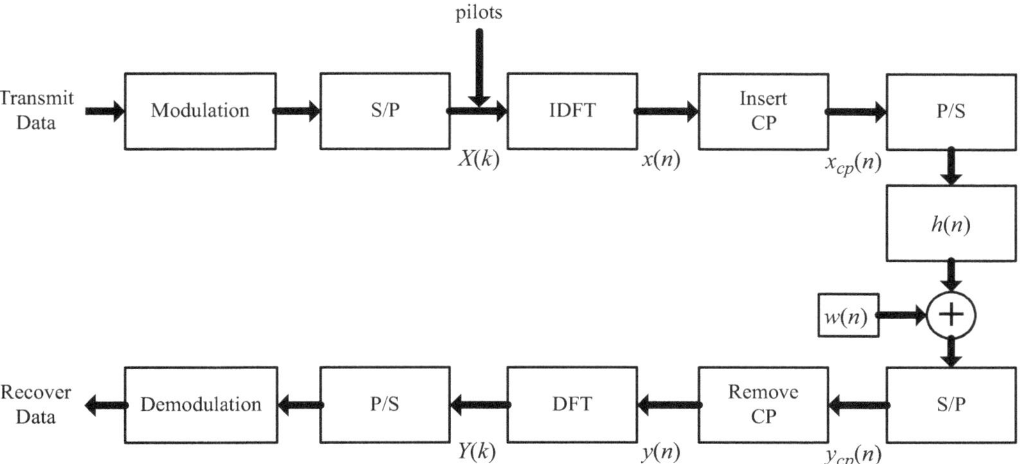

Figure 2.3 Block diagram of OFDM system.

where $\boldsymbol{\mu}_z$ and $\boldsymbol{\sigma}_z^2$ have the same dimensions as the latent vector $\mathbf{z}$. Given these outputs, the latent vector can be sampled from the approximate posterior probability by

$$\mathbf{z} = \boldsymbol{\mu}_z + \boldsymbol{\sigma}_z \odot \boldsymbol{\epsilon}, \tag{2.9}$$

where $\boldsymbol{\epsilon}$ is also a zero-mean, unit-variance, i.i.d. Gaussian vector. The encoder and decoder are jointly trained to maximize an approximation of the log-likelihood objective function, which is termed the evidence lower bound (ELBO) [63],

$$\max_{p,q} \frac{1}{N} \sum_i \log p(\mathbf{y}_i|\boldsymbol{\mu}_z(\mathbf{y}_i,\mathbf{v}_i) + \boldsymbol{\epsilon}\boldsymbol{\sigma}_z(\mathbf{y}_i,\mathbf{v}_i),\mathbf{v}_i) - D_{\mathrm{KL}}(q(\mathbf{z}|\mathbf{y}_i,\mathbf{v}_i)\|p(\mathbf{z})), \tag{2.10}$$

where $p(\mathbf{z})$ is the prior distribution of $\mathbf{z}$, which is often chosen to be a Gaussian distribution. N indicates the batch size in training.

Once trained, the CVAE-based models can be used conveniently. When the locations of UAVs and gNBs are known, the condition vector, $\mathbf{v}$, for each link can be provided. Random vectors $\mathbf{z}$ will then be sampled from the prior distribution $p(\mathbf{z})$. The decoder can then generate the parameters, $\mathbf{x}$, with the $\mathbf{v}$ and $\mathbf{z}$, following a distribution that closely approximates the actual one. In addition, modeling the small-scale dynamics of the channels is also feasible in this VAE-based framework. For example, Doppler shifts can be calculated when the velocities of the UAVs are given, based on which the CVAE-based model can derive the time-varying frequency response for each path.

2.2   Channel Estimation in OFDM: Traditional Methods

A block diagram of an OFDM communication system is shown in Figure 2.3. At the transmitter side, the input data stream is first modulated into complex symbols that are then passed through a serial-to-parallel (S/P) converter. The output is processed by an inverse discrete Fourier transform (IDFT) to generate time domain OFDM symbols $x(n)$:

$$x(n) = \frac{1}{\sqrt{N}} \sum_{k=0}^{N-1} X(k) \exp\left(j2\pi kn/N\right), 0 \leq n \leq N - 1, \tag{2.11}$$

where N is the number of subcarriers, or equivalently the size of the IDFT, and $X(k)$ is the modulated complex symbol stream, including training pilots, if any. To counter multipath fading, a cyclic prefix is added to the OFDM symbols, where the last N_{cp} samples of $x(n)$ are copied and put at the front of the OFDM symbol as the prefix, and the resulting time samples are passed through a parallel-to-serial (P/S) converter. The output then goes through a digital-to-analog converter to generate the baseband OFDM signal, to be upconverted to a carrier frequency and sent to the wireless channel.

After the signal reaches the receiver through the fading channel, it is downconverted to the baseband and then sampled and digitized by the analog-to-digital converter. These samples are serial-to-parallel converted, after which the prefix corresponding to the first N_{cp} samples is removed to yield the time samples $y(n)$:

$$y(n) = x(n) \otimes h(n) + z(n), 0 \leq n \leq N - 1, \tag{2.12}$$

where $\otimes$ denotes cyclic convolution, $h(n)$ is the discrete-time equivalent lowpass channel impulse response, and $z(n)$ is additive white Gaussian noise. These time samples $y(n)$ are converted to the frequency domain by performing a discrete Fourier transform (DFT), effectively leading to

$$Y(k) = X(k) * H(k) + Z(k), 0 \leq k \leq N - 1, \tag{2.13}$$

where $H(k)$ and $Z(k)$ are the frequency-domain samples of $h(n)$ and $z(n)$, respectively. The frequency samples $Y(k)$ are parallel-to-serial converted and passed through a demodulation stage to recover the transmitted data, using one-tap frequency equalization.

It is noted that the receiver mirrors all major processing blocks at the transmitter side, except at the demodulation phase, where knowledge of the channel $H(k)$, or equivalently $h(n)$, is needed for channel equalization. The process of using training pilots to estimate the $H(k)$, or $h(n)$, is exactly what a channel estimator does, and its performance largely determines the communication system's performance. In the following, we will discuss conventional channel estimation methods for OFDM systems in detail, paving the way for learning-based channel estimation to be covered in the subsequent sections of this chapter.

2.2.1 Channel Estimation for OFDM Systems

Depending on whether training pilots are used, channel estimation methods fall broadly into three categories: pilot-based, blind, and semi-blind estimation. Here

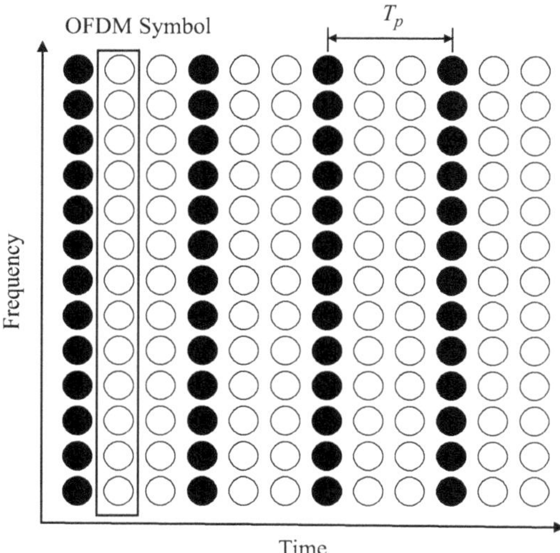

Figure 2.4 Block pilots for channel estimation in OFDM systems.

we focus on the pilot-based scheme for OFDM channel estimation, where a known sequence of training pilots is inserted into a portion of the subcarriers of some OFDM symbols. This is best visualized in an OFDM time–frequency grid, as illustrated in Figure 2.4, where we use black dots and empty circles to denote the pilot and data subcarriers in an OFDM symbol, respectively. The channels of these pilot subcarriers are estimated based on the received signal and known transmitted pilots. Interpolation algorithms can then be used to obtain the channels of the data subcarriers. We start with a brief introduction to three major types of pilot formats: block, comb, and lattice pilots.

Pilot Structure

Figure 2.4 shows the format of block pilots, where all subcarriers of some OFDM symbols, i.e., pilot symbols, are used as pilots whose channels can be estimated with relatively high accuracy. The intermediate symbols carry data, and their channels will be estimated through, e.g., time domain interpolation. For reliable performance, the pilot placement interval in time, T_p, is expected to be shorter than the channel coherence time, T_c, i.e., $T_p \leq T_c$. This could lead to very dense pilot placement in communication environments with high mobility, where the channels change rapidly in time. That said, block pilots work best for channels with high frequency selectivity due to their maximum frequency domain pilot placement. In addition to channel estimation based on interpolation, the decision-directed method is also used in practice; however, this is beyond our focus here and is hence ignored.

Figure 2.5 illustrates the placement of comb pilots. In contrast to the block pilot scheme, training pilots are now spread across all symbols in the time domain but

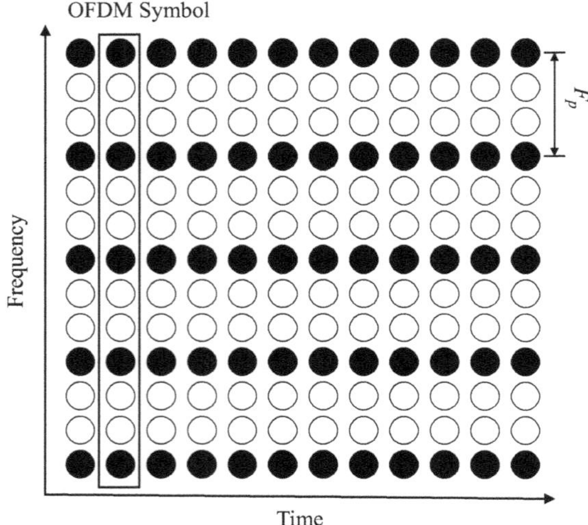

Figure 2.5 Comb pilots for channel estimation in OFDM systems.

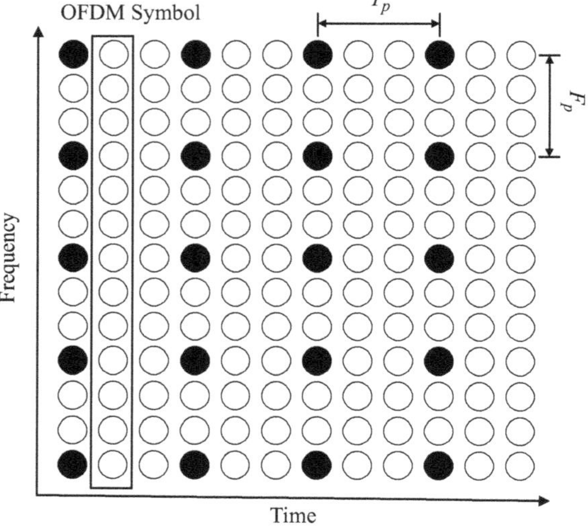

Figure 2.6 Lattice pilots for channel estimation in OFDM systems.

only on a few selected OFDM subcarriers periodically, resembling a "comb," hence the name. Whereas comb pilots perform well in tracking channels changing rapidly in time, extra care is needed in determining the pilot density in the frequency domain. Specifically, the pilot placement interval in frequency, F_p, must be less than the channel coherence bandwidth, B_c, to obtain reliable channel estimation performance, i.e., $F_p \leq B_c$.

The format of lattice pilots is shown in Figure 2.6, where pilots are scattered across the time–frequency grid. Let the period of inserting pilots along the time axis be T_p

and the period of inserting pilots along the frequency axis be F_p. Following similar arguments from the block and comb pilots, we would require $T_p \leq T_c$ and $F_p \leq B_c$ to be satisfied simultaneously in order to reliably track channel variations in both the time and frequency domains. The lattice pilot placement minimizes the use of training pilots and leaves more time–frequency resources for data delivery in OFDM systems. However, this also leads to more challenging channel estimation due to sparser pilot use and channel fading in both time and frequency.

LS and MMSE Channel Estimation

The most commonly used pilot-based channel estimation algorithms in OFDM systems are least squares (LS) and minimum mean square error (MMSE) methods [64, 65, 66]. Assuming the number of subcarriers carrying pilots in an OFDM symbol is N_P, we use $\mathbf{x}_P \in \mathbb{C}^{N_P \times 1}$, $\mathbf{y}_P \in \mathbb{C}^{N_P \times 1}$, and $\mathbf{h} \in \mathbb{C}^{N_P \times 1}$ to denote the transmitted pilot signal, received pilot signal, and frequency-domain channel vector over each pilot subcarrier, respectively. We also denote the noise vector as $\mathbf{z}_P \in \mathbb{C}^{N_P \times 1}$, whose elements are usually assumed to be i.i.d. complex Gaussian variables with mean zero and variance σ_z^2. For ease of illustration, we introduce a diagonal matrix $\mathbf{X}_P \in \mathbb{C}^{N_P \times N_P}$, whose diagonal elements equal each entry of the vector $\mathbf{x}_P$.

From (2.13),

$$\mathbf{y}_P = \mathbf{X}_P \mathbf{h} + \mathbf{z}_P. \tag{2.14}$$

The LS estimator requires no prior knowledge about the channel vector and directly minimizes

$$J(\hat{\mathbf{h}}) = \left\| \mathbf{y}_P - \mathbf{X}_P \hat{\mathbf{h}} \right\|^2, \tag{2.15}$$

which gives the LS estimator

$$\hat{\mathbf{h}}_{\mathrm{LS}} = \mathbf{X}_P^{-1} \mathbf{y}_P. \tag{2.16}$$

Since the matrix $\mathbf{X}_P$ is diagonal, the LS estimator reduces to a series of simple scalar divisions of the received signal over the transmitted pilots, making it appealing for practical implementation. Despite its simplicity, LS estimation may suffer from noise enhancement.

The MMSE channel estimation method improves on LS by taking channel statistics into account and exploiting the correlation between the channels of different subcarriers. In particular, the channel vector $\mathbf{h}$ is treated as a random variable with known statistics, e.g., a complex Gaussian random vector with mean zero and covariance $\mathbf{R_h} = \mathbb{E}[\mathbf{h}\mathbf{h}^H]$. Then the (linear) MMSE estimator is designed to minimize the mean squared error (MSE)

$$J(\hat{\mathbf{h}}) = \mathbb{E}[\|\hat{\mathbf{h}} - \mathbf{h}\|^2], \tag{2.17}$$

which gives the MMSE estimator

$$\hat{\mathbf{h}}_{\mathrm{MMSE}} = \mathbf{R_h} \left(\mathbf{R_h} + \frac{1}{\gamma} \mathbf{I} \right)^{-1} \hat{\mathbf{h}}_{\mathrm{LS}}, \tag{2.18}$$

where $\gamma = \sigma_x^2/\sigma_z^2$ is the signal-to-noise ratio (SNR), with σ_x^2 denoting the pilot signal power and σ_z^2 denoting the variance of the additive Gaussian noise. **I** represents an $N_P \times N_P$ identity matrix. Whereas the MMSE estimator improves the channel estimation performance noticeably, the price to pay is increased complexity due to an expensive matrix inversion, in addition to obtaining channel statistics.

Once the channels of the pilot subcarriers are estimated using (2.16) or (2.18), we can obtain the channels of the data subcarriers through interpolation in the time or frequency domains, or both, depending on the adopted pilot placement pattern. A rich body of literature has been dedicated to this area, where a multitude of one- or two-dimensional interpolators with different performance–complexity tradeoffs have been thoroughly investigated [66, 67, 68].

Example 2.2 Please derive the MMSE estimator (2.18) by minimizing the MSE between the estimated and true channel vectors given in (2.17).

Solution
From equation (2.16), we have $\hat{\mathbf{h}}_{\mathrm{LS}} = \mathbf{X}_P^{-1}\mathbf{y}_P \triangleq \tilde{\mathbf{h}}$. Next, we introduce a weight matrix $\mathbf{W}$ and define the MMSE channel estimate as $\hat{\mathbf{h}}_{\mathrm{MMSE}} \triangleq \mathbf{W}\tilde{\mathbf{h}}$. According to the orthogonality principle, to minimize the MSE in equation (2.17), the estimation error vector $\hat{\mathbf{h}}_{\mathrm{MMSE}} - \mathbf{h}$ must be orthogonal to $\tilde{\mathbf{h}}$. Therefore, we have

$$
\begin{aligned}
\mathbb{E}\{(\hat{\mathbf{h}}_{\mathrm{MMSE}} - \mathbf{h})\tilde{\mathbf{h}}^H\} &= \mathbb{E}\{(\mathbf{W}\tilde{\mathbf{h}} - \mathbf{h})\tilde{\mathbf{h}}^H\} \\
&= \mathbf{W}\mathbb{E}\{\tilde{\mathbf{h}}\tilde{\mathbf{h}}^H\} - \mathbb{E}\{\mathbf{h}\tilde{\mathbf{h}}^H\} \\
&= \mathbf{W}\mathbf{R}_{\tilde{\mathbf{h}}\tilde{\mathbf{h}}} - \mathbf{R}_{\mathbf{h}\tilde{\mathbf{h}}} = \mathbf{0},
\end{aligned}
\tag{2.19}
$$

where

$$
\begin{aligned}
\mathbf{R}_{\tilde{\mathbf{h}}\tilde{\mathbf{h}}} &= \mathbb{E}\{\tilde{\mathbf{h}}\tilde{\mathbf{h}}^H\} \\
&= \mathbb{E}\{(\mathbf{X}_P^{-1}\mathbf{y}_P)(\mathbf{X}_P^{-1}\mathbf{y}_P)^H\} \\
&= \mathbb{E}\{(\mathbf{h} + \mathbf{X}_P^{-1}\mathbf{z}_P)(\mathbf{h} + \mathbf{X}_P^{-1}\mathbf{z}_P)^H\} \\
&= \mathbb{E}\{\mathbf{h}\mathbf{h}^H\} + \mathbb{E}\{\mathbf{X}_P^{-1}\mathbf{z}_P\mathbf{z}_P^H(\mathbf{X}_P^{-1})^H\} \\
&= \mathbf{R}_{\mathbf{h}} + \frac{\sigma_z^2}{\sigma_x^2}\mathbf{I}
\end{aligned}
\tag{2.20}
$$

and

$$
\begin{aligned}
\mathbf{R}_{\mathbf{h}\tilde{\mathbf{h}}} &= \mathbb{E}\{\mathbf{h}\tilde{\mathbf{h}}^H\} \\
&= \mathbb{E}\{\mathbf{h}(\mathbf{X}_P^{-1}\mathbf{y}_P)^H\} \\
&= \mathbb{E}\{\mathbf{h}(\mathbf{h} + \mathbf{X}_P^{-1}\mathbf{z}_P)^H\} \\
&= \mathbb{E}\{\mathbf{h}\mathbf{h}^H + \mathbf{h}\mathbf{z}_P^H(\mathbf{X}_P^{-1})^H\} \\
&= \mathbb{E}\{\mathbf{h}\mathbf{h}^H\} = \mathbf{R}_{\mathbf{h}}.
\end{aligned}
\tag{2.21}
$$

Solving (2.19) for $\mathbf{W}$ yields

$$
\mathbf{W} = \mathbf{R}_{\mathbf{h}\tilde{\mathbf{h}}}\mathbf{R}_{\tilde{\mathbf{h}}\tilde{\mathbf{h}}}^{-1}.
\tag{2.22}
$$

Therefore, the MMSE channel estimate is derived as

$$\hat{\mathbf{h}}_{\text{MMSE}} = \mathbf{W}\tilde{\mathbf{h}} = \mathbf{R}_{\mathbf{h}\tilde{\mathbf{h}}}\mathbf{R}_{\tilde{\mathbf{h}}\tilde{\mathbf{h}}}^{-1}\hat{\mathbf{h}}_{\text{LS}} = \mathbf{R}_{\mathbf{h}}\left(\mathbf{R}_{\mathbf{h}} + \frac{1}{\gamma}\mathbf{I}\right)^{-1}\hat{\mathbf{h}}_{\text{LS}}. \tag{2.23}$$

2.2.2 Channel Estimation for MIMO-OFDM Systems

MIMO-OFDM has been the bedrock of modern wireless communication systems due to its capability to improve system capacity and cope with frequency-selective fading, as well as being implementation-friendly. In effect, MIMO operates over each parallel flat fading channel (subcarrier) that has been decomposed from the wideband frequency-selective fading channels by OFDM, as shown in Figure 2.7. Consider a MIMO-OFDM system with N_t transmit antennas, N_r receive antennas, and K subcarriers. The signal on the kth subcarrier received at the jth receive antenna is given by

$$Y^{(j)}(k) = \sum_{i=1}^{N_t} H^{(j,i)}(k)X^{(i)}(k) + W^{(j)}(k), k = 1, \cdots, K, \tag{2.24}$$

where $X^{(i)}(k)$ denotes the signal transmitted from the ith transmit antenna, $H^{(j,i)}(k)$ denotes the frequency-domain channel from the ith transmit antenna to the jth receive antenna, and $W^{(j)}(k)$ is the white Gaussian noise received at the jth receive antenna with mean zero and variance σ_w^2.

From (2.24), the signal at the receive antenna on each subcarrier is a superposition of signals from N_t transmit antennas on that same subcarrier. If we use the same pilot pattern as discussed in the last section – e.g., the lattice pilots in Figure 2.6, for all transmit antennas – we would end up having to estimate $N_t N_P$ unknowns with only N_P observations. Recall that N_P is the number of pilot subcarriers in an OFDM symbol. While this is challenging, one may still obtain a reasonable solution by increasing the pilot density in the frequency domain and leveraging the MMSE algorithm for channel estimation, which exploits the inherent correlation in channels of different subcarriers and across different transmit antennas. Another solution is to estimate the channel impulse response $h^{(j,i)}(n)$ in the time domain using the

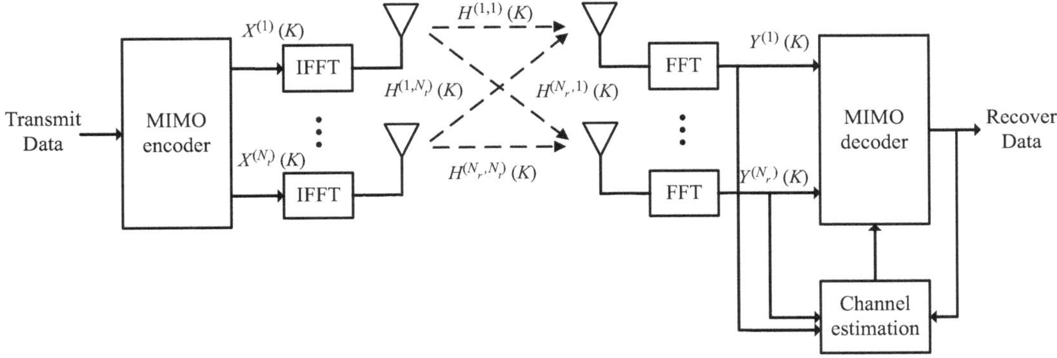

Figure 2.7 Illustration of MIMO-OFDM.

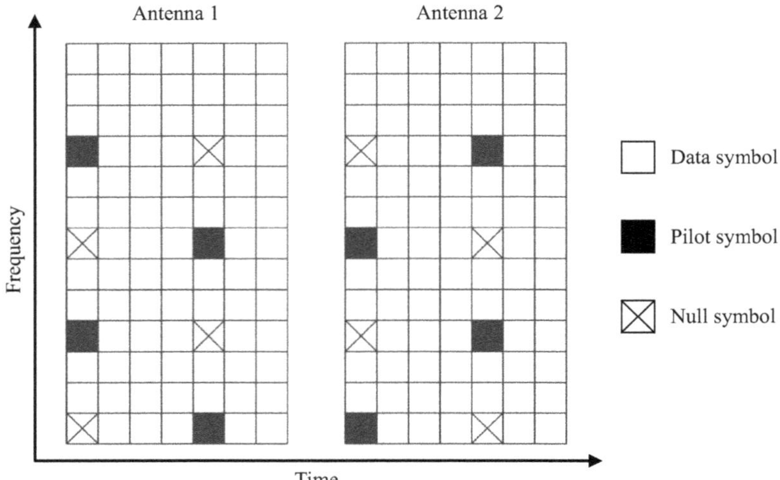

Figure 2.8 Pilot structure in the LTE system for two transmit antennas.

LS algorithm [69] since the significant channel taps, corresponding to the channel delay spread, are usually few in number, say L. Now we only need to estimate LN_P unknowns with the same N_P observations, making the problem much easier. Optimal training sequences for this estimation method that simplify the expensive matrix inversion by making it diagonal such that the computational complexity is significantly reduced are further developed in [70].

In commercial systems like Long-Term Evolution (LTE) and 5G cellular systems, pilot placement tends to be designed such that system implementation can be simplified. As a representative example, Figure 2.8 illustrates the pilot structure in LTE downlink transmission for two transmit antennas, where the pilots form a diamond shape as in [68], rather than a rectangular shape. In this case, different transmit antennas send pilot signals on non-overlapping subcarriers, thereby avoiding multi-antenna interference and effectively reducing the MIMO-OFDM system to a signal antenna OFDM system as discussed in the previous section. Then the LS and MMSE estimation methods in (2.16) and (2.18) can naturally be applied to estimate the channel for each transmit antenna, albeit with a reduced pilot density. Channels of data subcarriers can similarly be obtained through interpolation in the time and frequency domains.

It is noted that this pilot placement can be regarded as a special case of orthogonal pilot patterns for MIMO-OFDM. In particular, orthogonal pilot placement enables all transmit antennas to send signals on all pilot subcarriers but requires the transmitted pilot sequences to be orthogonal to each other. Each receive antenna then uses the signal received on all pilot subcarriers to estimate channels from the transmit antennas, possibly exploiting the property of pilot orthogonality. This is not much different from the other more sophisticated channel estimation approach we discussed before, which aims to estimate $N_t N_P$ unknown channels with signals received on N_P pilot subcarriers.

2.3 Learning-Based Channel Estimation: Data-Driven Methods

In this section, we study learning-based channel estimation methods that are primarily data driven. We first show how to use a simple fully connected deep neural network (DNN) for channel estimation in single-antenna, or single-input single-output (SISO), OFDM systems. We then present convolutional neural network (CNN)-based channel estimation for MIMO-OFDM that leverages space and frequency domain correlations.

2.3.1 DNN-Based Channel Estimation for SISO-OFDM Systems

Pilot-based channel estimation in OFDM systems estimates the unknown channel vector by periodically transmitting known training pilots. LS and MMSE estimation methods can be used to recover the channel vector based on the transmitted pilot, $\mathbf{x}_P$, and received signal, $\mathbf{y}_P$, according to (2.16) and (2.18), respectively. It is straightforward to show that with the same information, we can convert channel estimation to a supervised learning problem by treating $(\mathbf{x}_P, \mathbf{y}_P)$ as the input feature and the channel vector $\mathbf{h}$ as the training label. In particular, the training dataset can be constructed by generating a large number of channel realizations according to certain probability distributions and then passing the pilot signals through the generated channel to produce the output signal. The transmitted and received pilot signals and the channel vector for each pass are recorded as a training sample. The essence of a DNN-based channel estimator is then to learn a mapping from the transmitted and received pilot signal to the channel vector, and the mapping is optimized by adjusting parameters of the DNN to fit the dataset. This training stage is largely carried out offline, while in the online deployment phase, a forward pass through the DNN with the input of the current transmitted and received pilot signal yields the current channel estimate almost immediately.

An example DNN for channel estimation is proposed in [71], where the real and imaginary parts of the transmitted and received pilot signals, i.e., $\mathbf{x}_P$ and $\mathbf{y}_P$, are applied to a fully connected neural network, and the output, $\hat{\mathbf{h}}$, is the estimated channel vector. The neural network consists of two hidden layers with rectified linear unit (ReLU) activation. Training is performed by minimizing the quadratic loss between the estimated channel, $\hat{\mathbf{h}}$, and the true channel, $\mathbf{h}$, over a training set.

We show the MSE performance of the DNN-based channel estimation method compared against the linear MMSE (LMMSE) scheme in Figure 2.9. An OFDM system with 64 subcarriers and quadrature amplitude modulation (QAM) is simulated, where the first OFDM symbol contains quadrature phase shift keying (QPSK)-modulated pilot symbols over all subcarriers, and the second OFDM symbol places 64 data symbols with 64-QAM. The DNN is trained with SNRs assuming values of 5 to 40 dB. From the figure, the DNN-based method achieves a similar performance to the MMSE estimator that assumes the availability of the mean and covariance of the channel vector. Note that the gap between the two schemes in the low-SNR

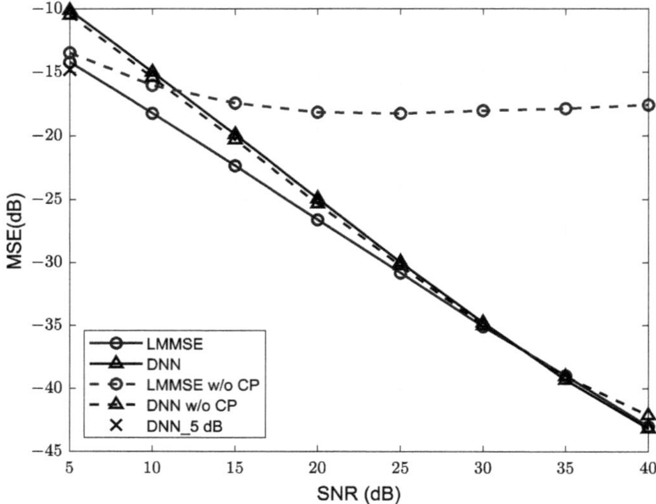

Figure 2.9 MSE performance of the DNN-based channel estimator and LMMSE in an OFDM system.

regime results from the discrepancy between the training and testing SNRs. To better illustrate the impact, we plot the performance at 5 dB (marked by "5 dB" in the figure), where such SNR discrepancy disappears and the DNN-based estimator even slightly outperforms the MMSE estimator. We also compare the performance of the two schemes when the cyclic prefix (CP) is removed in the OFDM system (marked by "w/o CP"), which inevitably causes inter-symbol interference. The DNN-based channel estimator exhibits robustness in such adverse settings, while LMMSE suffers significant performance loss.

2.3.2 CNN-Based Channel Estimation for MIMO-OFDM Systems

Featuring a limited number of scatters in the propagation environment, millimeter wave (mmWave) communications exhibit three kinds of channel correlations: spatial correlation among channels of different antennas, frequency correlation among channels at different subcarriers, and temporal correlation among channels in different time slots. These channel correlations are favorable for channel estimation; however, they cannot be fully exploited by traditional methods. In this part, we examine the use of CNNs to exploit such correlations to improve channel estimation performance.

Spatial-Frequency CNN-Based Channel Estimation

Consider a mmWave massive MIMO-OFDM system, where the transmitter has N_t antennas, and the receiver has N_r antennas. The frequency-domain channel of the kth subcarrier in OFDM, $\mathbf{H}_k \in \mathbb{C}^{N_r \times N_t}$, is given by

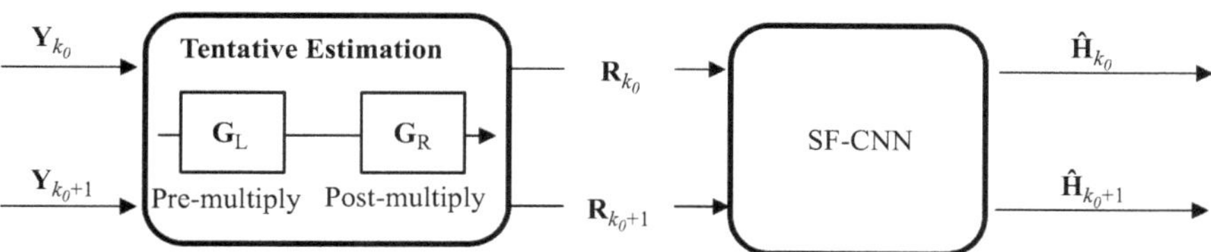

Figure 2.10 Spatial-frequency CNN-based channel estimation proposed in [72].

$$\mathbf{H}_k = \sqrt{\frac{N_t N_r}{L}} \sum_{l=1}^{L} \alpha_l e^{-j2\pi \tau_l f_s \frac{k}{K}} \mathbf{a}_r(\psi_l)\mathbf{a}_t^H(\phi_l), \tag{2.25}$$

where L is the number of paths, $\alpha_l \sim \mathcal{CN}(0,\sigma_\alpha^2)$ is the propagation gain of the lth path, and $\psi_l, \phi_l \in [0, 2\pi]$ denote the azimuth angles of arrival and departure (AoA/AoD) at the receiver and the transmitter, respectively. f_s is the sampling rate and K is the number of OFDM subcarriers. The corresponding response vectors for the uniform linear array can be written as

$$\mathbf{a}_r(\psi_l) = \frac{1}{\sqrt{N_r}}[1, e^{-j2\pi \frac{d}{\lambda}\sin(\psi_l)}, \ldots, e^{-j2\pi \frac{d}{\lambda}(N_r-1)\sin(\psi_l)}]^T \tag{2.26}$$

and

$$\mathbf{a}_t(\phi_l) = \frac{1}{\sqrt{N_t}}[1, e^{-j2\pi \frac{d}{\lambda}\sin(\phi_l)}, \ldots, e^{-j2\pi \frac{d}{\lambda}(N_t-1)\sin(\phi_l)}]^T, \tag{2.27}$$

where d is the distance between the adjacent antennas and λ denotes the carrier wavelength. The pilots are inserted in both frequency and time domains. The adjacent two subcarriers,[2] k_0 and $k_0 + 1$, have pilots inserted with the same length at the beginning of a coherence interval, and the rest of the time slots in each coherence interval are used for data transmission. Figure 2.10 illustrates the channel estimation procedure. The pilot signal matrix, $\mathbf{Y}_k$, corresponding to the kth subcarrier at the baseband, can be written as

$$\mathbf{Y}_k = \mathbf{W}_k^H \mathbf{H}_k \mathbf{F}_k \mathbf{X}_k + \tilde{\mathbf{N}}_k, \qquad k \in \{k_0, k_0 + 1\}, \tag{2.28}$$

where $\mathbf{W}_k$ and $\mathbf{F}_k$ are the combining and beamforming matrices respectively, and $\tilde{\mathbf{N}}_k$ is the combined additive white Gaussian noise. $\mathbf{X}_k$ in (2.28) is an $M_t \times M_t$ diagonal matrix with its diagonal elements denoting the transmit pilots, where M_t is the number of beamforming vectors used at the transmitter. Without loss of generality, it is assumed that $\mathbf{W}_k = \mathbf{W}$, $\mathbf{F}_k = \mathbf{F}$, and $\mathbf{X}_k = \sqrt{P}\mathbf{I}$, where P denotes the transmit power. Then (2.28) can be rewritten as

$$\mathbf{Y}_k = \sqrt{P}\mathbf{W}^H \mathbf{H}_k \mathbf{F} + \tilde{\mathbf{N}}_k, \qquad k \in \{k_0, k_0 + 1\}. \tag{2.29}$$

---

[2] To simplify the discussion, only two subcarriers are considered for exploiting the frequency correlation. Notwithstanding, the extension to more subcarriers is straightforward.

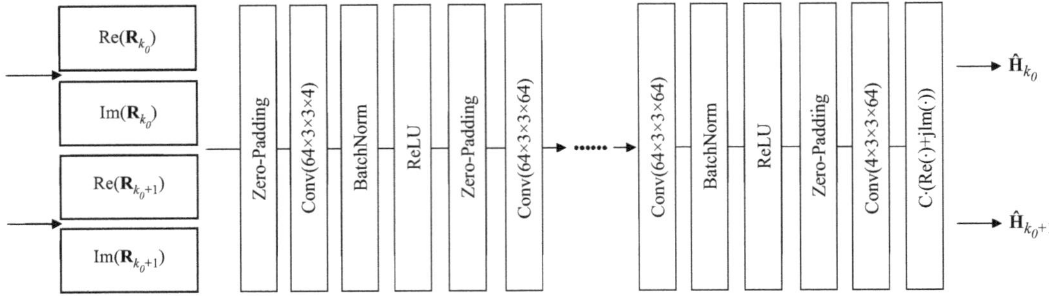

Figure 2.11 The structure of the SF-CNN [72].

Afterwards, $\mathbf{Y}_k$ goes through the tentative estimation module, which processes $\mathbf{Y}_k$ with $\mathbf{G}_L = (\mathbf{W}\mathbf{W}^H)^{-1}\mathbf{W}$ and $\mathbf{G}_R = \mathbf{F}^H(\mathbf{F}\mathbf{F}^H)^{-1}$ to produce a coarse estimation of $\mathbf{H}_k$, which is denoted by

$$\mathbf{R}_k = \sqrt{P}\mathbf{G}_L\mathbf{W}^H\mathbf{H}_k\mathbf{F}\mathbf{G}_R + \mathbf{G}_L\tilde{\mathbf{N}}_k\mathbf{G}_R. \tag{2.30}$$

Then, a spatial-frequency CNN (SF-CNN) exploits the channels' spatial and frequency correlations to refine these tentatively estimated channel matrices. As shown in Figure 2.10, the coarse estimations of channel matrices, $\mathbf{R}_{k_0}$ and $\mathbf{R}_{k_0+1}$, are fed into a SF-CNN. The structure of the SFT-CNN is illustrated in Figure 2.11. The SF-CNN receives two tentatively estimated channel matrices as input, and it separates their real and imaginary parts to obtain real-valued data of size $N_r \times N_t \times 4$. In the subsequent convolution layer, the data is processed by 64 filters of size $3 \times 3 \times 4$ with the ReLU activation function. Zero-padding is used to keep the feature matrix's size unchanged during the whole procedure. Afterward, a batch normalization layer is adopted to avoid gradient vanishing/explosion and prevent overfitting. The following several convolution layers use 64 filters of size $3 \times 3 \times 64$ to perform convolution with zero-padding, each of which is followed by a batch normalization layer and activated using ReLU. The output layer uses four $3 \times 3 \times 64$ filters to extract the real and imaginary parts of the scaled channel matrices at these two subcarriers, i.e., $\frac{\mathrm{Re}(\hat{\mathbf{H}}_{k_0})}{C}, \frac{\mathrm{Im}(\hat{\mathbf{H}}_{k_0})}{C}, \frac{\mathrm{Re}(\hat{\mathbf{H}}_{k_0+1})}{C}$, and $\frac{\mathrm{Im}(\hat{\mathbf{H}}_{k_0+1})}{C}$, where $C > 0$ is a scaling constant. After scaling up and combining the corresponding real and imaginary parts, the $N_r \times N_t$ estimated channels, $\hat{\mathbf{H}}_{k_0}$ and $\hat{\mathbf{H}}_{k_0+1}$, are obtained. The whole process can be represented as

$$\left\{\hat{\mathbf{H}}_{k_0}, \hat{\mathbf{H}}_{k_0+1}\right\} = f_\theta\left(\mathbf{R}_{k_0}, \mathbf{R}_{k_0+1}\right), \tag{2.31}$$

where θ denotes the parameters of the SF-CNN. The objective of the training for the SF-CNN is to minimize

$$\mathcal{L}(\theta) = \frac{1}{N_{\mathrm{tr}}C^2}\sum_{i=1}^{N_{\mathrm{tr}}}\sum_{q=1}^{2}\left\|\mathbf{H}_{k_0+q-1}^i - \hat{\mathbf{H}}_{k_0+q-1}^i\right\|_F^2, \tag{2.32}$$

where N_{tr} is the number of samples in the training set. Each sample is denoted by $(\mathbf{R}_i, \mathbf{H}_i)$, of which $\mathbf{R}_i \in \mathbb{C}^{N_r \times N_t \times 2}$ is the matrix composed of the coarse estimated channel matrices, and $\mathbf{H}_i = \frac{1}{C}[\mathbf{H}_{k_0}^i, \mathbf{H}_{k_0+1}^i]$ represents the scaled true channel matrices.

Spatial-Frequency-Temporal CNN-Based Channel Estimation

The SF-CNN approach can be modified to further incorporate channel temporal correlation. For time-varying channels, (2.25) becomes

$$\mathbf{H}_k(t) = \sqrt{\frac{N_t N_r}{L}} \sum_{l=1}^{L} \alpha_l e^{-j2\pi(\tau_l f_s \frac{k}{K} - v_l t)} \mathbf{a}_r(\psi_l)\mathbf{a}_t^H(\phi_l), \tag{2.33}$$

where v_l is the Doppler shift of the lth path. The received pilots after combining at the receiver in (2.29) become

$$\mathbf{Y}_k[n] = \sqrt{P}\mathbf{W}^H \mathbf{H}_k[n]\mathbf{F} + \tilde{\mathbf{N}}_k[n], \qquad k \in \{k_0, k_0 + 1\}, \tag{2.34}$$

where $\mathbf{Y}_k[n] = \mathbf{Y}_k(nT)$ is the discrete version of $\mathbf{Y}_k(t)$. T is the length of a time slot, which can be the duration of an OFDM symbol, as an example. Similar to SF-CNN, $\mathbf{Y}_k[n]$ goes through the tentative estimation module and generates tentatively estimated channel matrices, $\mathbf{R}_k[n]$, sequentially. Then, a spatial-frequency-temporal CNN (SFT-CNN) exploits the channels' spatial, frequency, and temporal correlations to refine these tentatively estimated channel matrices. As shown in Figure 2.12, two successive time slots, n_0 and $n_0 + 1$, are captured to conduct the channel estimation procedure. In the n_0th time slot, the coarse estimations of channel

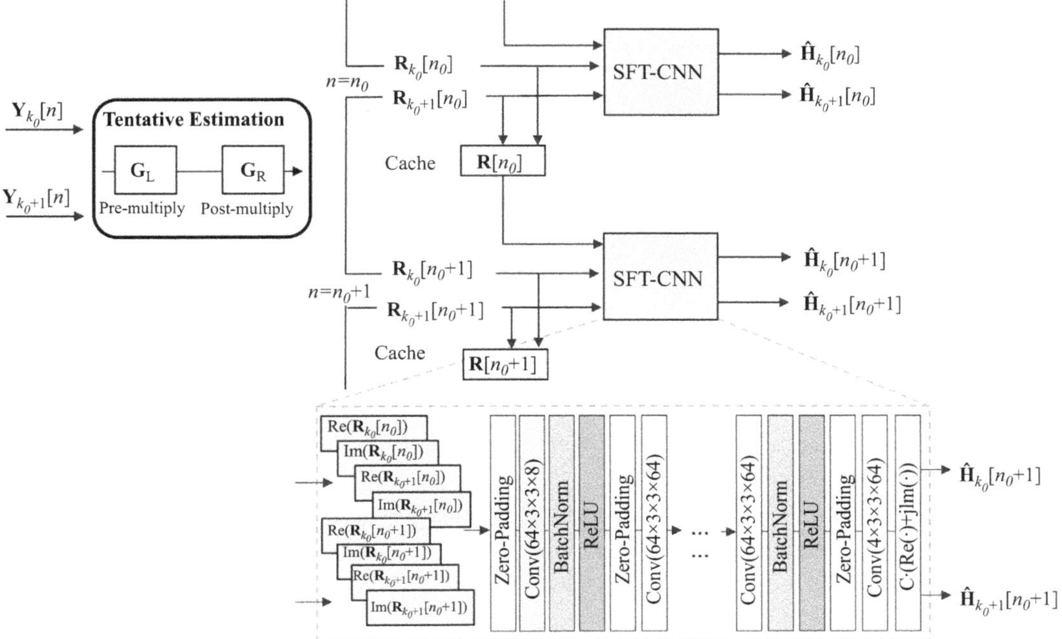

Figure 2.12 SFT-CNN for channel estimation, proposed in [72].

matrices, $\mathbf{R}_{k_0}[n_0]$ and $\mathbf{R}_{k_0+1}[n_0]$, are fed into an SFT-CNN. In addition, these two channel matrices will be stored in the cache for utilization at the next time slot. In the $(n_0 + 1)$th time slot, the SFT-CNN receives the current coarse estimations (i.e., $\mathbf{R}_{k_0}[n_0 + 1]$ and $\mathbf{R}_{k_0+1}[n_0 + 1]$), and the channel matrices from the cache to perform joint estimation. The cache is replaced by $\mathbf{R}_{k_0}[n_0 + 1]$ and $\mathbf{R}_{k_0+1}[n_0 + 1]$ simultaneously. The structure of SFT-CNN is similar to SF-CNN except that it has the additional input from the previous time slot. The whole channel estimation process can be represented as

$$\left\{ \hat{\mathbf{H}}_{k_0}[n_0 + 1], \hat{\mathbf{H}}_{k_0+1}[n_0 + 1] \right\}$$
$$= f_\theta \left(\mathbf{R}_{k_0}[n_0], \mathbf{R}_{k_0+1}[n_0], \mathbf{R}_{k_0}[n_0 + 1], \mathbf{R}_{k_0+1}[n_0 + 1] \right). \tag{2.35}$$

The objective for training the SFT-CNN is given by

$$\mathcal{L}(\theta) = \frac{1}{N_{\mathrm{tr}} C^2} \sum_{i=1}^{N_{\mathrm{tr}}} \sum_{q=1}^{2} \left\| \mathbf{H}_{k_0+q-1}^i[n_0 + 1] - \hat{\mathbf{H}}_{k_0+q-1}^i[n_0 + 1] \right\|_F^2, \tag{2.36}$$

where each sample is also denoted by $(\mathbf{R}_i, \mathbf{H}_i)$, of which $\mathbf{R}_i \in \mathbb{C}^{N_r \times N_t \times 4}$ is the matrix composed of the coarse estimated channel matrices in the n_0th and $(n_0 + 1)$th time slots, and $\mathbf{H}_i = \frac{1}{C}[\mathbf{H}_{k_0}^i[n_0 + 1], \mathbf{H}_{k_0+1}^i[n_0 + 1]]$ represents the scaled true channel matrices of the $(n_0 + 1)$th time slot.

Performance Comparison

We show the normalized MSE (NMSE) performance of the CNN-based channel estimation method against the MMSE schemes in Figure 2.13. When the covariance is computed from true channel realizations, we refer to the method as ideal MMSE. When the covariance is estimated from the LS solution, we refer to the method as non-ideal MMSE. NMSE is defined as

$$\mathrm{NMSE} = \mathbb{E}_{\mathbf{H}} \left\{ \frac{\left\| \mathbf{H} - \hat{\mathbf{H}} \right\|_F^2}{\| \mathbf{H} \|_F^2} \right\}, \tag{2.37}$$

where $\mathbf{H}$ and $\hat{\mathbf{H}}$ denote the true and estimated channel matrices, respectively. The clustered delay line model is used in simulation, where the number of main paths is set as $L = 3$, and an OFDM system with 64 subcarriers is adopted. The transmitter is equipped with a total of $N_t = 32$ antennas, while the receiver is equipped with $N_r = 16$ antennas.

The MMSE- and SFT-CNN-based channel estimation approaches utilize the channel information of the previous time slot, while the SF-CNN-based method does not. By comparing the curves of SFT-CNN and SF-CNN, the effect of temporal correlation on improving the NMSE performance can be clearly seen. Utilizing the additional channel information of adjacent subcarriers and time slots can improve the estimation accuracy remarkably. It can be found that with the same

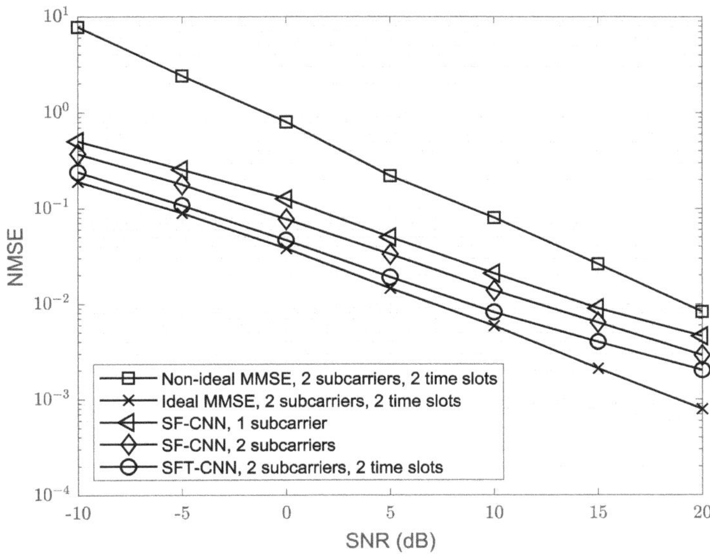

Figure 2.13 NMSE performance of CNN-based and existing methods [72].

channel correlation information, the SFT-CNN-based channel estimation method outperforms the non-ideal MMSE significantly in NMSE performance and is very close to that of the ideal MMSE estimator, which is quite difficult to implement in practical situations, at low and medium SNRs.

CNN-based channel estimation usually regards the channel matrix as a noisy image and extracts correlations in the spatial, frequency, and temporal domains through the convolution operation of CNN, which improves the accuracy of channel estimation. Furthermore, these correlations could be used by CNN-based methods to significantly reduce the spatial pilot overhead caused by massive antennas. More details can be found in [72].

2.3.3 Deep Generative Models for Channel Estimation

Achieving accurate estimation of high-dimensional wireless channels typically results in a significant amount of pilot overhead. There is an urgent need to incorporate prior knowledge of the channel structure into the estimation scheme to reduce this overhead. However, traditional handcrafted priors, such as sparsity and low-rank assumptions, are insufficient to capture the complicated features of real-world wireless channels. Deep generative modeling provides an efficient solution for characterizing wireless channels in a data-driven manner without assuming special structures for the channel distribution. This model-free approach can be particularly valuable as the carrier frequencies progressively increase, wherein traditional channel models lose efficacy. With the deep generative priors, the pilot overhead involved in high-dimensional channel estimation can be significantly reduced. In the following, we review two representative schemes using deep generative models for channel estimation: the GAN-based and the score-based generative model methods.

GAN-Based Channel Estimation

Consider a point-to-point, narrowband MIMO setup with N_t transmit and N_r receive antennas. The MIMO channel matrix characterizing the propagation of this scenario is given by $\mathbf{H} \in \mathbb{C}^{N_r \times N_t}$. At the ith slot allocated for channel estimation, a training beamformer $\mathbf{p}_i \in \mathbb{C}^{N_t \times 1}$ is utilized to transmit a complex-valued symbol s. The beamformer is also referred to as the pilot vector throughout this subsection. Let $\mathbf{W} \in \mathbb{C}^{N_r \times N_r}$ denote the combining matrix at the receiver. The received signal at the ith slot is given by

$$\mathbf{y}_i = \mathbf{W}^H(\mathbf{H}\mathbf{p}_i s + \mathbf{n}_i), \tag{2.38}$$

where $\mathbf{n}_i \sim \mathcal{CN}(\mathbf{0}, \sigma_n^2\mathbf{I})$ represents the AWGN vector with variance σ_n^2 per element. For simplification of the analysis, we suppose that $s = 1$ and $\mathbf{W} = \mathbf{I}$. The channel estimation stage contains a total of N_P time slots, where the N_P pilot vectors constitute a pilot matrix $\mathbf{P} = [\mathbf{p}_1, \ldots, \mathbf{p}_{N_P}]$. Assuming that the MIMO channel, $\mathbf{H}$, remains constant across these time slots, the received signal $\mathbf{Y} \in \mathbb{C}^{N_r \times N_P}$ within this period can be written as

$$\mathbf{Y} = \mathbf{H}\mathbf{P} + \mathbf{N}, \tag{2.39}$$

where $\mathbf{N} = [\mathbf{n}_1, \ldots, \mathbf{n}_{N_P}]$. Vectorization, denoted by $\mathrm{vec}(\cdot)$, can be performed to derive a compact expression as follows:

$$\mathbf{y} = \mathbf{A}\mathbf{h} + \mathbf{n}, \tag{2.40}$$

where $\mathbf{y} = \mathrm{vec}(\mathbf{Y}) \in \mathbb{C}^{N_r N_P \times 1}$, $\mathbf{A} = \mathbf{P}^T \otimes \mathbf{I}$, $\mathbf{h} = \mathrm{vec}(\mathbf{H}) \in \mathbb{C}^{N_r N_t \times 1}$, and $\mathbf{n} = \mathrm{vec}(\mathbf{N}) \in \mathbb{C}^{N_r N_P \times 1}$. Typically, given $\mathbf{y}$ and $\mathbf{A}$, accurate recovery of $\mathbf{h}$ requires $N_P \geq N_t$ to prevent the problem from becoming underdetermined. However, this leads to substantial pilot overhead for large-scale MIMO systems with numerous antennas. Therefore, the adoption of deep generative priors is helpful to facilitate channel estimation with limited pilot placement.

To use a GAN for channel estimation, we first train the GAN using a collection of real channel data samples $\mathbf{h}$ to learn the underlying channel structure and distribution. By using GAN-based Channel modeling approaches described in Section 2.1.2, specifically adversarial training, the GAN's generator G learns to synthesize channel samples from the desired distribution given low-dimensional noise vectors $\mathbf{z} \in \mathbb{R}^{d \times 1}$ as input.[3]

After successful learning of the channel distribution, the generator G can be utilized in the framework presented in Figure 2.14 for channel estimation. This framework follows the paradigm of compressed sensing using generative models [74] by assuming that the target channel lies in the span of the pre-trained generator.

---

[3] Note that in this subsection, G denotes a function that maps from $\mathbb{R}^{d \times 1}$ to $\mathbb{C}^{N_r N_t \times 1}$.

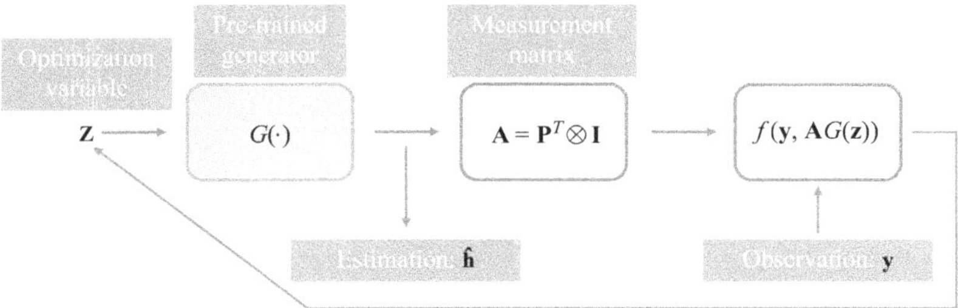

Figure 2.14 GAN-based channel estimation proposed in [73].

Specifically, the generator takes a latent vector $\mathbf{z}$, which is to be optimized, as input, and produces the estimated channel. Given the observation $\mathbf{y}$, gradient descent is performed over the input variable, $\mathbf{z}$, to find a low-dimensional representation $\mathbf{z}^*$ of the true channel, $\mathbf{h}$, such that a pre-defined loss function $f(\mathbf{y}, \mathbf{A}G(\mathbf{z}))$ is minimized. A practical choice for f is the ℓ_2 loss, leading to the following optimization problem:

$$\mathbf{z}^* = \arg\min_{\mathbf{z}\in\mathbb{R}^{d\times 1}} \|\mathbf{y} - \mathbf{A}G(\mathbf{z})\|_2^2. \tag{2.41}$$

After the optimization, the estimated channel is simply given by

$$\hat{\mathbf{h}}_{\mathrm{GAN}} = G(\mathbf{z}^*). \tag{2.42}$$

The rationale of this strategy can be explained as follows: Equation (2.42) ensures the prior consistency of the estimate, producing plausible samples that align with the underlying channel distribution, while the gradient descent optimization enhances the pilot consistency of the estimate. In addition, it is proved in [74] that $\hat{\mathbf{h}}_{\mathrm{GAN}} = G(\mathbf{z}^*)$ will be among the nearest points to the true channel within the span of G given mild assumptions, as long as gradient descent obtains an approximate solution to (2.41). To further enhance the reconstructed performance, an additional regularization term is introduced in [73], imposing a Gaussian assumption on $\mathbf{z}$ and adapting (2.41) as follows:

$$\mathbf{z}^* = \arg\min_{\mathbf{z}\in\mathbb{R}^{d\times 1}} \|\mathbf{y} - \mathbf{A}G(\mathbf{z})\|_2^2 + \lambda_{\mathrm{reg}}\|\mathbf{z}\|_2^2, \tag{2.43}$$

where λ_{reg} is a regularization parameter determined by line search.

Score-Based Generative Models for Channel Estimation

The score-based generative models introduced in Section 1.2.2 provide a novel approach to achieving high-fidelity channel estimation with limited pilot overhead, showcasing state-of-the-art performance [75]. Furthermore, this approach demonstrates exceptional robustness to test-time distributional shifts, thereby being practically effective in highly dynamic wireless environments.

Channel estimation using score-based generative models can be divided into the offline training stage and the online inference stage, as shown in Figure 2.15. The training stage learns the inherent channel distribution, $p_h(\mathbf{h})$, from a set of channel

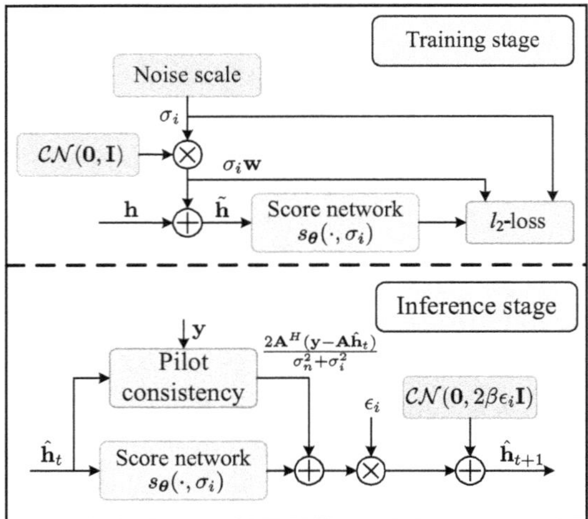

Figure 2.15 Score-based generative models for channel estimation proposed in [75].

realizations by using a score network to approximate the score of the underlying distribution, i.e., $\nabla_{\mathbf{h}} \log p_h(\mathbf{h})$. As described in Section 1.2.2, the training of the score network is achieved by multi-scale noise-perturbed score matching. Consider the perturbation model as $\tilde{\mathbf{h}} = \mathbf{h} + \sigma_i \mathbf{w}$, where σ_i denotes the noise scale, and $\mathbf{w} \sim \mathcal{CN}(\mathbf{0}, \mathbf{I})$ denotes the Gaussian noise added to the channel data $\mathbf{h} \sim p_h(\mathbf{h})$. Given a set of L noise scales $\{\sigma_i\}_{i=1}^{L}$, the expected loss for training a noise-conditional score network $s_{\boldsymbol{\theta}}(\tilde{\mathbf{h}}, \sigma_i)$ with parameters $\boldsymbol{\theta}$ is given by

$$\mathcal{L}_{\text{score}}(\boldsymbol{\theta}) = \mathbb{E}_{i, \mathbf{h} \sim p_h, \mathbf{w} \sim p_w} \left[\sigma_i^2 \left\| s_{\boldsymbol{\theta}}(\mathbf{h} + \sigma_i \mathbf{w}, \sigma_i) + \frac{\mathbf{w}}{\sigma_i^2} \right\|_2^2 \right], \tag{2.44}$$

where the weighting of each ℓ_2 norm term with σ_i^2 compensates for the infinite magnitude of the scaled noise, $\mathbf{w}/\sigma_i^2$, for small σ_i, thereby stabilizing training. In practice, a mini-batch version of (2.44) is employed to update $\boldsymbol{\theta}$ using back-propagation. At each training step, the data sample within a mini-batch is perturbed by noise at a randomly selected noise scale from $\{\sigma_i\}_{i=1}^{L}$. The network is trained to predict the scaled negative noise, which indicates the direction away from the perturbed sample and toward the original data sample, representing the score of the noise-conditional distribution $p_{\sigma_i}(\tilde{\mathbf{h}}|\mathbf{h})$. It is proved in [37] that the network minimizing (2.44) satisfies $s_{\boldsymbol{\theta}}(\tilde{\mathbf{h}}, \sigma_i) = \nabla_{\tilde{\mathbf{h}}} \log p_{\sigma_i}(\tilde{\mathbf{h}})$ almost surely. Herein, $p_{\sigma_i}(\tilde{\mathbf{h}}) = \int p_{\sigma_i}(\tilde{\mathbf{h}}|\mathbf{h})p_h(\mathbf{h})d\mathbf{h}$ denotes the perturbed channel distribution, which approaches the true channel distribution, $p_h(\mathbf{h})$, when the noise scale, σ_i, is sufficiently small. For the implementation of the score network, $s_{\boldsymbol{\theta}}$, the RefineNet architecture [76] is employed. This architecture comprises multiple RefineNet blocks operating at different resolutions in parallel, with shortcut connections between them. Further details can be found in [75, 76]. This architecture facilitates learning the structural characteristics of the channel and estimating the score function. A notable feature of this score-based model learning is that no information on the measurement model (2.40) is involved, and the model is only responsible for learning the channel prior, differentiating this approach from numerous supervised learning-based channel estimators. Hence, this approach will be robust across different measurement setups, such as varying pilot counts and SNRs, during the inference stage.

The inference stage utilizes the pre-trained score-based model and the received pilot observations to perform channel estimation. The proposed scheme resorts to sampling from the posterior distribution, $p_{h|y}(\mathbf{h}|\mathbf{y})$. Initialized with a random noise sample, Langevin dynamics for posterior sampling takes the following update rule:

$$\hat{\mathbf{h}} \leftarrow \hat{\mathbf{h}} + \epsilon \nabla_{\mathbf{h}} \log p_{h|y}(\hat{\mathbf{h}}|\mathbf{y}) + \sqrt{2\epsilon}\mathbf{w}, \quad \mathbf{w} \sim \mathcal{CN}(\mathbf{0}, \mathbf{I}), \tag{2.45}$$

where ϵ is the step size. The posterior score $\nabla_{\mathbf{h}} \log p_{h|y}(\hat{\mathbf{h}}|\mathbf{y})$ involved in (2.45) guides the sampling to high-density regions of the posterior distribution. The additional noise term, $\sqrt{2\epsilon}\mathbf{w}$, enhances sample diversity and prevents the update from getting trapped in local minima. By applying Bayes' rule, the posterior score can be decomposed as

$$\nabla_{\mathbf{h}} \log p_{h|y}(\mathbf{h}|\mathbf{y}) = \nabla_{\mathbf{h}} \log p_h(\mathbf{h}) + \nabla_{\mathbf{h}} \log p_{y|h}(\mathbf{y}|\mathbf{h}), \tag{2.46}$$

where the prior score $\nabla_{\mathbf{h}} \log p_h(\mathbf{h})$ structures the channel estimate according to the underlying channel distribution, $p_h(\mathbf{h})$, while the likelihood score, $\nabla_{\mathbf{h}} \log p_{y|h}(\mathbf{y}|\mathbf{h})$, enhances consistency with received pilots. This likelihood score can be computed in closed form based on the pilot measurements and noise statistics as follows:

$$\nabla_{\mathbf{h}} \log p_{y|h}(\mathbf{y}|\mathbf{h}) = \frac{2\mathbf{A}^H(\mathbf{y} - \mathbf{Ah})}{\sigma_n^2}. \tag{2.47}$$

Note that the closed-form expression of the prior score is unavailable, and only the noise-perturbed version, $\nabla_{\tilde{\mathbf{h}}} \log p_{\sigma_i}(\tilde{\mathbf{h}})$, can be approximated by the pre-trained score network, s_θ. Hence, we sample from the noise-perturbed posterior distribution, $p_{\sigma_i}(\tilde{\mathbf{h}}|\mathbf{y})$, with gradually decreased noise scales, which finally approaches the true posterior distribution, $p_{h|y}(\mathbf{h}|\mathbf{y})$. To this end, annealed Langevin dynamics introduced in Section 1.2.2 is used, proceeding as follows at the ith noise scale ($i = 1, \cdots, L$):

$$\hat{\mathbf{h}}_{t+1} = \hat{\mathbf{h}}_t + \epsilon_i \left(\nabla_{\tilde{\mathbf{h}}} \log p_{\sigma_i}(\hat{\mathbf{h}}_t) + \nabla_{\tilde{\mathbf{h}}} \log p_{\sigma_i}(\mathbf{y}|\hat{\mathbf{h}}_t) \right)$$

$$+ \sqrt{2\beta\epsilon_i}\mathbf{w}_t, \quad \mathbf{w}_t \sim \mathcal{CN}(\mathbf{0}, \mathbf{I}), \quad t = 1, \cdots, T, \tag{2.48}$$

where $\epsilon_i = \epsilon_0 \cdot \sigma_i^2/\sigma_L^2$ is the annealed step size, and T is the number of sampling steps at each noise scale. The scalars ϵ_0 and β are hyperparameters. Since $\nabla_{\tilde{\mathbf{h}}} \log p_{\sigma_i}(\hat{\mathbf{h}}_t) \approx s_\theta(\hat{\mathbf{h}}_t, \sigma_i)$, the only unknown term in (2.48) is $\nabla_{\tilde{\mathbf{h}}} \log p_{\sigma_i}(\mathbf{y}|\hat{\mathbf{h}}_t)$. This term differs from the likelihood score in (2.47) and renders the score, $\nabla_{\tilde{\mathbf{h}}} \log p_{\sigma_i}(\mathbf{y}|\hat{\mathbf{h}}_t)$, intractable to compute. To resolve the challenge, it is suggested in [75, 77] to heuristically add an annealing term to (2.47) and compute the score as follows:

$$\nabla_{\tilde{\mathbf{h}}} \log p_{\sigma_i}(\mathbf{y}|\hat{\mathbf{h}}_t) = \frac{2\mathbf{A}^H(\mathbf{y} - \mathbf{A}\hat{\mathbf{h}}_t)}{\sigma_n^2 + \sigma_i^2}. \tag{2.49}$$

This results in the final channel estimation procedure outlined in Algorithm 2.1.

Algorithm 2.1 MIMO Channel Estimation Using Score-Based Generative Models

---

Require: Measurement matrix $\mathbf{A} = \mathbf{P}^T \otimes \mathbf{I}$, received pilots $\mathbf{y}$, received noise variance σ_n^2, pre-trained score network s_θ, inference noise scales $\{\sigma_i\}_{i=1}^L$, hyperparameters ϵ_0, β, T.

Initialize $\hat{\mathbf{h}}_1 \sim \mathcal{CN}(\mathbf{0}, \mathbf{I})$.

for $i = 1$ to L **do**

   Set annealed step size $\epsilon_i = \epsilon_0 \cdot \sigma_i^2 / \sigma_L^2$.

   for $t = 1$ to T **do**

      Generate perturbation noise $\mathbf{w}_t \sim \mathcal{CN}(\mathbf{0}, \mathbf{I})$.

$$\hat{\mathbf{h}}_{t+1} = \hat{\mathbf{h}}_t + \epsilon_i \left(s_\theta(\hat{\mathbf{h}}_t, \sigma_i) + \frac{2\mathbf{A}^H(\mathbf{y} - \mathbf{A}\hat{\mathbf{h}}_t)}{\sigma_n^2 + \sigma_i^2} \right) + \sqrt{2\beta\epsilon_i}\,\mathbf{w}_t.$$

   $\hat{\mathbf{h}}_1 = \hat{\mathbf{h}}_{T+1}$.

Ensure: Estimated channel $\hat{\mathbf{h}} = \hat{\mathbf{h}}_{T+1}$.

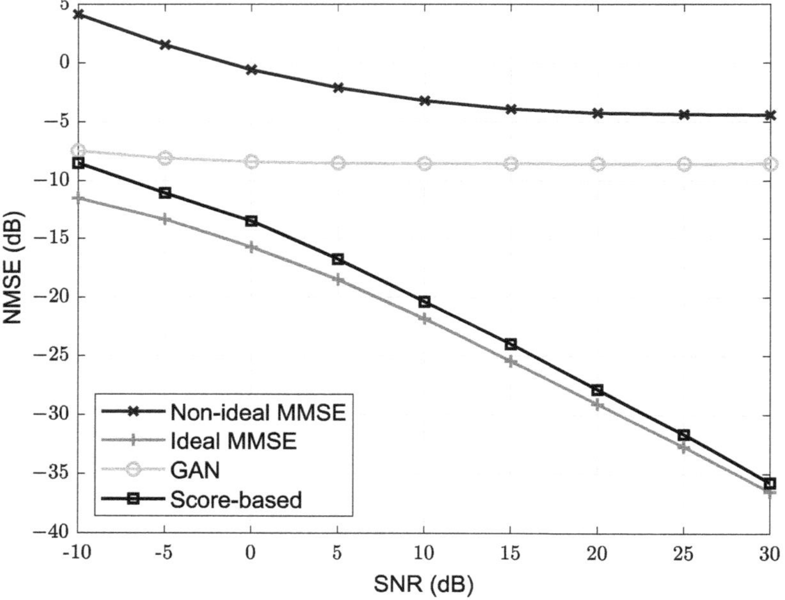

Figure 2.16 NMSE performance of deep generative model-based channel estimation methods.

Performance Comparison

We examine the NMSE performance of deep generative model-based channel estimation approaches in Figure 2.16. A MIMO system with $N_t = 64$ and $N_r = 16$ is considered. Uniform linear arrays with half-wavelength antenna spacings are employed at both the transmitter and the receiver. Channel realizations are generated according to the clustered delay line model specified in the 3rd generation partnership project (3GPP) 38.901 technical report [78]. The carrier frequency is set as 40 GHz. The number of pilot vectors is set as $N_P = 38 \approx 0.6N_t$. Each element of the pilot vector is randomly drawn from the QPSK lattice. A total of 100 independent channel realizations are utilized for performance evaluation. The network setups and training details can be found in the original paper [73, 75].

We compare the GAN- and score-based models against the conventional MMSE methods introduced in Section 2.3.2. The GAN-based channel estimator exhibits significant performance gains over the non-ideal MMSE. However, this estimator has a substantial gap compared to the ideal MMSE benchmark. In contrast, the channel estimator using the powerful score-based generative model closely approaches the ideal MMSE baseline, showcasing the capability of score-based models in accurately modeling the distribution of wireless channels and the effectiveness of the posterior sampling scheme.

2.4 Learning-Based Channel Estimation: Model-Driven Methods

Large-scale MIMO channels are known to exhibit sparsity due to limited scattering. This is particularly true for mmWave communications, where only a very small number of dominant multipaths exist, resulting from the propagation characteristics at the high frequency. Compressed sensing, as an effective model-based approach to deal with sparsity in signal processing, is exploited in this section to enhance learning-based channel estimation. We begin with a brief introduction to the basic concepts and prototypical algorithms in the field of compressed sensing. We then explore two examples, i.e., learned denoising approximate message passing (LDAMP) [13] and learnable iterative shrinkage thresholding algorithm-based channel estimator (LISTA-CE) [79], to show how learning-based channel estimation can be enhanced by expert knowledge in compressed sensing. Finally, some other model-driven learning methods for channel estimation are introduced briefly.

2.4.1 Basics of Compressed Sensing

Compressed sensing, or compressive sampling, is a new sensing or sampling framework that captures and reconstructs compressible signals with far fewer samples than are required by Shannon's celebrated sampling theorem. It employs a small number of non-adaptive linear projections to capture useful information about the signal, which are then used to perform perfect signal recovery via an optimization process. The ground of the compressed sensing theory is generally believed to have been laid by the three foundational papers [80, 81, 82] published in 2006. In this part, we provide a short introduction to the basic concept of compressed sensing and refer interested readers to the three pioneering papers, excellent tutorial papers [83, 84], and textbook [85] for a more in-depth understanding of the subject.

The sensing problem in compressed sensing is to recover the signal $\mathbf{x} \in \mathbb{R}^N$ from the linear measurement in the form of

$$\mathbf{y} = \mathbf{\Phi}\mathbf{x}, \tag{2.50}$$

in the case where the measurement is free from noise, or

$$\mathbf{y} = \mathbf{\Phi}\mathbf{x} + \mathbf{n}, \tag{2.51}$$

in the case where the measurement is noisy, where $\mathbf{\Phi} \in \mathbb{C}^{M \times N}$ is a linear sensing matrix and $\mathbf{n} \in \mathbb{C}^N$ is a noise vector. Typically, the number of measurements, M, is much smaller than the signal dimension, N; hence the name compressed sensing. The process of recovering $\mathbf{x} \in \mathbb{R}^N$ from $\mathbf{y} \in \mathbb{R}^M$ may appear ill-posed in general when $M < N$. However, the problem is shown to be well defined when the signal $\mathbf{x}$ is sparse and the measurement matrix $\mathbf{\Phi}$ is properly chosen according to the compressed sensing theory.

Suppose that $\mathbf{x}$ is K-sparse, i.e., the number of non-zero entries in $\mathbf{x}$ is no more than K. In the case of noise-free measurement, the compressed sensing theory shows that the signal $\mathbf{x}$ of length N can be reconstructed with overwhelming probability via solving the optimizing problem

$$\min_{\mathbf{x} \in \mathbb{R}^N} \|\mathbf{x}\|_1 \quad \text{subject to} \quad \mathbf{y} = \mathbf{\Phi}\mathbf{x}, \tag{2.52}$$

where $\|\mathbf{x}\|_1$ is the ℓ_1 norm of $\mathbf{x}$, with $\mathbf{\Phi}$ constructed by sampling i.i.d. entries from a Gaussian probability density function with zero mean and variance $1/N$, and the number of measurements, M, satisfies

$$M \geq cK \log(N/K), \tag{2.53}$$

where c is a small constant [80, 81, 83].

In case of noisy measurements, or if the signal of interest is not exactly sparse but only approximately sparse, i.e., a large number of entries in $\mathbf{x}$ are close to (but not precisely) zero, the compressed sensing theory has demonstrated that robust signal recovery is achievable by solving the ℓ_1 minimization problem

$$\min_{\mathbf{s} \in \mathbb{R}^N} \|\mathbf{x}\|_1 \quad \text{subject to} \quad \|\mathbf{\Phi}\mathbf{x} - \mathbf{y}\|_2 \leq \epsilon, \tag{2.54}$$

where ϵ bounds the amount of noise in the data.

An important notion called the restricted isometry property (RIP), introduced in [86], is key to studying the general robustness of compressed sensing. If the sensing matrix, $\mathbf{\Phi}$, satisfies the RIP property, then the signal recovery is accurate up to the noise level and the closeness of the (approximately sparse) signal to its K-sparse counterpart that is obtained by setting all but the K largest entries to zero. It is shown that the sensing matrix $\mathbf{\Phi}$ obtained via sampling from the Gaussian distribution with (2.53) satisfied obeys the RIP property with overwhelming probability [83]. This supports the conclusion of accurate signal recovery via solving (2.52).

In practice, the signal, $\mathbf{x} \in \mathbb{R}^N$, may not be sparse in the original domain where it is acquired but can be expressed in a sparse or approximately sparse form in a transform domain, e.g., the Fourier or wavelet domains. In this case, we can express the signal with sparse or approximately sparse coefficients $\mathbf{s}$ in a proper orthonormal basis

$$\mathbf{x} = \mathbf{\Psi}\mathbf{s}, \tag{2.55}$$

where $\boldsymbol{\Psi} \in \mathbb{R}^{N \times N}$ contains N basis (i.e., orthogonal and unit) vectors of size $N \times 1$. Then substituting (2.55) back in (2.50) or (2.51) yields an equivalent sensing matrix $\boldsymbol{\Theta} = \boldsymbol{\Phi}\boldsymbol{\Psi}$. The compressed sensing theory shows that the sensing matrix, $\boldsymbol{\Phi}$, is universal in the sense that $\boldsymbol{\Theta} = \boldsymbol{\Phi}\boldsymbol{\Psi}$ obeys the RIP property with overwhelming probability regardless of the choice of orthonormal basis $\boldsymbol{\Psi}$, provided that the entries of $\boldsymbol{\Phi}$ are drawn from the Gaussian distribution and (2.53) is satisfied.

Finally, we note that the problem in (2.52) can be recast as a linear program known as basis pursuit (BP) [87], and the problem in (2.54) is also a convex optimization problem that can be formulated as second-order cone programming (SOCP) [88]. To solve these problems, a wide variety of algorithms have been proposed in the literature that range widely in computational complexity and empirical effectiveness. Examples of these algorithms include the iterative shrinkage thresholding algorithm (ISTA) [89, 90], approximate message passing (AMP) [91], and orthogonal matching pursuit (OMP) [92, 93]. Next, we present the details of the ISTA and AMP algorithms.

Iterative Shrinkage Thresholding Algorithm

ISTA is an iterative algorithm for solving the following ℓ_1 regularized least-squares problem:

$$\widehat{\mathbf{x}} = \arg\min_{\mathbf{x}} \frac{1}{2}\|\mathbf{y} - \mathbf{A}\mathbf{x}\|_2^2 + \lambda\|\mathbf{x}\|_1, \tag{2.56}$$

where $\lambda > 0$ is a regularization parameter, providing a tradeoff between the sparsity of $\widehat{\mathbf{x}}$ and the measurement fidelity. This problem covers a wide range of applications in signal processing and wireless communications, such as channel estimation.

ISTA solves (2.56) by breaking the problem into a sequence of proximal problems. Defining the first term in (2.56) as $f(\mathbf{x}) = \frac{1}{2}\|\mathbf{y} - \mathbf{A}\mathbf{x}\|_2^2$, the minimization of this quadratic term can be achieved by the gradient descentt iteration:

$$\widehat{\mathbf{x}}_{t+1} = \widehat{\mathbf{x}}_t - \beta_t \nabla f\left(\widehat{\mathbf{x}}_t\right), \tag{2.57}$$

where $\widehat{\mathbf{x}}_t$ is the estimate at the tth iteration and β_t is a suitable step size. $\nabla f\left(\cdot\right)$ denotes the gradient of $f\left(\cdot\right)$, and we have $\nabla f\left(\widehat{\mathbf{x}}_t\right) = -\mathbf{A}^H(\mathbf{y} - \mathbf{A}\widehat{\mathbf{x}}_t)$. The gradient descentt iteration can be equivalently expressed as a proximal regularization of the linear approximation of $f\left(\cdot\right)$ at $\widehat{\mathbf{x}}_t$ [90]:

$$\widehat{\mathbf{x}}_{t+1} = \arg\min_{\mathbf{x}} f(\widehat{\mathbf{x}}_t) + \langle \mathbf{x} - \widehat{\mathbf{x}}_t, \nabla f\left(\widehat{\mathbf{x}}_t\right)\rangle + \frac{1}{2\beta_t}\|\mathbf{x} - \widehat{\mathbf{x}}_t\|_2^2$$

$$= \arg\min_{\mathbf{x}} \frac{1}{2\beta_t}\|\mathbf{x} - (\widehat{\mathbf{x}}_t - \beta_t \nabla f(\widehat{\mathbf{x}}_t))\|_2^2. \tag{2.58}$$

Adapting the idea of (2.58) to the non-smooth ℓ_1 regularization problem (2.56) leads to a sequence of proximal problems as follows:

$$\widehat{\mathbf{x}}_{t+1} = \arg\min_{\mathbf{x}} \frac{1}{2\beta_t}\|\mathbf{x} - (\widehat{\mathbf{x}}_t - \beta_t \nabla f(\widehat{\mathbf{x}}_t))\|_2^2 + \lambda\|\mathbf{x}\|_1. \tag{2.59}$$

Since the ℓ_1-norm in the above equation is separable, the computation of $\widehat{\mathbf{x}}_{t+1}$ can be transformed into component-wise minimization problems, and we can derive the following update rule by simple algebra:

$$\widehat{\mathbf{x}}_{t+1} = \eta(\widehat{\mathbf{x}}_t - \beta_t \nabla f(\widehat{\mathbf{x}}_t); \beta_t \lambda). \tag{2.60}$$

$\eta(\cdot)$ in (2.60) is a component-wise shrinkage operator (denoiser) and is defined as

$$\eta(x; \theta) = \text{sign}(x) \max\{|x| - \theta, 0\}, \tag{2.61}$$

where θ is a threshold. Defining $\mathbf{v}_t = \mathbf{y} - \mathbf{A}\widehat{\mathbf{x}}_t$ and substituting $\nabla f(\widehat{\mathbf{x}}_t) = -\mathbf{A}^H(\mathbf{y} - \mathbf{A}\widehat{\mathbf{x}}_t)$ into (2.60), the ISTA can be summarized as iterating between the following two steps:

$$\mathbf{v}_t = \mathbf{y} - \mathbf{A}\widehat{\mathbf{x}}_t, \tag{2.62}$$

$$\widehat{\mathbf{x}}_{t+1} = \eta\left(\widehat{\mathbf{x}}_t + \beta_t \mathbf{A}^H \mathbf{v}_t; \beta_t \lambda\right). \tag{2.63}$$

A typical requirement to ensure the convergence of $\widehat{\mathbf{x}}_t$ to a minimizer of (2.56) is that the step size, β_t, should be within the range $(0, 1/\|\mathbf{A}^H \mathbf{A}\|)$ [90], where $\|\cdot\|$ denotes the spectral norm of a matrix.

Approximate Message Passing

Inspired by approximate belief propagation in the context of code-division multiple-access (CDMA) detection [94], the AMP algorithm is proposed for signal recovery in compressed sensing [91]. AMP is also an iterative algorithm with two steps, given by

$$\mathbf{v}_t = \mathbf{y} - \mathbf{A}\widehat{\mathbf{x}}_t + \frac{N}{M}\mathbf{v}_{t-1}\langle\eta'(\widehat{\mathbf{x}}_{t-1} + \mathbf{A}^H \mathbf{v}_{t-1}; \lambda + \gamma_{t-1})\rangle, \tag{2.64}$$

$$\widehat{\mathbf{x}}_{t+1} = \eta\left(\widehat{\mathbf{x}}_t + \mathbf{A}^H \mathbf{v}_t; \lambda + \gamma_t\right), \tag{2.65}$$

where $\eta'(\cdot)$ is the derivative of the shrinkage operator, $\eta(\cdot)$, $\langle\cdot\rangle$ is the empirical mean of a vector given by $\langle\mathbf{x}\rangle = N^{-1}\sum_{i=1}^{N} x_i$, and the threshold level γ_t is iteratively computed as follows:

$$\gamma_t = \frac{N}{M}(\lambda + \gamma_{t-1})\langle\eta'(\widehat{\mathbf{x}}_{t-1} + \mathbf{A}^H \mathbf{v}_{t-1}; \lambda + \gamma_{t-1})\rangle. \tag{2.66}$$

The critical difference between AMP and ISTA lies in the last term of equation (2.64), known as the Onsager correction [91]. The term is closely related to the Thouless–Anderson–Palmer (TAP) formula used in the spin glass model in statistical physics [95], which realizes the asymptotic Gaussianity of estimation error before the denoising step performed by the shrinkage operator in (2.65). For a detailed derivation of the Onsager term and the update rules presented in equations (2.64) to (2.66), we direct readers to reference [96], where a detailed derivation is provided based on the standard sum-product belief propagation approach.

AMP shows improved performance as compared to ISTA in terms of solving the optimization problem in (2.56). Specifically, AMP achieves an attractive sparsity-undersampling tradeoff comparable to that of the computationally expensive linear programming-based reconstruction methods [91], unattainable by the ISTA method. Moreover, AMP has been proven to achieve optimal recovery performance in scenarios with i.i.d. zero-mean sub-Gaussian measurements when the compression ratio M/N surpasses the threshold established for the belief propagation algorithm [97].

Example 2.3 This example illustrates how the ISTA and AMP algorithms are employed to address channel estimation problems in mmWave communications. Figure 2.17 illustrates a typical lens-based mmWave massive MIMO system, where the base station (BS) is equipped with a three-dimensional (3D) lens featuring energy focusing capability. Additionally, a matching $U \times V$ antenna array is placed on the focal surface of the lens. This lens antenna array functions as a spatial DFT matrix, which transforms the spatial channel into beamspace by concentrating signals from different directions onto different antennas. The UV antennas are connected to N_{RF} radio frequency (RF) chains through a $N_{\mathrm{RF}} \times UV$ selection network, denoted by a matrix $\mathbf{W} \in \mathbb{R}^{N_{\mathrm{RF}} \times UV}$. Here we assume $N_{\mathrm{RF}} \ll UV$ since the mmWave channel is sparse in beamspace. Hence, only a small number of dominant beams are selected, and the hardware cost and power consumption of the RF chains are significantly reduced. Moreover, the elements of $\mathbf{W}$ are randomly chosen from the set $\frac{1}{\sqrt{UV}}\{-1, +1\}$, i.e., with identical normalized amplitudes, which can be realized by using one-bit phase shifters. For the sake of convenience, a single-user mmWave system is considered. Extending it to support multiple users is straightforward by utilizing orthogonal pilot signals for distinct users.

The widely used Saleh–Valenzuela channel model for mmWave communications is adopted, where the beamspace channel is defined as $\mathbf{H} \in \mathbb{R}^{U \times V}$.[4] The channel vector, $\mathbf{h} \in \mathbb{R}^{UV \times 1}$, is obtained by vectorizing $\mathbf{H}$. Let s represent the training symbol

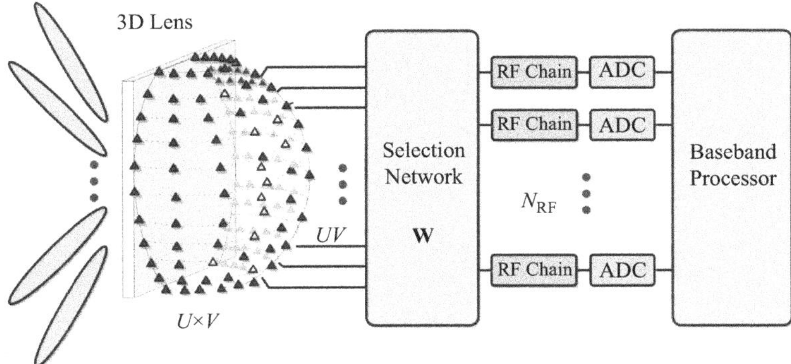

Figure 2.17 Block diagram of the base station in a 3D lens antenna array-based mmWave MIMO system [13].

---

[4] Note that a simplified real-valued channel is considered since neural networks generally operate in the real-valued domain. Extensions to complex channel matrices are straightforward by transforming the complex-valued $\mathbf{H}$ into its real-valued counterpart.

(pilot) sent by the user in the uplink training phase, and the received signal vector, $\mathbf{y} \in \mathbb{R}^{UV \times 1}$, at the BS is given by

$$\mathbf{y} = \mathbf{h}s + \mathbf{n}, \tag{2.67}$$

where $\mathbf{n} \sim \mathcal{N}(\mathbf{0}, \sigma_n^2 \mathbf{I})$ is a Gaussian noise vector with variance σ_n^2. Given a selection network $\mathbf{W}$ at the receiver, the received signal $\mathbf{r}$ from the RF chain can be expressed as

$$\mathbf{r} = \mathbf{W}\mathbf{y} = \mathbf{W}\mathbf{h} + \bar{\mathbf{n}}, \tag{2.68}$$

where $\bar{\mathbf{n}} = \mathbf{W}\mathbf{n}$ is the equivalent noise after the selection network at the receiver and is assumed to follow $\mathcal{N}(\mathbf{0}, \sigma_n^2 \mathbf{I})$ as each entry in matrix $\mathbf{W}$ has normalized amplitudes. Furthermore, the pilot signal is assumed to be known at the receiver side and is set as $s = 1$ for convenience. Equation (2.51) and equation (2.68) are similar. The selection network, $\mathbf{W}$, can be regarded as the sensing matrix in (2.51). Therefore, the introduced ISTA and AMP algorithms can be utilized to solve the channel estimation problem.

The NMSE is chosen to be the performance metric, defined as

$$\text{NMSE} = \mathbb{E} \left\{ \frac{\|\hat{\mathbf{h}} - \mathbf{h}\|_2^2}{\|\hat{\mathbf{h}}\|_2^2} \right\}. \tag{2.69}$$

A four-path mmWave channel is considered in the simulation. The size of the 3D lens is $U = V = 32$.

In the context of this investigation, we adopt a compression ratio of one-half, i.e. $N_{\text{RF}} = UV/2 = 512$, alongside an SNR of 20 dB as our simulation parameters. Figure 2.18 illustrates the NMSE performance of both the ISTA and the AMP algorithms with respect to number of iterations. Owing to the relatively high SNR, both algorithms converge to similar NMSE values. Nevertheless, a discernible distinction

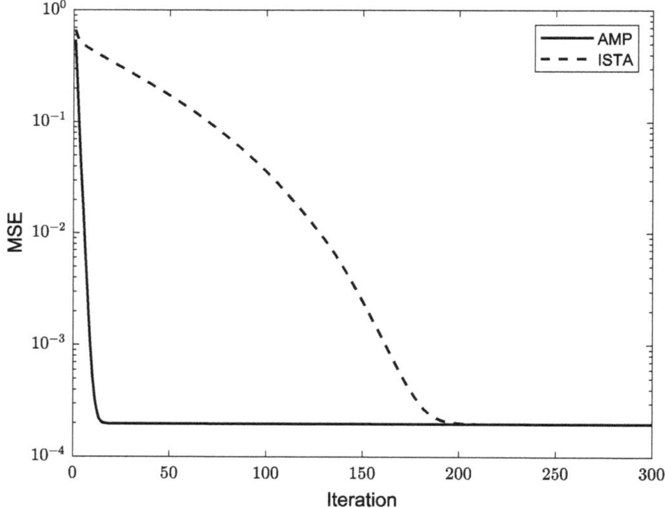

Figure 2.18 NMSE performance of ISTA and AMP versus iterations.

emerges regarding the convergence speed between AMP and ISTA. Specifically, AMP displays a significantly expedited convergence, requiring only approximately 10% of the iterations of ISTA to attain the desired NMSE of about -20 dB. This marked disparity in convergence speed is attributed to the incorporation of the Onsager correction term as delineated in equation (2.64).

2.4.2 LDAMP-Based Channel Estimation

While ISTA and AMP demonstrate commendable performance in the task of reconstructing the compressed channel, their efficacy often hinges upon the fine-tuning of numerous hyperparameters. In this section, we introduce the novel framework known as learned denoising-based approximate message passing, representing a pioneering endeavor in combining deep learning techniques with traditional compressed sensing-based approaches for channel estimation. This innovative approach seeks to harness the power of machine learning to mitigate the intricacies associated with parameter tuning, thereby enhancing the practical applicability and performance of compressed sensing-based channel estimation methods. In this approach, the channel matrix is regarded as a two-dimensional (2D) natural image, and the denoising convolutional neural network (DnCNN) [98] is incorporated into the AMP algorithm. According to the Saleh–Valenzuela channel model, the elements of the beamspace channel vector are not independent. They are correlated through the antenna array response matrix. This characteristic is highly similar to a 2D natural image, i.e., the channel is sparse, and the changes between adjacent elements are subtle. Furthermore, by comparing equation (2.51) with equation (2.68), the similarity between the channel estimation problem and the signal reconstruction problem can be identified. Therefore, the LDAMP network, initially developed for image recovery, can be applied to exploit the correlation in beamspace channel estimation. In this approach, we present the LDAMP-based method to estimate the beamspace channel vector $\mathbf{h}$ from the received signal $\mathbf{r}$ and the selection network $\mathbf{W}$ in (2.68), leveraging the power of deep learning to enhance the performance of compressed sensing-based channel estimation.

The LDAMP neural network consists of L cascaded layers, each having the same structure. As illustrated in Figure 2.19, each layer of the LDAMP network contains a high-performance image denoiser $D_{\hat{\sigma}^l}(\cdot)$ to recover the channel from its noisy observation, a divergence estimator $\mathrm{div}D_{\hat{\sigma}^l}(\cdot)$ for the denoiser, and tied weights. Considering that the elements of the beamspace channel vector are correlated through the antenna array response matrix, the denoising $D_{\hat{\sigma}^l}(\cdot)$ is performed by the DnCNN instead of $\eta(\cdot)$ in equation (2.65). DnCNN can exploit the correlation between the adjacent elements in the channel matrix to recover the beamspace channel from its noisy estimation. The divergence estimator, $\mathrm{div}D_{\hat{\sigma}^l}(\cdot)$, is employed to construct the Onsager correction term similar to that introduced in the AMP algorithm, i.e., (2.64). The computation of $\mathrm{div}D_{\hat{\sigma}^l}(\cdot)$ is realized by a Monte Carlo approximation.

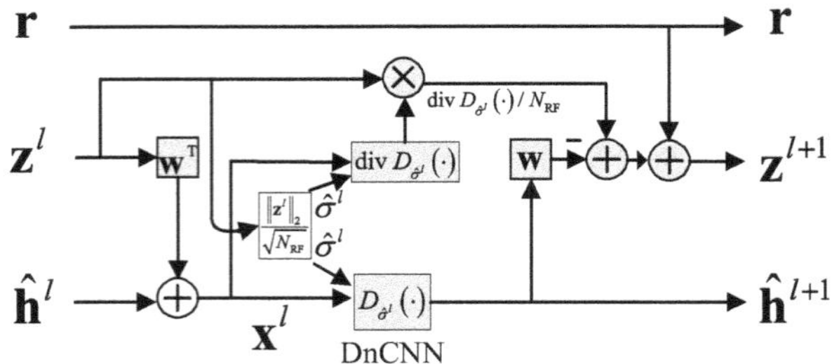

Figure 2.19 Architecture of the lth layer of the LDAMP network [13].

For the lth layer of the LDAMP neural network, the channel is estimated as follows:

$$\mathbf{z}^{l+1} = \mathbf{r} - \mathbf{W}\hat{\mathbf{h}}^{l+1} + \frac{1}{N_{\mathrm{RF}}}\mathbf{z}^l \mathrm{div}D_{\hat{\sigma}^l}(\hat{\mathbf{h}}^l + \mathbf{W}^T\mathbf{z}^l), \qquad (2.70)$$

$$\hat{\mathbf{h}}^{l+1} = D_{\hat{\sigma}^l}(\hat{\mathbf{h}}^l + \mathbf{W}^T\mathbf{z}^l), \qquad (2.71)$$

where $\hat{\mathbf{h}}^l$ is the input channel estimate of the lth layer, $\mathbf{z}^l$ represents the input residual vector of the lth layer, and $\hat{\sigma}^l$ represents an estimate of the standard deviation of the equivalent noise input to the denoiser, which is given by $\hat{\sigma}^l = \|\mathbf{z}^l\|_2/\sqrt{N_{\mathrm{RF}}}$. Note that the update rule in (2.70) and (2.71) can be viewed as an advanced instance of the AMP algorithm discussed in the previous subsection.

The input of the denoiser, $\mathbf{x}^l = \hat{\mathbf{h}}^l + \mathbf{W}^T\mathbf{z}^l$, can be regarded as a noisy observation of the true channel vector:

$$\mathbf{x}^l = \mathbf{h} + \hat{\mathbf{n}}^l, \qquad (2.72)$$

where the equivalent noise is $\hat{\mathbf{n}}^l = (\hat{\mathbf{h}}^l - \mathbf{h} + \mathbf{W}^T\mathbf{z}^l) \sim \mathcal{N}(\mathbf{0}, (\hat{\sigma}^l)^2\mathbf{I})$. The role of the denoiser $D_{\hat{\sigma}^l}(\cdot)$ is to recover channel $\mathbf{h}$ from noisy channel $\mathbf{x}^l$ by removing equivalent noise $\hat{\mathbf{n}}^l$. The equivalent noise variance $(\hat{\sigma}^l)^2$ depends on the estimated residual vector $\mathbf{z}^l$. As the number of network layers increases, $(\hat{\sigma}^l)^2$ diminishes and finally converges to a finite value. Furthermore, the Onsager correction term, $N_{\mathrm{RF}}^{-1}\mathbf{z}^l \mathrm{div}D_{\hat{\sigma}^l}(\hat{\mathbf{h}}^l + \mathbf{W}^T\mathbf{z}^l)$, removes bias from the intermediate solutions, rendering the equivalent noise $\hat{\mathbf{n}}^l$ uncorrelated with the true channel $\hat{\mathbf{h}}^l$, as required by typical image denoisers.

The divergence of denoiser $D_{\hat{\sigma}^l}(\cdot)$ with respect to its input $\mathbf{h} \in \mathbb{R}^{UV \times 1}$ is defined as

$$\mathrm{div}D_{\hat{\sigma}^l} = \sum_{k=1}^{UV} \frac{\partial D_{\hat{\sigma}^l k}(\mathbf{h})}{\partial h_k}, \qquad (2.73)$$

where $D_{\hat{\sigma}^l k}(\mathbf{h})$ and h_k denote the kth element of $D_{\hat{\sigma}^l}(\mathbf{h})$ and $\mathbf{h}$, respectively. Since (2.73) is analytically tractable only for linear denoisers or element-wise non-linear denoisers, not applicable to the DnCNN denoiser used in LDAMP, the following Monte Carlo approximation [99] is utilized to compute the divergence $\mathrm{div}D_{\hat{\sigma}^l}(\cdot)$. Given a denoiser $D_{\hat{\sigma}^l}(\cdot)$, and using an i.i.d. random vector $\mathbf{b} \sim \mathcal{N}(\mathbf{0}, \mathbf{I})$, the divergence can be estimated by

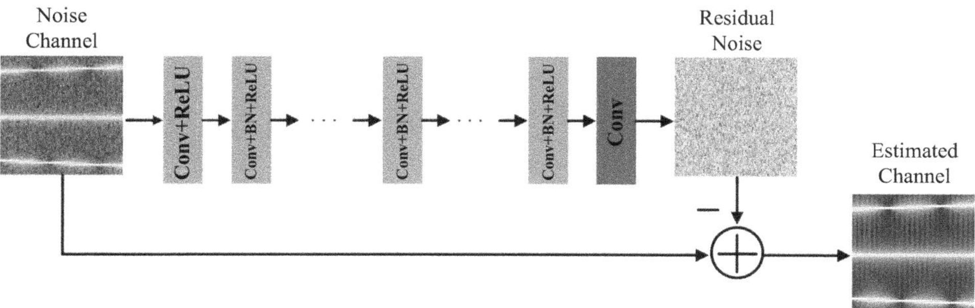

Figure 2.20 Network architecture of the DnCNN denoiser [13].

$$\mathrm{div}D_{\hat{\sigma}^l} = \lim_{\epsilon \to 0} \mathbb{E}_{\mathbf{b}} \left\{ \mathbf{b}^T \left(\frac{D_{\hat{\sigma}^l}(\mathbf{h} + \epsilon\mathbf{b}) - D_{\hat{\sigma}^l}(\mathbf{h})}{\epsilon} \right) \right\} \tag{2.74}$$

$$\approx \frac{1}{\epsilon} \mathbf{b}^T (D_{\hat{\sigma}^l}(\mathbf{h} + \epsilon\mathbf{b}) - D_{\hat{\sigma}^l}(\mathbf{h})), \tag{2.75}$$

where ϵ is set to an extremely small number to mimic the limit in (2.74).

The DnCNN denoiser used in the LDAMP network plays a key role in channel estimation. The DnCNN, introduced in [98], is capable of addressing Gaussian denoising problems with an unknown noise level and outperforms other denoising techniques in terms of accuracy and computational efficiency. Figure 2.20 illustrates the network architecture of the DnCNN denoiser. The network consists of 20 convolutional layers. The first convolutional layer uses 64 different filters of size $3 \times 3 \times 1$, followed by a ReLU activation. In the subsequent 18 convolutional layers, each layer utilizes 64 different filters of size $3 \times 3 \times 64$, followed by batch normalization and a ReLU. The final convolutional layer uses a single $3 \times 3 \times 64$ filter to reconstruct the channel. Instead of directly learning the mapping from noisy images to denoised images, the residual noise is chosen to be learned, known as residual learning [100]. To illustrate this approach, three pseudo-color images are plotted in Figure 2.20, representing the noisy channel, the residual noise, and the estimated channel, respectively. The network takes the noisy channel $\mathbf{h} + \sigma\mathbf{z}$ as input and generates the residual noise $\hat{\mathbf{z}}$ instead of the estimated channel $\hat{\mathbf{h}}$ as output. This approach allows the network to focus on removing highly structured components from the input, instead of unstructured noise. Consequently, residual learning improves both the training efficiency and accuracy of the network.

We show the performance of the LDAMP network as compared with the support detection (SD) algorithm [101], the sparse informative parameter estimator-based sparse analysis AMP for imaging (SCAMPI) algorithm [102], and the denoising-based BM3D-AMP algorithm [103] in Figure 2.21. A four-path mmWave channel is considered in the simulation. The size of the 3D lens is $U = V = 64$. The number of layers in the LDAMP network is set to 10.

Figure 2.21 shows that both the BM3D-AMP algorithm and the LDAMP network exhibit superior performance compared to the SD and SCAMPI algorithms due to the efficacy of the denoiser. Moreover, the LDAMP network outperforms

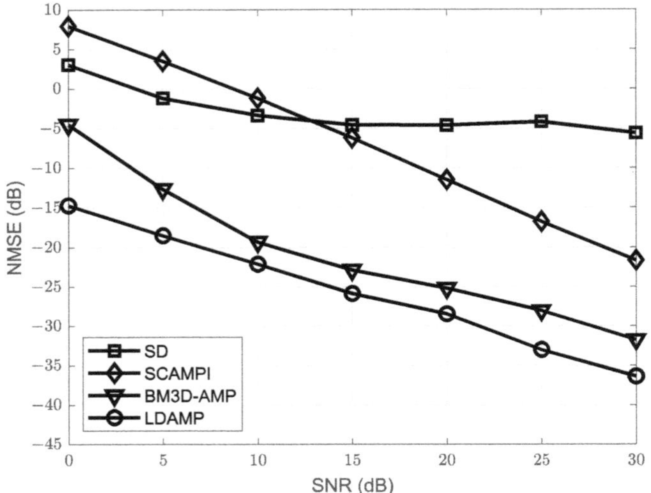

Figure 2.21 Comparison of NMSE performance between the LDAMP network and other methods with $N_{\mathrm{RF}}/(UV) = 0.1$.

state-of-the-art BM3D-AMP by a wide margin in NMSE. This improvement is mainly attributed to the incorporation of DnCNN, which effectively harnesses channel correlations to reconstruct the channel.

LDAMP is the first work that introduces model-driven neural network-based methods into channel estimation. This network inherits the superiority of iterative signal recovery algorithms and deep learning techniques, presenting excellent performance and demonstrating the power of model-driven deep learning in addressing channel estimation problems. LISTA-CE is another effort along this line of thought and will be discussed next.

2.4.3 LISTA-CE

LISTA-CE is designed based on ISTA [88]. Most state-of-the-art methods use the Fourier transform for the sparse transformation. However, in the broadband mmWave communication system, the Fourier transform matrix often cannot achieve the desired sparse transformation due to the beam squint effect. LISTA-CE is designed by using deep learning to design such sparse transformation matrices and then exploit the sparsity of mmWave channels for channel estimation.

LISTA-CE is designed considering a lens-based mmWave massive MIMO system, similar to the one considered by LDAMP. The main difference is that LISTA-CE considers an OFDM system. Thus the received signal vector $\mathbf{y}_{m,q} \in \mathbb{C}^{N_{\mathrm{RF}} \times 1}$ at the BS can be written as

$$\mathbf{y}_{m,q} = \mathbf{W}_q \tilde{\mathbf{h}}_m s_{m,q} + \mathbf{W}_q \mathbf{n}_{m,q}, \tag{2.76}$$

where $s_{m,q}$ is the pilot transmitted at subcarrier m in time slot q for $m = 1, 2, \ldots, M_{\mathrm{c}}$ and $q = 1, 2, \ldots, Q$, $\mathbf{n}_{m,q} \sim \mathcal{N}_{\mathbb{C}}\left(\mathbf{0}, \sigma_n^2 \mathbf{I}\right)$ represents a Gaussian noise vector, and

$\mathbf{W}_q \in \mathbb{C}^{N_{\mathrm{RF}} \times N}$ is the adaptive selection network that is fixed for different subcarriers. Note that M_c represents the number of subcarriers and N is the dimension of the antenna array. Setting $s_{m,q} = 1$ for convenience, the received signal, $\bar{\mathbf{y}}_m$, in Q time steps is given by

$$\bar{\mathbf{y}}_m = [\mathbf{y}_{m,1}^T, \ldots, \mathbf{y}_{m,Q}^T]^T = \bar{\mathbf{W}}\tilde{\mathbf{h}}_m + \mathbf{W}\mathbf{n}_m^{\mathrm{eq}}, \tag{2.77}$$

where $\bar{\mathbf{W}} = [\mathbf{W}_1^T, \mathbf{W}_2^T, \ldots, \mathbf{W}_Q^T]^T \in \mathbb{C}^{QN_{\mathrm{RF}} \times N}$ and $\mathbf{n}_m^{\mathrm{eq}} = [\mathbf{n}_{m,1}^T, \ldots, \mathbf{n}_{m,Q}^T]^T$.

By stacking M_c beamspace channel vectors into a matrix and transforming the complex values into real values, we have the following signal recovery model:

$$\mathbf{Y} = \bar{\mathbf{W}}\mathbf{H} + \mathbf{N}, \tag{2.78}$$

where $\mathbf{Y} = [\mathrm{Re}(\bar{\mathbf{y}}_1, \bar{\mathbf{y}}_2, \ldots, \bar{\mathbf{y}}_{M_c}), \mathrm{Im}(\bar{\mathbf{y}}_1, \bar{\mathbf{y}}_2, \ldots, \bar{\mathbf{y}}_{M_c})] \in \mathbb{R}^{QN_{\mathrm{RF}} \times 2M_c}$ is the received signal in the real-valued domain, $\mathbf{H} = [\mathrm{Re}(\tilde{\mathbf{h}}_1, \tilde{\mathbf{h}}_2, \ldots, \tilde{\mathbf{h}}_{M_c}), \mathrm{Im}(\tilde{\mathbf{h}}_1, \tilde{\mathbf{h}}_2, \ldots, \tilde{\mathbf{h}}_{M_c})] \in \mathbb{R}^{N \times 2M_c}$, and $\mathbf{N} = [\mathrm{Re}(\mathbf{n}_1^{\mathrm{eq}}, \mathbf{n}_2^{\mathrm{eq}}, \ldots, \mathbf{n}_{M_c}^{\mathrm{eq}}), \mathrm{Im}(\mathbf{n}_1^{\mathrm{eq}}, \mathbf{n}_2^{\mathrm{eq}}, \ldots, \mathbf{n}_{M_c}^{\mathrm{eq}})]$. Considering the correlation between different subcarriers, the channel matrix can be regarded as a natural 2D image. Therefore, many compressed image restoration methods are readily available for beamspace channel estimation, facilitating the development of model-driven deep learning-based channel estimation networks.

On the basis of traditional ISTA methods for channel estimation, the network structure of LISTA-CE is proposed. Substituting the variables in the channel estimation problem into the ISTA iterative formula, we can rewrite equations 2.62 and (2.63) as

$$\mathbf{R}^{(t)} = \hat{\mathbf{H}}^{(t-1)} - \rho\bar{\mathbf{W}}^T\left(\bar{\mathbf{W}}\hat{\mathbf{H}}^{(t-1)} - \mathbf{Y}\right), \tag{2.79}$$

$$\hat{\mathbf{H}}^{(t)} = \mathbf{\Psi}^T\left(\eta\left(\mathbf{\Psi}\mathbf{R}^{(t)}; \theta^{(t)}\right)\right), \tag{2.80}$$

where $\mathbf{R}^{(t)}$ can be regarded as the noisy signal and (2.80) is used to perform denoising on $\mathbf{R}^{(t)}$, $\mathbf{\Psi}$ is the transformation matrix that makes $\mathbf{\Psi}\mathbf{R}^{(t)}$ sparse in some transform domain, t is the iteration index, and ρ is the iteration step size. The denoiser $\eta(x; \theta)$ is the same as the one in (2.61).

Although ISTA is an efficient algorithm for recovering compressed signals, many parameters, such as λ and ρ, require manual selection. Furthermore, choosing a suitable transformation matrix $\mathbf{\Psi}$ is challenging. Usually, the DFT matrix is utilized, but it is insufficient, especially for a wideband mmWave communication system with the beam squint effect. For example, the beam-delay domain is not completely sparse in the transform domain, which motivates the use of deep learning to learn the sparse transformation matrix $\mathbf{\Psi}$.

To this end, the LISTA-CE network is developed. The structure of LISTA-CE is illustrated in Figure 2.22, which is a revised version of ISTA formed by adding learnable parameters. Compared with traditional ISTA, which only transforms the channel in the frequency domain, the channel is transformed in both the spatial and frequency domains to enhance sparsity. Therefore, the network consists of T cascaded layers, each having the same structure with four modules: module $\mathbf{R}^{(t)}$, module $\hat{\mathbf{H}}^{(t)}$, module $\mathbf{R}'^{(t)}$, and module $\hat{\mathbf{H}}'^{(t)}$. The former two, module $\mathbf{R}^{(t)}$ and module

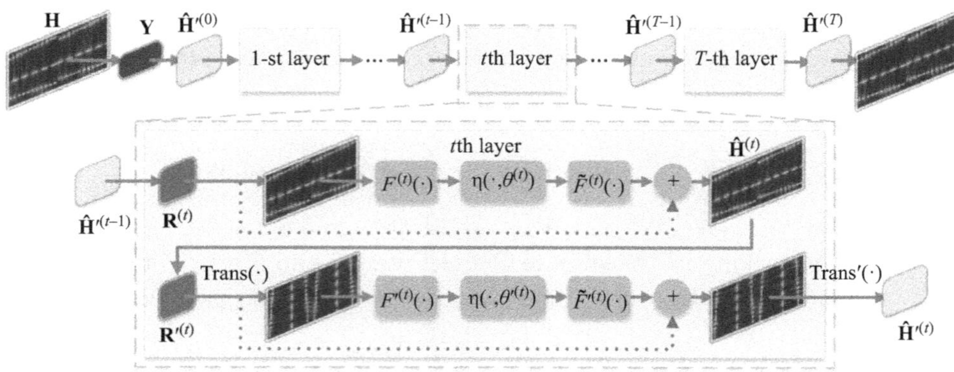

Figure 2.22 Illustration of the LISTA-CE framework [79].

$\hat{\mathbf{H}}^{(t)}$, perform channel transformation and denoising in the frequency domain, while the other two do so in the spatial domain. The input to LISTA-CE is the received signal $\mathbf{Y}$, and the final output $\hat{\mathbf{H}}'^{(T)}$ is an estimate of $\mathbf{H}$, with $\hat{\mathbf{H}}'^{(0)}$ initialized with zero. The four modules are introduced in detail as follows.

- **Module $\mathbf{R}^{(t)}$:** Module $\mathbf{R}^{(t)}$ can be written as

$$\mathbf{R}^{(t)} = \hat{\mathbf{H}}^{(t-1)} - \rho^{(t)}\bar{\mathbf{W}}^T\left(\bar{\mathbf{W}}\hat{\mathbf{H}}'^{(t-1)} - \mathbf{Y}\right), \tag{2.81}$$

which is similar to equation (2.79). $\hat{\mathbf{H}}^{(t-1)}$ and $\hat{\mathbf{H}}'^{(t-1)}$ are calculated in modules $\hat{\mathbf{H}}^{(t-1)}$ and $\hat{\mathbf{H}}'^{(t-1)}$, respectively, which will be introduced later. Note that step size $\rho^{(t)}$ in (2.81) is learned from data and not shared for each layer to further improve the performance.

- **Module $\hat{\mathbf{H}}^{(t)}$:** As stated before, one of the major difficulties in ISTA is to choose the right sparsifying transformation matrix. In response, we propose to use deep learning to learn this transformation, as given by

$$\mathcal{F}^{(t)}(\mathbf{R}^{(t)}) = \mathbf{B}^{(t)} \cdot \text{ReLU}\left(\mathbf{A}^{(t)} \cdot \left(\mathbf{R}^{(t)}\right)^T\right), \tag{2.82}$$

where $\mathbf{A}^{(t)} \in \mathbb{R}^{w_1 \times 2M_c}$ and $\mathbf{B}^{(t)} \in \mathbb{R}^{w_2 \times w_1}$ are two learnable matrices. In addition, $\widetilde{\mathcal{F}}^{(t)}(\cdot)$ signifies the transformation of the signal from the sparse domain to the spatial-frequency domain. The network structure of $\widetilde{\mathcal{F}}^{(t)}(\cdot)$ is symmetric to that of $\mathcal{F}^{(t)}(\cdot)$, which is composed of two fully connected layers. Therefore, the transformation in (2.80) for ISTA is improved to become

$$\hat{\mathbf{H}}^{(t)} = \widetilde{\mathcal{F}}^{(t)}\left(\eta\left(\mathcal{F}^{(t)}\left(\mathbf{R}^{(t)}\right); \theta^{(t)}\right)\right), \tag{2.83}$$

where $\theta^{(t)}$ represents the threshold of the denoiser in the tth layer and is also trainable.

To further speed up the convergence, LISTA-CE considers the concept of residual learning [100]. The dotted line in Figure 2.22 represents the direct connection for residual learning, which enables the network to concentrate on learning the residual noise of the channel. This modification prevents gradient vanishing and significantly accelerates the convergence of the training process. Equation (2.83) can therefore be rewritten as

$$\hat{\mathbf{H}}^{(t)} = \mathbf{R}^{(t)} + \widetilde{\mathcal{F}}^{(t)} \left(\eta \left(\mathcal{F}^{(t)} \left(\mathbf{R}^{(t)} \right) ; \theta^{(t)} \right) \right). \tag{2.84}$$

- **Module $\mathbf{R}'^{(t)}$**: module $\mathbf{R}'^{(t)}$ is similar to module $\mathbf{R}^{(t)}$, except that it takes $\hat{\mathbf{H}}^{(t)}$ as input and $\rho'^{(t)}$ as step size, that is,

$$\mathbf{R}'^{(t)} = \hat{\mathbf{H}}^{(t)} - \rho'^{(t)} \bar{\mathbf{W}}^T \left(\bar{\mathbf{W}} \hat{\mathbf{H}}^{(t)} - \mathbf{Y} \right). \tag{2.85}$$

- **Module $\hat{\mathbf{H}}'^{(t)}$**: In order to perform channel transformation and denoising in the beam domain, module $\hat{\mathbf{H}}'^{(t)}$ first transposes $\mathbf{R}'^{(t)}$, because the left multiplication operation represents the transformation of the matrix columns. Denote the transpose operations as $\mathrm{Trans}(\mathbf{R}'^{(t)}) = [\mathbf{R}'^{(t)}(:,1:M_{\mathrm{c}})^T, \mathbf{R}'^{(t)}(:,M_{\mathrm{c}}+1:2M_{\mathrm{c}})^T]$ and the inverse operation $\mathrm{Trans}'(\cdot)$. Then, $\hat{\mathbf{H}}'^{(t)}$ can be expressed as

$$\hat{\mathbf{H}}'^{(t)} = \mathbf{R}'^{(t)} + \mathrm{Trans}' \left(\widetilde{\mathcal{F}}'^{(t)} \left(\eta \left(\mathcal{F}'^{(t)} \left(\mathrm{Trans} \left(\mathbf{R}'^{(t)} \right) \right) ; \theta'^{(t)} \right) \right) \right). \tag{2.86}$$

The structures of $\mathcal{F}'^{(t)}(\cdot)$ and $\widetilde{\mathcal{F}}'^{(t)}(\cdot)$ are similar to $\mathcal{F}^{(t)}(\cdot)$ and $\widetilde{\mathcal{F}}^{(t)}(\cdot)$, respectively, but with different weights.

From the above discussion, the learnable variables in the LISTA-CE network are $\Theta = \{\rho^{(t)}, \rho'^{(t)}, \theta^{(t)}, \theta'^{(t)}, \mathcal{F}^{(t)}(\cdot), \widetilde{\mathcal{F}}^{(t)}(\cdot), \mathcal{F}'^{(t)}(\cdot), \widetilde{\mathcal{F}}'^{(t)}(\cdot)\}_{t=1}^{T}$.

Numerical results are provided to show the performance of LISTA-CE for wideband beamspace channel estimation. NMSE is used as the performance metric, which is defined as

$$\mathrm{NMSE} = \mathbb{E} \left\{ \|\hat{\mathbf{H}}'^{(T)} - \mathbf{H}\|_2^2 / \|\mathbf{H}\|_2^2 \right\}. \tag{2.87}$$

Figure 2.23 shows that LISTA-CE outperforms other compressed sensing-based algorithms, including ISTA and OMP. In the ISTA approach, the DFT is employed to convert the frequency domain channel into the delay domain. In contrast, LISTA-CE learns a transformation that results in a much sparser representation than the DFT, leading to superior performance.

In conclusion, LISTA-CE leverages the inherent sparsity of the mmWave channel to reduce both the complexity and the number of learnable parameters. Simulation results demonstrate that LISTA-CE significantly surpasses other compressed sensing-based algorithms. Further details can be found in [79].

2.4.4 Other Model-Driven Methods

The success of LDAMP demonstrates the feasibility of introducing deep learning into channel estimation. In recent years, more and more researchers have embraced this idea, leading to two primary directions of research. The first is to improve the LDAMP algorithm, and the second is to combine deep learning with other channel estimation algorithms based on compressed sensing. We next briefly introduce several representative works.

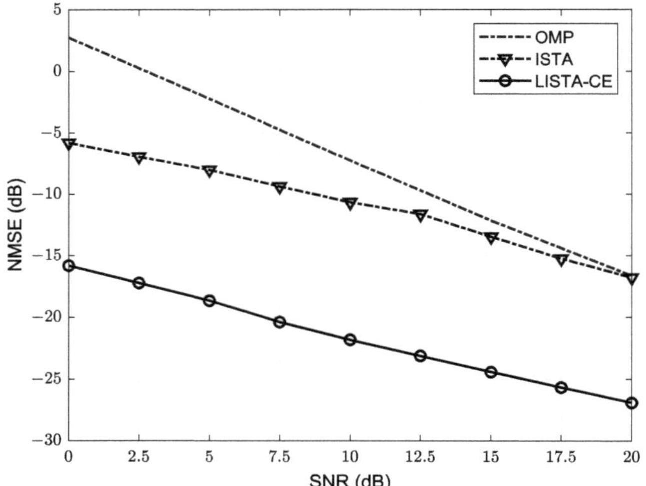

Figure 2.23 NMSE analysis of the LISTA-CE and other channel estimation algorithms.

An AMP-Based Network with Deep Residual Learning [104]

Inspired by LDAMP, an AMP-based network with deep residual learning, termed LampResNet, is proposed by combining the AMP algorithm with residual learning to solve the channel estimation problem of a beamspace mmWave massive MIMO system. LampResNet consists of a cascaded Learned-AMP (LAMP) network and a ResNet. The former performs initial channel estimation and obtains a two-dimensional channel matrix as the input to the latter part, while the latter further optimizes the channel matrix and performs denoising. In contrast to from the DnCNN, ResNet is a denoiser that contains multiple concatenated denoising blocks, and each denoiser contains three convolutional layers. Simulation results show that LampResNet outperforms LDAMP with lower computational complexity.

Gaussian Mixture LAMP [105]

Gaussian mixture LAMP (GM-LAMP) assumes that the elements in the beam domain channel matrix follow the Gaussian distribution. With this prior information, a new shrinkage function is derived and used to refine the AMP algorithm. The original shrinkage function in the LAMP network is replaced by the derived Gaussian mixture shrinkage function. Extensive simulation results are provided to demonstrate the gain of GM-LAMP.

Learned Denoising-Based Generalized Expectation Consistent Algorithm [106]

Most of the current research is based on the AMP algorithm, which requires offline training before online deployment. When the training environment is quite different from the test environment, the performance of the network tends to decline significantly. However, it is challenging to obtain an accurate channel for training the neural network in wireless communication. A learned denoising-based generalized expectation consistent (GEC) algorithm is proposed by combining the DnCNN denoiser with the GEC algorithm. At the same time, Stein's unbiased risk estimator loss is proposed, enabling the training of the neural networks using solely the received signal, without the need for precise channel information. Simulation results show that such a neural network using unsupervised learning can obtain comparable performance to a neural network using supervised learning with accurate channel information at high SNR, and the network can be trained in practical deployment to achieve good performance.

2.5 Learning Channel Compression and Reconstruction

Massive MIMO significantly improves both the spectral and energy efficiency of communication systems by equipping a large number of antennas at the BS. To achieve such performance gains, having high-quality CSI at the BS is important as it dictates the beamforming directions toward each piece of user equipment (UE) in the downlink transmission.

In time-division duplexing (TDD) systems, the uplink and downlink operate on the same frequency band, and channel reciprocity holds, in which the BS can obtain the downlink CSI based on the uplink through proper calibration. However, in frequency-division duplexing (FDD) systems, the uplink and downlink operate on different frequency bands, and reciprocity no longer holds. In order to acquire downlink CSI, the BS transmits pilots for users to estimate the channel, which will report the estimated CSI to the BS. In massive MIMO systems, the huge number of antennas at the BS leads to a high CSI dimension, making the feedback overhead unaffordable. Therefore, the feedback methods based on codebook [107] and compressed sensing [108, 109] have been introduced to reduce CSI feedback overhead. However, the feedback overhead of these methods increases linearly with the number of antennas, posing challenges in massive MIMO systems. Since FDD systems are widely used, how to efficiently feed back downlink CSI becomes a key issue.

In this section, we shall first introduce these traditional methods of CSI compression and reconstruction, and then discuss in detail how to leverage deep learning techniques to improve the performance.

2.5.1 Conventional Feedback Methods

Codebook-Based CSI Feedback

Consider a single-cell massive MIMO-OFDM system with N_c subcarriers, where the BS is equipped with a uniform linear array with N_t transmit antennas, and the UE

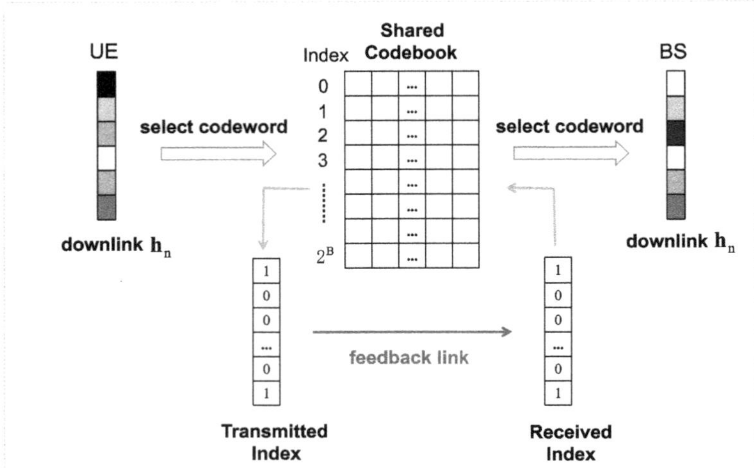

Figure 2.24 Architecture of codebook-based CSI feedback.

is equipped with a single receive antenna. The received signal over the nth subcarrier at the UE is

$$y_n = \mathbf{h}_n^H \mathbf{v}_n x_n + w_n, \quad n = 1, 2, \cdots, N_c, \tag{2.88}$$

where $\mathbf{h}_n \in \mathbb{C}^{N_t \times 1}$, $x_n \in \mathbb{C}$, $\mathbf{v}_n \in \mathbb{C}^{N_t \times 1}$, and $w_n \in \mathbb{C}$ denote the channel vector, transmit data symbol, pre-coding vector, and additive Gaussian white noise, respectively. Therefore, the downlink CSI matrix in the spatial-frequency domain can be obtained by

$$\tilde{\mathbf{H}} = \begin{bmatrix} \mathbf{h}_1 & \mathbf{h}_2 & \cdots & \mathbf{h}_{N_c} \end{bmatrix} \in \mathbb{C}^{N_t \times N_c}. \tag{2.89}$$

In FDD systems, the UE should feed $\tilde{\mathbf{H}}$ back to the BS. The total number of feedback parameters is $N = 2 \times N_t \times N_c$, which will be prohibitively high for massive MIMO systems.

As shown in Figure 2.24, a predefined codebook is shared by the BS and UE in the codebook-based CSI feedback. The UE first selects the codeword closest to the downlink CSI and feeds the index of the selected codeword to the BS. Then, the BS obtains the codeword based on the received index. Taking random vector quantization (RVQ) as an example, the codeword for the downlink CSI can be obtained by

$$\hat{\mathbf{h}}_n = \arg\max_{\mathbf{C}_i \in \mathcal{C}} \left\| \mathbf{h}_n^H \mathbf{C}_i \right\|, \tag{2.90}$$

where $\mathbf{C}_i$ and $\mathcal{C}$ are the codeword and the shared codebook, respectively. Assuming that the size of the codebook is $2^B = |\mathcal{C}|$, the number of feedback bits is reduced to B.

Many codebook design methods have been proposed to improve feedback accuracy, which unfortunately is still insufficient to meet the stringent requirement of massive MIMO systems. Moreover, the incurred feedback overhead and the codeword search complexity become increasingly high with growing antenna sizes, which together make the codebook-based feedback methods untenable in massive MIMO communication systems.

Compressed Sensing-Based CSI Feedback

We have seen in Section 2.4 how compressed sensing can be leveraged to estimate high-dimensional channels with many fewer observations than are required in conventional signal processing methods. The key underlying assumption is that the high-dimensional channel matrix is (approximately) sparse in an appropriate domain, which is indeed the case in massive MIMO systems due to limited scattering in the propagation environment. Such sparsity can be further exploited in the CSI feedback task to reduce the feedback overhead. The CSI matrix, $\tilde{\mathbf{H}}$, can be converted into the angular-delay domain matrix, $\mathbf{H}$, by a two-dimensional discrete Fourier transform (2D-DFT) as

$$\mathbf{H} = \mathbf{F}_{\mathrm{a}}\tilde{\mathbf{H}}\mathbf{F}_{\mathrm{d}}, \tag{2.91}$$

where $\mathbf{F}_{\mathrm{a}} \in \mathbb{C}^{N_t \times N_t}$ and $\mathbf{F}_{\mathrm{d}} \in \mathbb{C}^{N_c \times N_c}$ are the DFT matrices. The CSI compression at the UE can be realized by

$$\mathbf{L} = \Phi\mathbf{H}, \tag{2.92}$$

where $\Phi \in \mathbb{C}^{M \times N_t}$ and $\mathbf{L}$ stand for the sensing matrix and the compressed measurement matrix, respectively. The reconstruction of the CSI can be formulated as an optimization problem as follows:

$$\min \|\hat{\mathbf{H}}\|_0, \quad \text{s.t.} \quad \mathbf{L} = \Phi\hat{\mathbf{H}}, \tag{2.93}$$

where $\hat{\mathbf{H}}$ denotes the signal to be reconstructed and $\|\cdot\|_0$ is the ℓ_0-norm. This problem can be readily solved by the methods introduced in Section 2.4.1.

However, conventional compressed sensing algorithms rely heavily on the assumption that the CSI matrix is sparse and do not make full use of channel structural characteristics. In addition, existing compressed sensing reconstruction algorithms, as discussed in Section 2.4.1, are often iterative, resulting in low reconstruction speed.

2.5.2 Deep Learning-Based CSI Feedback

As shown in Figure 1.10 in Chapter 1, the autoencoder consists of an encoder and a decoder, forming a neural network model capable of data compression and decompression. The input data is first compressed into a low-dimensional embedding by the encoder. Then, the decoder performs reconstruction to generate an output similar to the input. The autoencoder is trained in an unsupervised approach to force the output to be the same as the input. With appropriate dimension and sparsity constraints, the autoencoder can learn data projections that are more meaningful than principal component analysis (PCA) or other techniques. So far, autoencoders have been applied to data denoising and dimension reduction in data visualization.

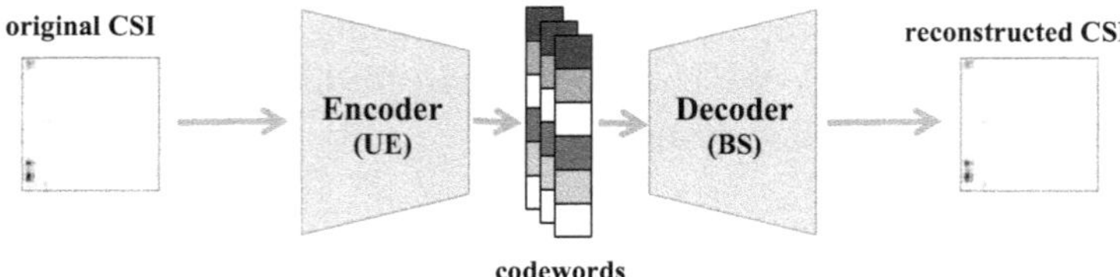

Figure 2.25 CSI feedback based on an autoencoder.

Given the advantages of autoencoders in data compression, researchers have achieved many advances in image compression and reconstruction based on autoencoders. A classical image autoencoder maps an image to a latent vector space through an encoder and then reconstructs the original image through a decoder. The autoencoder is then trained with target data that is the same as the input image.

On the basis of the above work, the autoencoder is introduced into CSI feedback. In general, since the deep learning framework cannot handle complex numbers, the CSI matrices are separated into two channels that represent the real part and imaginary part, respectively. The CSI matrices can be viewed as images to be processed by the autoencoder networks. Specifically, the encoder at the UE converts the CSI matrices into low-dimensional codewords. Then, the UE sends the compressed CSI matrices back to the BS through the feedback link, and the decoder at the BS reconstructs the original CSI matrices from the received low-dimensional codewords. On this basis, the mechanism of deep learning-based CSI feedback can be illustrated in Figure 2.25.

Example 2.4 Inspired by the significant breakthrough of CNNs in image compression and reconstruction, CsiNet [14] is proposed to address CSI feedback in massive MIMO systems. The structure of CsiNet is shown in Figure 2.26, where "Conv" is the convolutional layer, "Dense" is the fully-connected layer, and "Reshape" is the layer that remodels the structure of the data without changing its total number of elements. The numbers in the convolutional layers represent the convolution kernel size, the numbers in the remodeling and fully connected layers represent the output data dimensions, and the numbers marked above each layer represent the number of generated channels.

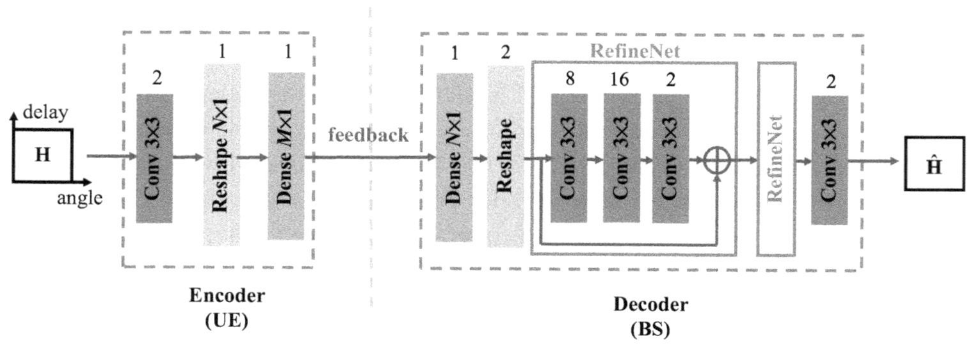

Figure 2.26 Overall architecture of CsiNet.

Specifically, the encoder consists of a convolutional layer and a fully-connected layer. First, the convolutional layer extracts CSI features from the real and imaginary parts of the input CSI matrix to generate two-channel feature maps. After this, the feature maps are combined into a vector. The vector is then input into the fully connected layer, which converts the N-dimensional CSI matrix into an M-dimensional codeword, where M is greater than N. The decoder consists of a fully connected layer, two RefineNet units, and a convolutional layer. The first fully connected layer decompresses the received codeword as an initial estimation of the CSI matrix. The reconstructed CSI matrix is then improved by two RefineNet units. Each RefineNet unit is composed of three convolutional layers. These three layers generate feature maps of 8, 16, and 2 channels, respectively. The input of the first layer and the output of the last layer are added as the output of the whole unit, which effectively alleviates gradient vanishing [100]. Leaky rectified linear unit (LeakyReLU) is used as the activation function for all convolutional layers except the last one. The final convolutional layer uses a sigmoid activation function to normalize the output CSI matrix. All convolutional layers use the "same padding" method to ensure that the output size is the same as the original CSI matrix. In addition, to make the training easier, batch normalization is applied after each convolutional layer, and the output values of each layer are normalized to $(0, 1)$, which can make the training process more robust.

The COST 2100 channel model [110] is used to generate the dataset. Detailed simulation settings can be found in [14]. In the indoor scenario, the number of antennas at the BS is set to $N_t = 32$, and the UE is equipped with a single antenna. The downlink operates in the central frequency band of 5.3 GHz, and the number of subcarriers is $N_c = 1024$. In the outdoor scenario, the central frequency of the system is 300 MHz, and other settings are consistent with the indoor scenario. Before training, a 2D-DFT is used to convert generated CSIs into sparse matrices. Due to the limited channel delay spread, the majority of non-zero values are concentrated in the first $N_c'(N_c' << N_c = 1024)$ columns of $\mathbf{H}$. Therefore, the first $N_c' = 32$ columns of $\mathbf{H}$ are retained and the remaining columns are removed. For simplicity and continuity of notation, the truncated matrix of dimensions $N_t \times N_c'$ is still denoted by $\mathbf{H}$. As a result, the compression size can be reduced to $N = 2N_c'N_t = 2048$.

An end-to-end learning method is used to train CsiNet. The objective function optimized in training is the MSE. The adaptive moment estimation (Adam) optimization algorithm is used for training. The learning rate, the epochs, and the batch size are set to 0.001, 1000, and 200, respectively. The NMSE is employed for evaluating the performance of CsiNet.

CsiNet is compared with three compressed sensing-based methods (LASSO [89], TVAL3 [111], and BM3D-AMP [103]), and the simulation results are shown in Table 2.2. The "CR" defined as M/N in Table 2.2 stands for the compression ratio. CsiNet achieves the lowest NMSE for all compression ratios, significantly surpassing the compressed sensing-based methods. The high performance of CsiNet is mainly attributed to feature extraction. In addition, the reconstruction algorithm is non-iterative and several orders of magnitude faster than the iterative compressed sensing-based algorithms.

Table 2.2 Performance of CsiNet and compressed sensing-based methods in NMSE (dB).

	Indoor			
CR	1/4	1/16	1/32	1/64
LASSO	−7.59	−2.72	−1.03	−0.14
BM3D-AMP	−4.33	0.26	24.72	0.22
TVAL3	−14.87	−2.61	−0.27	0.63
CsiNet	**−17.36**	**−8.65**	**−6.24**	**−5.84**
	Outdoor			
CR	1/4	1/16	1/32	1/64
LASSO	−5.08	−1.01	−0.24	−0.06
BM3D-AMP	−1.33	0.55	22.66	25.45
TVAL3	−6.90	−0.43	0.46	0.76
CsiNet	**−8.75**	**−4.51**	**−2.81**	**−1.93**

Time Correlation-Aided CSI Feedback

CsiNet shows much better reconstruction performance than the traditional compressed sensing-based algorithms. However, as the compression ratio decreases, the reconstruction accuracy of CsiNet drops dramatically. This phenomenon stems from the fact that CsiNet primarily focuses on CSI features within the angular-delay domain, while reconstructing the CSI matrices in different time slots independently and disregarding the correlation in the time domain.

To improve the performance of CSI feedback, CsiNet-long short-term memory (LSTM) in [112] simultaneously utilizes the features in the angular-delay domain and the correlation in the time domain. Recall that the coherence time of the channel is T_c. For convenience of expression, notation $\mathbf{H}$ is replaced by $\mathbf{H}_t$, signifying the CSI matrix at time t in the spatial-frequency domain. We further denote the feedback interval as δt. The T adjacent instantaneous CSI matrices in the angular-delay domain formulate a channel group, denoted by $\{\mathbf{H}_t''\}_{t=1}^{T} = \{\mathbf{H}_1'', \ldots, \mathbf{H}_t'', \ldots, \mathbf{H}_T''\}$, where $\mathbf{H}_t''$ is derived by retaining the first $N_c'(N_c' < N_c)$ columns of $\mathbf{H}_t' = \mathbf{F}_a\mathbf{H}_t\mathbf{F}_d$. These T CSI matrices are correlated if δt and T satisfy the following conditions:

$$0 \leqslant \delta t \cdot T \leqslant T_c. \tag{2.94}$$

Therefore, the correlation between current CSI matrices and previous CSI matrices can be exploited to further reduce the feedback overhead. The procedure of CSI feedback with CsiNet-LSTM is elaborated as follows. First, the UE uses an encoder to compress a group of CSI matrices $\{\mathbf{H}_t''\}_{t=1}^{T}$ into low-dimensional codewords,

$$\mathbf{s}_t = f_{en}\left(\mathbf{H}_t''\right), \tag{2.95}$$

where $f_{en}(\cdot)$ represents the preprocessing and encoding function. Then, a decoder with memory function is used at the BS to extract the correlation from the CSI

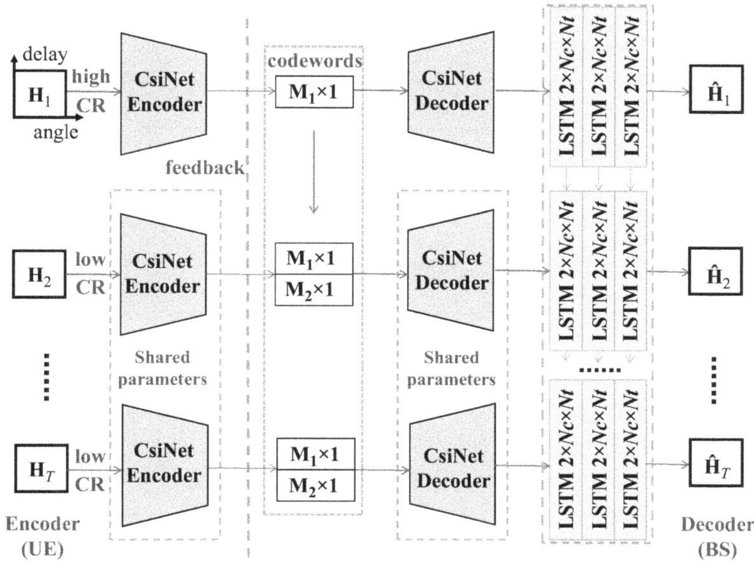

Figure 2.27 Overall architecture of CsiNet-LSTM [112].

matrices $\mathbf{H}_1'', \mathbf{H}_2'', \ldots, \mathbf{H}_{t-1}''$ reconstructed at the previous time. Finally, these reconstructed CSI matrices are combined with codewords $\mathbf{s}_t$ to reconstruct the current CSI matrix $\mathbf{H}_t''$. This process can be expressed as follows:

$$\mathbf{H}_t'' = f_{\mathrm{de}}\left(\mathbf{s}_t, \mathbf{H}_1'', \mathbf{H}_2'', \ldots, \mathbf{H}_{t-1}''\right), \tag{2.96}$$

where $f_{\mathrm{de}}(\cdot)$ represents the decoding and postprocessing function.

CsiNet-LSTM can simultaneously complete the reconstruction of a channel group with T CSI matrices. As shown in Figure 2.27, the encoder and decoder of CsiNet-LSTM have the same structures as CsiNet. When compressing the CSI matrices, CsiNet-LSTM adopts two different compression ratios. The first CsiNet module uses a high compression ratio M_1/N to preserve sufficient structural information of the first CSI matrix for subsequent high-resolution recovery. The remaining $T-1$ CSI matrices are compressed in a more aggressive compression ratio M_2/N. The reconstruction of the first CSI matrix is obtained by feeding the compressed codeword directly into the CsiNet decoder. When reconstructing the other $T-1$ CSI matrices, the first high compression ratio codeword, M_1, is connected in series to the front of all the low compression ratio codewords M_2. Then the connected codewords are input into the CsiNet decoders to make full use of the channel correlation information for decoding. The LSTM in CsiNet-LSTM consists of three LSTM cells with a hidden unit number of $2 \times N_c' \times N_t$, and the number of time steps is set to T. We input T initial reconstructed CSI matrices into the LSTM sequentially. After each time step, the correlation information is updated to capture the change in the CSI over time.

CsiNet-LSTM is based on the observation that the correlation between CSI matrices is similar to the correlation between frames in video signals [113]. At present,

Table 2.3 Performance of CsiNet-LSTM and other methods in NMSE (dB).

		Indoors	
CsiNet	−10.59	−7.35	−6.09
CsiNet-LSTM	**−23.06**	**−22.33**	**−21.24**
		Outdoors	
CsiNet	−3.60	−2.14	−1.65
CsiNet-LSTM	**−9.86**	**−9.18**	**−8.83**

video signals are often processed by recurrent CNNs, where the structure information of each frame is extracted by CNNs, and the correlation information between frames is extracted by recurrent neural networks. Accordingly, CsiNet-LSTM first uses the CNNs to extract the features of CSI matrices in the angular-delay domain and obtain the initial reconstructed CSI matrices. LSTM networks are used to extract the correlation of CSI matrices in the time domain to improve upon the initial reconstructed CSI, given their good performance in dealing with sequence problems. $\mathbf{H}_1''$ can be reconstructed accurately due to its higher compression ratio. At each time step, the LSTM implicitly learns the temporal correlation information from previous CSI matrices and integrates it with the CSI matrix to be reconstructed from the input of the current time step. With the help of the temporal correlation information learned by the LSTM, the $T - 1$ CSI matrices can also achieve reconstruction accuracy similar to or even better than that achieved by the first CSI matrix. Thus, the CsiNet-LSTM improves the tradeoff between the quality of CSI reconstruction and the compression ratio.

The COST 2100 channel model is used to generate simulation data. The CSI feedback interval $\delta t = 0.04$ s, and the channel group size is $T = 10$. The reconstruction performance of CsiNet-LSTM is tested with different compression ratios. In CsiNet-LSTM, the first CSI matrix is compressed using a compression ratio of 1/4. The comparison results are summarized in Table 2.3. CsiNet-LSTM has the lowest NMSE at all compression ratios, surpassing the reconstruction performance of CsiNet. This advantage is particularly noticeable in the outdoor scenario with a low compression ratio. The integration of LSTM networks within CsiNet-LSTM enables the model to capture temporal dependencies and patterns in the CSI data that are crucial for precise reconstruction, especially in dynamic and challenging outdoor conditions.

2.5.3 Other Feedback Methods

Following the pioneering work of [14] that initially applied deep learning to CSI feedback, researchers have undertaken numerous investigations to further reduce feedback overhead and improve recovery accuracy. These studies are divided into four categories: non-ideal feedback, bitstream generation, processing of complex models, and joint design with other modules.

First, most existing algorithms ignore the interference and non-linear effects in feedback links. In fact, the CSI reconstruction accuracy is inevitably reduced in the presence of feedback noise. Therefore, an independent denoising network called DNNet is designed in [114] to improve feedback performance. Second, the UE usually directly feeds compressed codewords with continuous values back to the BS, which is impractical in digital communication systems with limited bandwidth. Therefore, sending codewords quantized in bitstreams is necessary for practical deployment. Considering the influence of quantization distortion, a bit-level optimization design of neural networks called JCResNet is proposed in [115]. JCResNet substantially improves the reconstruction accuracy in terms of bit-level quantization performance. Third, most new feedback networks are very onerous in terms of memory and computing requirements. In practical applications, the encoder needs to be lightweight because the hardware resources of the UE are limited. Therefore, some representative neural network compression and acceleration techniques, such as knowledge distillation, efficient neural network design, pruning, quantization, and low-rank approximation are further incorporated into the CSI feedback network in [116]. Experiments show that these techniques can achieve the same or even higher reconstruction accuracy as the original model with reduced complexity. Finally, the existing works only focus on improving the channel feedback accuracy but ignore the corresponding precoding performance. Therefore, an implicit CSI feedback framework is proposed in [117]. This framework combines feedback and pre-coding modules for optimization to maximize pre-coding performance gain. In addition, the framework is extended to multi-cell systems, where the feedback of the CSI with interference is also needed for designing pre-coding vectors.

2.6 Exercises

Exercise 2.1 Consider the set of empirical measurements of P_r/P_t, i.e., the ratio of received to transmitted power given in Table 2.4 at 900 MHz.

Find the path loss exponent γ and the shadowing variance $\sigma^2_{\phi_{dB}}$ in the simplified path loss and shadowing model in (2.1) that best fit the data in terms of minimizing the MSE, assuming $d_0 = 1$ m and K is determined from the free-space path loss model at this d_0 given by

Table 2.4 Path loss measurements.

Distance from transmitter	P_r/P_t
5 m	-60 dB
30 m	-80 dB
60 m	-105 dB
110 m	-115 dB
500 m	-135 dB

Table 2.5 ITU indoor office environment power delay profiles [52].

Delay in ns	Fractional power
0	0.61722
50	0.30934
110	0.06172
170	0.00978
290	0.00155
310	0.00039

$$10 \log_{10} K = 20 \log_{10} \frac{\lambda}{4\pi d_0}. \tag{2.97}$$

Exercise 2.2   Consider the power delay profiles for indoor office environments given in Table 2.5.

Find the delay spread of the channel and the maximum symbol rate such that a linearly modulated signal does not experience non-negligible inter-symbol interference when transmitted through the channel. Assume that this happens when the symbol duration is no less than 10 times the delay spread.

Exercise 2.3   Section 2.1.2 introduces how we can use CVAE to model the channels with the necessary condition vector $\mathbf{v}$. In fact, the goal of Channel modeling based on VAE is to maximize the log-likelihood objective, $\log p(\mathbf{y}_i|\mathbf{v}_i)$.

(a) Derive the ELBO as presented in (2.10).
(b) Clarify the reasons why we express the latent vector $\mathbf{z}$ as (2.9). (Hint: This procedure is called the reparameterization trick.)

Exercise 2.4   QUAsi Deterministic RadIo channel GenerAtor (QuaDRiGa), is a powerful toolbox for generating realistic radio channel impulse responses for system-level simulations of mobile radio networks. QuaDRiGa was mainly developed to enable the modeling of MIMO channels for specific network configurations, such as indoor, satellite, or heterogeneous configurations. In addition, QuaDRiGa contains a collection of features created in SCM(e) and WINNER channel models along with novel modeling approaches that provide features to enable quasi-deterministic multi-link tracking of user (receiver) movements in changing environments. The code can be found at https://github.com/le-liang/wcmlbook/tree/main/ch2/Exercise_2.4.

(a) Use QuaDRiGa to generate real MIMO channels for one specific configuration.
(b) Use the GAN-based framework to model the MIMO channel you generated in 2.4(a).

Exercise 2.5   DeepMIMO is a framework for generating large-scale MIMO datasets based on an accurate three-dimensional ray-tracing method. They can be found in www.deepmimo.net. For example, the O1 Scenario contains the channels of an outdoor scenario, in which there are many buildings and roads, constructing a relatively realistic communication environment. Choose one scenario in the DeepMIMO dataset, and use the GAN-based framework to model the channel.

Exercise 2.6 For an OFDM system with a delay spread $\tau_s = 50$ ns, if the pilot symbols are arranged in a comb-type pattern, clarify what requirement the pilot placement frequency interval must satisfy to achieve reliable channel estimation. Moreover, is this arrangement better suited for a fast-fading channel or a slow-fading channel?

Exercise 2.7 Section 2.3.1 introduces data-driven SISO-OFDM channel estimation. Please reproduce the simulation results in Figure 2.9. Reference code can be found at https://github.com/le-liang/wcmlbook/tree/main/ch2/Exercise_2.7.

(a) Implement channel estimators using both DNN-based and LMMSE methods for a SISO channel estimation problem. The OFDM system has 64 subcarriers. The first OFDM symbol contains 64 QPSK-modulated pilot symbols, while the second consists of 64 data symbols with 64-QAM modulation. Calculate the MSE performance for SNRs ranging from 5 to 40 dB in 5 dB increments to reproduce the solid line results.
(b) To demonstrate inter-symbol interference, repeat the above experiment but remove the CP from the OFDM system configuration. This should reproduce the dashed line results.

Exercise 2.8 Demonstrate how the optimization problem in (2.52) can be reformulated as a linear program when the data is real-valued. What if the data is complex-valued?

Exercise 2.9 In Example 2.3 involving channel estimation within a lens-based mmWave massive MIMO system, the compression ratio is set to 0.5, and the SNR is set as 20 dB. How will the performance of the algorithm be affected if the compression ratio and SNR change? Verify your conjecture through simulation.

Exercise 2.10 Derive the step size β_t for ISTA that ensures $\widehat{\mathbf{x}}_t$ converges to a minimizer of (2.56) according to reference [90]. Additionally, demonstrate why the introduction of the Onsager correction term enables AMP to converge significantly faster than ISTA.

Exercise 2.11 Use Python along with the deep learning frameworks (e.g., Tensorflow or Pytorch) to complete the following programming tasks. The relevant code for LDAMP can be found at https://github.com/le-liang/wcmlbook/tree/main/ch2/Exercise_2.11.

(a) Compare the performance of LDAMP and LISTA-CE under different channel sparsity levels, evaluating their estimation accuracy (e.g., MSE) and convergence speed.
(b) Adjust the parameters in LDAMP, such as the denoising network architecture, learning rate, and training iterations, and analyze their impact on channel

estimation performance. Similarly, adjust the parameters in LISTA-CE, and assess their effects on the accuracy of channel estimation.

Exercise 2.12   In real-world scenarios, the performance of deep learning-based techniques for channel estimation may deteriorate significantly when there is a mismatch between the training and testing environments. A typical case is the mismatch between indoor and outdoor propagation scenarios. Perform an analysis of the robustness of the LDAMP or LISTA-CE algorithms in the presence of this mismatch based on the QuaDRiGa toolbox introduced in Exercise 2.4. If their robustness is found to be inadequate, explore strategies to enhance their resilience and maintain performance consistency even in the face of environmental discrepancies.

Exercise 2.13   Discuss the advantages and disadvantages of the codebook-based and deep learning-based CSI feedback methods introduced in Section 2.5. Please refer to the 3GPP standardization document technical specification 38.214 [118] for a detailed description of the codebook in 5G NR.

Exercise 2.14   Please transform the channel matrix derived in Exercise 2.4 from the spatial-frequency domain to the angular-delay domain and confirm its sparsity. Additionally, please demonstrate why the channel matrix exhibits sparsity in the angular-delay domain.

Exercise 2.15   How does CsiNet perform on other channel datasets? Please use Python and TensorFlow to complete the following programming tasks. Reference code can be found at https://github.com/le-liang/wcmlbook/tree/main/ch2/Exercise_2.15.

(a) Please adopt the COST 2100 channel model to generate more than five different channel datasets referring to [14], such as changing the distribution of users.
(b) Evaluate the CSI reconstruction NMSE of the trained CsiNet model on each of these datasets.
(c) Please mix the above different channel datasets and use them to train CsiNet. Then, compare the reconstruction performance with that in (b). Based on the results, consider how to improve the generalization of the feedback methods for complex channels in practical systems.

3   Learning Receiver Design: Signal Detection and Channel Decoding

This chapter addresses communication receiver design using machine learning techniques. In particular, we focus on two key receiver modules, i.e., signal detection and channel decoding, which are crucial for reliable communication in wireless networks. Conventional signal detection and channel decoding design typically relies heavily on mathematical models, which may not fully capture the non-ideal effects present in real-world wireless systems. This can lead to suboptimal performance, especially in scenarios with complicated channel conditions and interference. To overcome these limitations, we illustrate how deep learning approaches can be used as a data-driven solution to address the non-ideal effects without relying on predefined mathematical models. Specifically, we discuss the use of deep learning models for signal detection and channel decoding. These models are trained on large datasets of channel measurements and the corresponding transmitted signals, allowing them to learn the complex relationships between the input and output data. In addition to the data-driven approaches, we also explore the use of model-based deep learning methods. These methods combine the advantages of both the conventional model-based approach and the data-driven deep learning approach. They leverage prior knowledge of the wireless channel and the communication system to design deep neural networks that can improve performance beyond either approach alone. We highlight the potential benefits of these model-driven deep learning approaches, including improved performance, flexibility, and adaptability to changing channel conditions.

3.1   Learning-Based OFDM Signal Detection

In this section, we investigate the signal detection problem in orthogonal frequency-division multiplexing (OFDM) systems. The receiver signal processing in conventional OFDM systems is typically implemented in a module-by-module manner, which may not achieve globally optimal performance. To tackle this issue, the use of data-driven learning techniques can be employed to improve symbol detection in OFDM systems. One such approach is the fully connected deep neural network (FC-DNN) proposed in [15]. Compared with the traditional OFDM receiver, the FC-DNN receiver can handle difficult scenarios such as cyclic prefix (CP)-free or pilotless transmissions.

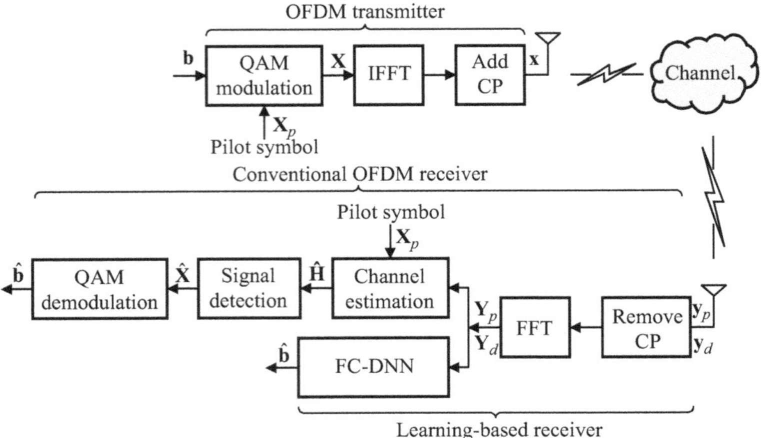

Figure 3.1 Block diagram of an OFDM system with a traditional receiver and a learning-based receiver.

3.1.1 Signal Detection in an OFDM System

The block diagram of an OFDM system with both a traditional receiver and a learning-based receiver is shown in Figure 3.1. The transmitter first modulates the input information bits **b** into quadrature amplitude modulation (QAM) symbols. Then, pilot symbols are inserted to form the transmitted OFDM frame, which consists of two blocks. The first block contains the pilot symbols denoted by $\mathbf{X}_p$, while the second block contains the transmitted data denoted by $\mathbf{X}_d$. The symbol vectors are converted from the frequency domain to the time domain using the inverse fast Fourier transform (IFFT). A CP with N_g time samples is inserted before transmission in the channel to mitigate inter-symbol interference. The channel is assumed to be time-invariant during the transmission of one frame but changes over different frames.

The frequency-domain received sample $Y(k)$ can be expressed as (2.13) based on the discussions in Section 2.2 of Chapter 2. The data detection at the conventional OFDM receiver is carried out by exploiting the estimated channel $\hat{\mathbf{H}}$ in the frequency domain. With OFDM, the transmitted frequency-domain data symbol can be obtained by simple least squares (LS) equalization in each subcarrier as follows:

$$\hat{X}(k) = \frac{Y(k)}{\hat{H}(k)}, \tag{3.1}$$

where $\hat{H}(k)$ is the estimated channel frequency response at the kth subcarrier. The detected symbol $\hat{X}(k)$ is then demodulated to obtain the estimated bitstream $\hat{\mathbf{b}}$.

3.1.2 Deep Neural Network-Based Signal Detection in an OFDM System

Conventional signal detection in OFDM systems depends heavily on channel state information (CSI), and the channel estimator and signal detector are designed separately. The incredible success of deep learning in a wide range of domains has shown its potential to address these limitations by shifting from traditional model-based approaches to data-driven learning methods. The core concept of this data-driven learning paradigm is to replace the conventional modular design with data-driven pipelines. Specifically, deep neural networks (DNNs) can be trained with extensive, well-labeled datasets to seamlessly establish the mapping from inputs to the predictive outputs, circumventing the reliance on analytical modeling or approximations that are intractable in complex scenarios. Moreover, data-driven techniques demonstrate a remarkable capability to discern the salient features from the observed data, which may not be fully captured by conventional mathematical models. We have discussed learning-based channel estimation in Chapter 2. Here we move forward to introduce a joint channel estimation and signal detection model based on the data-driven learning approaches in Example 3.1.

Example 3.1 Figure 3.2 presents the architecture of the DNN used for joint channel estimation and signal detection for the OFDM system illustrated in Figure 3.1. The network takes the received signals as inputs and recovers the transmitted data in an end-to-end manner. Specifically, the input of the network includes the received pilot and data vectors, which are denoted by $\mathbf{Y}_p$ and $\mathbf{Y}_d$, respectively. The two vectors are initially reshaped from complex to real values and then delivered to the input layer of the network. The size of the input layer is proportional to the total number of subcarriers with pilots and data. For example, 64 subcarriers are allocated for pilots and data each in [15]. Thus, the size of the input layer is $64 \times 2 = 128$ for both the pilot and data symbols in Figure 3.2 (containing real and imaginary parts). Then, the input data passes through three hidden layers with the numbers of neurons being 500, 256, and 120, respectively. The size of the output layer is set as 16, which means that every 16 bits of the transmitted data are grouped and recovered by an independent FC-DNN model. Therefore, a total of 8 identical DNNs can work in parallel to predict the 128 transmitted bits under the quadrature phase shift keying (QPSK) modulation scheme. Such a design is beneficial for reducing the complexity and solving the difficulty in training the network because the number of neurons in the output layer decreases from 128 to 16, and high data recovery precision is easier to achieve.

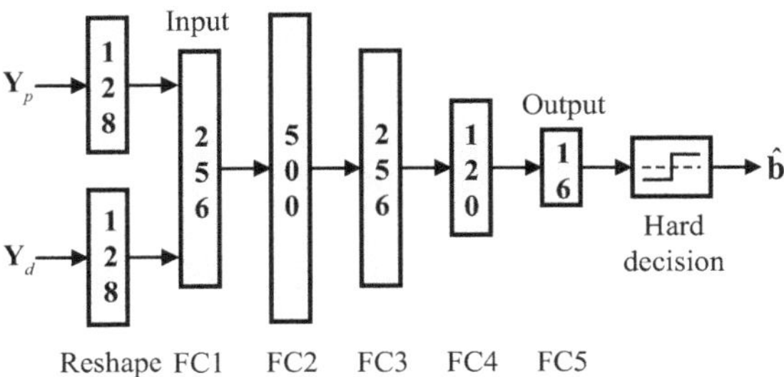

Figure 3.2 The architecture of the FC-DNN.

Two types of activation functions are utilized in the model. All layers except the output layer use the rectifier linear unit (ReLU) function, i.e., $f_{\mathrm{Re}}(a) = \max(0, a)$, and the output layer uses the sigmoid function, i.e., $\sigma(a) = \frac{1}{1+e^{-a}}$. The sigmoid function maps the output into the interval $[0, 1]$, which can be viewed as a soft decision of the raw input binary bits. Finally, hard decisions can be carried out based on the soft output to derive the estimated bits $\hat{\mathbf{b}}$.

The implementation of the FC-DNN receiver includes two stages, i.e., the offline training and the online deployment stages. During the offline training, random training data samples are generated by simulation on the channels with certain statistical properties developed by experts in the wireless communications domain. In particular, random information sequences are first generated and modulated to obtain the transmitted symbols, which together with the pilot symbols constitute an OFDM frame. Notably, the pilot sequence should be fixed during the training and deployment stages, because the network learns how to deal with channel distortions and accurately recover the transmitted data (without estimating the channel response) based on the pilot sequence in the training stage. In other words, a latent mapping conditioned on the pilot between the received signal and the transmitted data is learned through training. Hence, this constant pilot sequence is critical to keeping the network operating properly and robustly. The current channel realization is randomly selected from the specified channel model, and the received signal, which undergoes the current channel together with the additive noise, is collected. Therefore, each training data sample contains the raw input binary data, which functions as the label for supervised learning, and the corresponding received OFDM signal as the input of the network. In terms of the loss function during training, since the label is binary data, the detection problem can be formulated as a binary classification problem, and quadratic or cross-entropy loss can be used to evaluate the network. In [15], the model is trained by minimizing the quadratic loss, i.e., ℓ_2 loss, between the output prediction $\hat{\mathbf{b}}$ of the network and the label $\mathbf{b}$:

$$\ell(\hat{b}, b) = \frac{1}{K} \sum_{k=1}^{K} (\hat{b}(k) - b(k))^2, \tag{3.2}$$

where $\hat{b}(k)$ and $b(k)$ are the kth element of the prediction and the label, respectively. After sufficient training, the coefficients and weights in the model are fixed, and the FC-DNN is deployed online for forward inference. This process can be realized efficiently since the complexity only lies in several matrix multiplications.

Figure 3.3 compares the bit error rate (BER) performance of the FC-DNN with conventional methods in an OFDM wireless communication system and investigates the impact of the pilot numbers and CP on performance. Traditional receivers use LS or minimum mean square error (MMSE) for channel estimation and LS for symbol detection and are named after their corresponding channel estimation methods for short. First, Figure 3.3a shows the performance comparison with different numbers of pilots. When 64 pilots are used, the same as the total number of subcarriers, the FC-DNN achieves much better performance than the LS method and is almost equal to the MMSE method with perfect knowledge of the channels' second-order statistics. Moreover, when the number of pilot subcarriers is reduced to eight,[1] the performance of the LS and MMSE methods severely degrades when the SNR exceeds 20 dB. However, the FC-DNN can continue to reduce its BER as the SNR increases and remarkably outperform the traditional methods in the high-SNR regime. This result indicates that the deep learning-based method is more robust than traditional methods when fewer training pilots are used.

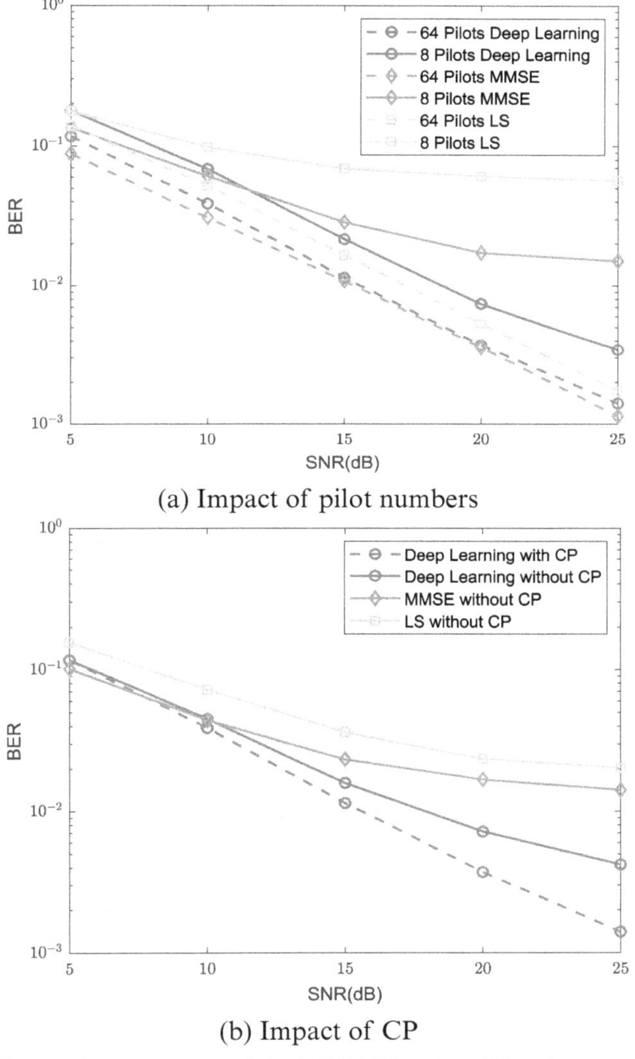

(a) Impact of pilot numbers

(b) Impact of CP

Figure 3.3 BER curves of the FC-DNN and traditional methods. An OFDM system with 64 subcarriers is considered, and the impact of non-linear distortion is investigated.

---

[1] The first OFDM block consists of 8 pilot subcarriers and 56 data subcarriers, and the input and output of the FC-DNN remain unchanged.

Second, the impact of CP-free transmission on different methods is demonstrated in Figure 3.3b. The CP is critical for transforming the linear convolution in wireless channels to circular convolution and combating inter-symbol interference. Symbol detection becomes challenging when no CP is inserted before transmission in the wireless channel, which explains the performance slippage of all the methods investigated in Figure 3.3b. However, the FC-DNN still has advantages over traditional methods, whose BER curves tend to saturate when the SNR is above 15 dB. This result reveals the robustness of the deep learning method against CP removal.

From the figure, we can conclude that the data-driven learning method for symbol detection demonstrates strong power under non-linear scenarios with insufficient pilots or CP removal, where the conventional receiver seriously degrades and saturates at a low- or medium-SNR regime because neither MMSE nor LS can accurately estimate the channel, and the coherent data detection is no longer effective.

The FC-DNN completes the symbol detection and demodulation process without explicitly estimating the channel. This is achieved by merging several modules at the receiver into one black box and leveraging advanced deep learning techniques to optimize reception performance globally. After sufficient training, the network can learn to use the received pilot and data sequence to compensate for the impact exerted by the multipath channel and recover the raw input data. Moreover, initial experiments verify the generalization ability and robustness of the DNN to various mismatches and non-ideal scenarios. This pioneering work demonstrates that deep learning is a promising tool for signal detection problems in receiver design.

3.2 Learning-Based MIMO Detection: Approximate Inference

In this section, we will first formulate the signal detection problem in multiple-input multiple-output (MIMO) systems. Then we present multiple approximate inference algorithms for MIMO detection, including both deterministic and stochastic methods.

3.2.1 Signal Detection in MIMO

Consider a MIMO system equipped with N_t transmit and N_r receive antennas, as shown in Figure 3.4. The transmitted signals can be stacked into a vector $\tilde{\mathbf{x}} \in \tilde{\mathcal{A}}^{N_t}$, where $\tilde{\mathcal{A}}$ is a discrete finite alphabet. Let $\tilde{\mathbf{H}} \in \mathbb{C}^{N_r \times N_t}$ denote the complex channel gain matrix, whose (j, i)th element is the channel gain between the ith transmit antenna and jth receive antenna, with $i = 1, 2, \ldots, N_t$ and $j = 1, 2, \ldots, N_r$. Then, the received signal $\tilde{\mathbf{y}} \in \mathbb{C}^{N_r}$ can be written as

$$\tilde{\mathbf{y}} = \tilde{\mathbf{H}}\tilde{\mathbf{x}} + \tilde{\mathbf{w}}, \tag{3.3}$$

where $\tilde{\mathbf{w}} \in \mathbb{C}^{N_r}$ denotes the receive antenna noise vector.

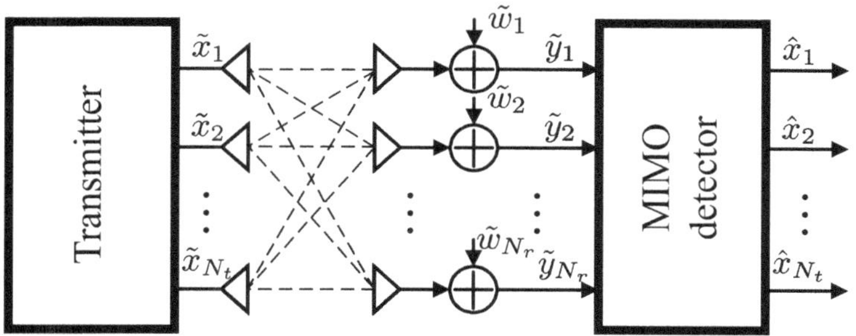

Figure 3.4 Schematic of the MIMO detection model.

To simplify the analysis, the following assumptions are usually made:

- A rectangular $\tilde{M}$-QAM constellation set is used for transmissions with $\tilde{M} = |\tilde{\mathcal{A}}|$, and power normalization is imposed on the transmitted symbols, that is, $\mathbb{E}[|x_i|^2] = \tilde{E}_s = 1, \forall i = 1, 2, \ldots, N_t$.
- The columns of $\tilde{\mathbf{H}}$ are normalized, that is, $\mathbb{E}[\|\mathbf{h}_i\|^2] = 1$, for all $i = 1, 2, \ldots, N_t$.
- The noise vector $\tilde{\mathbf{w}}$ follows the circular symmetric complex Gaussian distribution, $\mathcal{CN}(\mathbf{0}, \tilde{\sigma}_w^2 \mathbf{I}_{N_r})$, where $\tilde{\sigma}_w^2$ is the noise variance.

In addition, the complex-valued MIMO model defined in (3.3) has an equivalent real-valued form, expressed as

$$\mathbf{y} = \mathbf{H}\mathbf{x} + \mathbf{w}, \tag{3.4}$$

where

$$\mathbf{x} = \left[\mathrm{Re}\{\tilde{\mathbf{x}}\}^T, \mathrm{Im}\{\tilde{\mathbf{x}}\}^T\right]^T \in \mathcal{A}^K, \mathbf{y} = \left[\mathrm{Re}\{\tilde{\mathbf{y}}\}^T, \mathrm{Im}\{\tilde{\mathbf{y}}\}^T\right]^T \in \mathbb{R}^N,$$

$$\mathbf{w} = \left[\mathrm{Re}\{\tilde{\mathbf{w}}\}^T, \mathrm{Im}\{\tilde{\mathbf{w}}\}^T\right]^T \in \mathbb{R}^N, \mathbf{H} = \begin{bmatrix} \mathrm{Re}\{\tilde{\mathbf{H}}\} & -\mathrm{Im}\{\tilde{\mathbf{H}}\} \\ \mathrm{Im}\{\tilde{\mathbf{H}}\} & \mathrm{Re}\{\tilde{\mathbf{H}}\} \end{bmatrix} \in \mathbb{R}^{N \times K},$$

with $N = 2N_r$, $K = 2N_t$, and $\mathcal{A} = \{a_1, \ldots, a_M\}$ being the real-valued constellation with $|\mathcal{A}| = M = \sqrt{\tilde{M}}$. The average power per transmit antenna is $E_s = \tilde{E}_s/2 = 1/2$, and the noise covariance matrix is $\sigma_w^2 \mathbf{I}_N$ with $\sigma_w^2 = \tilde{\sigma}_w^2/2$.

Provided with the aforementioned system model, the MIMO symbol detection problem[2] is to recover the transmitted symbol vector $\mathbf{x}$ based on the received signals $\mathbf{y}$ and the estimated channel matrix $\mathbf{H}$. The optimal detector in the sense of minimum error probability is based on the maximum a posteriori (MAP) criterion considering the system model of (3.4),

$$\hat{\mathbf{x}}_{\mathrm{MAP}} = \arg\max_{\mathbf{x} \in \mathcal{A}^K} p(\mathbf{x}|\mathbf{y}). \tag{3.5}$$

Using Bayes' theorem and assuming that the transmitted signals are equiprobable, the posterior distribution can be expressed as

---

[2] We focus our attention on coherent symbol-by-symbol MIMO detection in this section.

$$p(\mathbf{x}|\mathbf{y}) = \frac{p(\mathbf{y}|\mathbf{x})p(\mathbf{x})}{p(\mathbf{y})} \propto \underbrace{\exp\left(-\frac{1}{2\sigma_w^2}\|\mathbf{y} - \mathbf{H}\mathbf{x}\|^2\right)}_{p(\mathbf{y}|\mathbf{x})} \underbrace{\prod_{k=1}^{K} p(x_k)}_{p(\mathbf{x})}, \qquad (3.6)$$

and the MAP criterion reduces to

$$\hat{\mathbf{x}}_{\mathrm{MLD}} = \arg\max_{\mathbf{x}\in\mathcal{A}^K} p(\mathbf{y}|\mathbf{x},\mathbf{H}) = \arg\min_{\mathbf{x}\in\mathcal{A}^K} \|\mathbf{y} - \mathbf{H}\mathbf{x}\|^2. \qquad (3.7)$$

Equation (3.7) provides a way to obtain the maximum likelihood detector (MLD). However, this method is often impractical due to its complexity, which grows exponentially with the number of decision variables. Linear detection methods offer significant complexity reduction, with one of the most prominent being the linear MMSE detector given by

$$\hat{\mathbf{x}}_{\mathrm{MMSE}} = (\mathbf{H}^T\mathbf{H} + \sigma_w^2\mathbf{I}_K)^{-1}\mathbf{H}^T\mathbf{y}. \qquad (3.8)$$

However, linear detectors are widely known to be suboptimal and incapable of achieving a desirable performance–complexity tradeoff. To address this issue, there have been many research efforts to develop highly efficient approximation schemes based on Bayesian inference. Among them, two major categories of approximate inference methods have emerged: deterministic and stochastic approximations. Deterministic methods assume a specific factorization or parametric form (e.g., Gaussian) to obtain analytical approximations of the posterior distribution [91, 119]. These methods are computationally efficient and provide a closed-form approximation. In contrast, stochastic methods are based on numerical sampling and can be computationally expensive. However, they provide more accurate approximations of the posterior distribution and are more flexible in modeling complex distributions.

In the following, we will discuss a number of deterministic methods for estimating $\mathbf{x}$ in (3.4). These include approximate message passing (AMP) [91], which has been shown to be near-optimal under large-scale MIMO Rayleigh fading channels [120]; orthogonal AMP (OAMP) [121], which is applicable to more general channel models; and the expectation propagation (EP) method [119]. Additionally, we will introduce the stochastic sampling-based Markov chain Monte Carlo (MCMC) method [122, 123]. These methods have been widely used in MIMO detection and have shown enhanced performance compared to conventional detectors, e.g., LS and MMSE.

3.2.2 AMP-Based Detector

AMP [91], introduced in Chapter 2, is a variant of the loopy belief propagation (BP) algorithm and can be viewed as an approximate inference method used for signal recovery. The main feature of AMP is the utilization of an Onsager correction term [95] to mitigate the influence of previous estimation errors on the current estimation errors caused by the independent and identically distributed (i.i.d.) dense observation matrix. The iterative process of the AMP-based MIMO detector can be characterized as the iteration between a linear estimator (LE) and a non-linear estimator (NLE) [121]:

$$\text{LE:} \quad \mathbf{r}_t = \hat{\mathbf{x}}_t + \mathbf{H}^T \left(\mathbf{y} - \mathbf{H}\hat{\mathbf{x}}_t \right) + \mathbf{r}_t^{\text{Onsager}}, \tag{3.9}$$

$$\text{NLE:} \quad \hat{\mathbf{x}}_{t+1} = \eta_t \left(\mathbf{r}_t \right), \tag{3.10}$$

where the subscript t is the iteration index, η_t is a component-wise Lipschitz continuous function, and the Onsager correction term $\mathbf{r}_t^{\text{Onsager}}$ is given by

$$\mathbf{r}_t^{\text{Onsager}} = \frac{K}{N} \left(\frac{1}{K} \sum_{j=1}^{K} \eta'_{t-1} \left(r_{t-1}^{j} \right) \right) \left(\mathbf{r}_{t-1} - \hat{\mathbf{x}}_{t-1} \right), \tag{3.11}$$

where r_t^{j} is the jth element of $\mathbf{r}_t$, and $\eta'_t(\cdot)$ is the derivative of $\eta_t(\cdot)$. A typical choice of $\eta_t(\cdot)$ is the MMSE denoiser given by[3]

$$\eta_t(r) = \mathbb{E}[\hat{x}|r, \tau] = \sum_{a_m \in \mathcal{A}} a_m \frac{\mathcal{N} \left(\hat{x} = a_m : r, \tau^2 \right)}{\sum_{a_m \in \mathcal{A}} \mathcal{N} \left(\hat{x} = a_m : r, \tau^2 \right)}, \tag{3.12}$$

where $\mathcal{N} \left(\hat{x} : r, \tau^2 \right)$ denotes a Gaussian distribution for a random variable $\hat{x}$ with mean r and variance τ^2, and τ is the standard deviation of the equivalent input noise of the NLE, which is ensured to be Gaussian distributed by the Onsager correction term.

The analysis of convergence in [124] indicates that AMP is Bayes-optimal when the measurement matrix has a zero-mean i.i.d. sub-Gaussian distribution, provided that the compression ratio $\delta = N/K$ is greater than a specific value. However, if the channel matrix deviates from the strong assumption of being i.i.d. sub-Gaussian, AMP may become unstable. To expand the applicability of AMP to a wider range of problems, a new algorithm, OAMP, has been proposed, which will be discussed in detail in the following subsection.

3.2.3 OAMP-Based Detector

OAMP is a variation of AMP that allows for unitarily invariant channel matrices instead of the assumption of i.i.d. Gaussian channels. Similar to AMP, the mechanism of OAMP [121] can be characterized by the interactive iteration between the LE and NLE:

$$\text{LE:} \quad \mathbf{r}_t = \hat{\mathbf{x}}_t + \mathbf{W}_t \left(\mathbf{y} - \mathbf{H}\hat{\mathbf{x}}_t \right), \tag{3.13}$$

$$\text{NLE:} \quad \hat{\mathbf{x}}_{t+1} = \eta_t^{\text{df}} \left(\mathbf{r}_t \right), \tag{3.14}$$

where $\mathbf{W}_t$ is the de-correlated matrix, which satisfies $\text{tr}(\mathbf{I} - \mathbf{W}_t\mathbf{H}) = 0$, and $\eta_t^{\text{df}}(\cdot)$ is an element-wise divergence-free function. How to properly choose these two terms is discussed next.

---

[3] A uniform prior distribution over the QAM constellation $\mathcal{A}$ is assumed for the transmit symbol.

1. The de-correlated matrix $\mathbf{W}_t$ in the LE is given by

$$\mathbf{W}_t = \zeta_t \hat{\mathbf{W}}_t, \tag{3.15}$$

where the de-correlated coefficient $\zeta_t = \frac{K}{\mathrm{tr}(\hat{\mathbf{W}}_t \mathbf{H})}$ is critical for maintaining the orthogonality between estimation errors. Meanwhile, several common choices for $\hat{\mathbf{W}}_t$ are as follows:

- matched filter (MF):

$$\hat{\mathbf{W}}^{\mathrm{MF}} = \mathbf{H}^T, \tag{3.16}$$

- pseudo-inverse (PINV):

$$\hat{\mathbf{W}}^{\mathrm{PINV}} = \begin{cases} \mathbf{H}^T(\mathbf{H}\mathbf{H}^T)^{-1}, & N \leq K, \\ (\mathbf{H}^T\mathbf{H})^{-1}\mathbf{H}^T, & N > K, \end{cases} \tag{3.17}$$

- LMMSE:

$$\hat{\mathbf{W}}^{\mathrm{LMMSE}} = \mathbf{H}^T \left(\mathbf{H}\mathbf{H}^T + \frac{\sigma_w^2}{v^2}\mathbf{I} \right)^{-1}, \tag{3.18}$$

where v^2 is the error variance of the NLE.

Among the three choices, the LMMSE structure $\hat{\mathbf{W}}^{\mathrm{LMMSE}}$ brings up the Bayes-optimal OAMP.

2. The divergence-free η^{df} in the NLE satisfies $\mathbb{E}[\eta^{\mathrm{df}}(r)'] = 0$, which exempts OAMP from the Onsager correction term. One way to construct η^{df} is given by

$$\eta^{\mathrm{df}}(\mathbf{r}) = C \left(\hat{\eta}(\mathbf{r}) - \left(\frac{1}{K}\sum_{j=1}^{K} \hat{\eta}'(r_j) \right) \mathbf{r} \right), \tag{3.19}$$

where $\hat{\eta}$ is an arbitrary function, and C is an arbitrary constant [121].

It can be proved that the properties of being de-correlated and divergence-free ensure the orthogonality of the output error terms for the LE and NLE. Two error vectors $\mathbf{e}_t = \mathbf{r}_t - \mathbf{x}$ and $\mathbf{f}_t = \hat{\mathbf{x}}_t - \mathbf{x}$ are also introduced to evaluate the accuracy of the estimators. Then, the error variance estimators are defined as

$$\tau_t^2 = \frac{1}{K}\mathbb{E}[\|\mathbf{e}_t\|^2], \quad v_{t+1}^2 = \frac{1}{K}\mathbb{E}[\|\mathbf{f}_t\|^2]. \tag{3.20}$$

These error variances can be used as parameters for constructing the de-correlated matrix $\mathbf{W}$ and divergence-free η^{df} in OAMP and thus minimizing the MSE in the iterations [121].

The Bayes-optimal OAMP has an inherent disadvantage in that an LMMSE estimate is involved in every iteration, which requires a matrix inversion and restricts the use of OAMP in massive MIMO systems. Moreover, the Bayes-optimality of OAMP is derived under the assumptions of large-scale systems and a unitarily invariant sensing matrix. As a result, the OAMP detector severely deteriorates in practical finite-dimensional MIMO systems.

3.2.4 EP-Based Detector

EP [125] provides an alternative form of Bayesian inference, which is aimed at approximating the desired distribution by a function within the exponential distributions family.[4] The exponential distributions represent a family whose probability density function (PDF) can be expressed as

$$q(\mathbf{x}|\boldsymbol{\theta}) \propto h(\mathbf{x}) \exp(\boldsymbol{\theta}^T \mathbf{u}(\mathbf{x})), \tag{3.21}$$

where $\boldsymbol{\theta}$ and $\mathbf{u}(\mathbf{x})$ are the parameter vector and the sufficient statistics of the distribution, respectively. For instance, $\mathbf{u}(\mathbf{x})$ is $\{x_i, x_i x_j\}_{i,j=1}^{d}$ for the d-dimensional multivariate Gaussian distribution. In EP, we aim to minimize the Kullback–Leibler divergence between the true posterior distribution and the approximating distribution $q(\mathbf{x}|\boldsymbol{\theta})$, given by

$$\mathrm{KL}(p\|q) = -\int p(\mathbf{x}|\mathbf{y}) \ln \frac{q(\mathbf{x}|\boldsymbol{\theta})}{p(\mathbf{x}|\mathbf{y})} \, d\mathbf{x}. \tag{3.22}$$

Within the exponential family, minimizing the Kullback–Leibler divergence in the framework of the approximate inference method is equivalent to

$$\mathbb{E}_{q(\mathbf{x}|\boldsymbol{\theta}^*)}[\mathbf{u}(\mathbf{x})] = \mathbb{E}_{p(\mathbf{x}|\mathbf{y})}[\mathbf{u}(\mathbf{x})]. \tag{3.23}$$

Equation (3.23) is known as the moment matching condition, which is used to achieve the resemblance between $q(\mathbf{x})$ and $p(\mathbf{x}|\mathbf{y})$.[5] In the EP framework, this condition is satisfied in an iterative manner by updating the parameters $\boldsymbol{\theta}^*$.

In MIMO detection problems, (3.23) is achieved by replacing the non-Gaussian factors in (3.6) with unnormalized Gaussians, which belong to the exponential family, to approximate the posterior belief, given by [119]:

$$\begin{aligned}
q(\mathbf{x}) &\propto \mathcal{N}\left(\mathbf{y} : \mathbf{Hx}, \sigma_w^2 \mathbf{I}_N\right) \cdot \prod_{k=1}^{K} \exp\left(\gamma_k x_k - \frac{1}{2}\lambda_k x_k^2\right) \\
&\propto \mathcal{N}\left(\mathbf{x} : \mathbf{H}^{\dagger}\mathbf{y}, \sigma_w^2 \left(\mathbf{H}^T\mathbf{H}\right)^{-1}\right) \cdot \mathcal{N}\left(\mathbf{x} : \boldsymbol{\lambda}^{-1}\boldsymbol{\gamma}, \boldsymbol{\lambda}^{-1}\right) \\
&\propto \mathcal{N}\left(\mathbf{x} : \boldsymbol{\mu}, \boldsymbol{\Sigma}\right),
\end{aligned} \tag{3.24}$$

where $\gamma_k \in \mathbb{R}$ and $\lambda_k \in \mathbb{R}^+$, $\boldsymbol{\gamma} = [\gamma_1, \ldots, \gamma_K]^T$, and $\boldsymbol{\lambda} = \mathrm{diag}([\lambda_1, \ldots, \lambda_K])$ is a diagonal matrix with $[\lambda_1, \ldots, \lambda_K]$ being the diagonal elements. Using the Gaussian product lemma,[6] the mean vector $\boldsymbol{\mu}$ and the covariance matrix $\boldsymbol{\Sigma}$ of $q(\mathbf{x})$ can be expressed as

---

[4] For a comprehensive review of the exponential families and their properties, we refer the reader to [126].

[5] To keep the notation uncluttered, we do not explicitly state the conditional probability $q(\mathbf{x}|\boldsymbol{\theta}^*)$ and replace it by $q(\mathbf{x})$ in what follows.

[6] The product of two Gaussians results in another Gaussian [127]

$$\mathcal{N}(\mathbf{x} : \mathbf{a}, \mathbf{A}) \cdot \mathcal{N}(\mathbf{x} : \mathbf{b}, \mathbf{B}) \propto \mathcal{N}\left(\mathbf{x} : \left(\mathbf{A}^{-1} + \mathbf{B}^{-1}\right)^{-1}\left(\mathbf{A}^{-1}\mathbf{a} + \mathbf{B}^{-1}\mathbf{b}\right), \left(\mathbf{A}^{-1} + \mathbf{B}^{-1}\right)^{-1}\right).$$

$$\Sigma = \left(\sigma_w^{-2}\mathbf{H}^T\mathbf{H} + \lambda\right)^{-1}, \tag{3.25}$$

$$\mu = \Sigma\left(\sigma_w^{-2}\mathbf{H}^T\mathbf{y} + \gamma\right). \tag{3.26}$$

The EP algorithm iteratively updates the pairs $(\gamma_k, \lambda_k), k = 1, \ldots, K$ to satisfy the moment matching condition, that is,

$$\mu = \mathbb{E}_{p(\mathbf{x}|\mathbf{y})}[\mathbf{x}], \tag{3.27}$$

$$\Sigma = \mathrm{Cov}_{p(\mathbf{x}|\mathbf{y})}[\mathbf{x}], \tag{3.28}$$

in the context of MIMO detection problems. The parameters γ_k and λ_k can be initialized as $\gamma_k^{(0)} = 0$ and $\lambda_k^{(0)} = 1/E_s$ for all k, and then the pairs $(\gamma_k^{(t)}, \lambda_k^{(t)})$ for all k are updated in parallel at each EP iteration t. Letting, $q^{(t)}(\mathbf{x})$ denote the approximation to $q(\mathbf{x})$ in (3.24) at iteration t, the pair $(\gamma_k^{(t)}, \lambda_k^{(t)})$ can be updated as follows.

- Update the cavity marginal distribution $q^{(t)\backslash k}(x_k)$, denoted by

$$q^{(t)\backslash k}(x_k) = \frac{q^{(t)}(x_k)}{\exp\left(\gamma_k^{(t-1)}x_k - \frac{1}{2}\lambda_k^{(t-1)}x_k^2\right)} \propto \mathcal{N}\left(x_k : x_{e,k}^{(t)}, v_{e,k}^{(t)}\right), \tag{3.29}$$

where $x_{e,k}^{(t)}$ and $v_{e,k}^{(t)}$ denote the elements of the mean and covariance[7] of the marginal distribution, respectively, and can be expressed as

$$v_{e,k}^{(t)} = \frac{\Sigma_k^{(t)}}{1 - \Sigma_k^{(t)}\lambda_k^{(t-1)}}, \tag{3.30}$$

$$x_{e,k}^{(t)} = v_{e,k}^{(t)}\left(\frac{\mu_k^{(t)}}{\Sigma_k^{(t)}} - \gamma_k^{(t-1)}\right). \tag{3.31}$$

- Calculate the mean $\hat{x}_k^{(t)}$ and variance $v_k^{(t)}$ with the estimated posterior distribution $\hat{p}^{(t)}(x_k|\mathbf{y}) \propto q^{(t)\backslash k}(x_k)p(x_k)$.
- Update the pair $(\gamma_k^{(t)}, \lambda_k^{(t)})$ so that

$$q^{(t)\backslash k}(x_k)\exp\left(\gamma_k^{(t)}x_k - \frac{1}{2}\lambda_k^{(t)}x_k^2\right) \propto \hat{p}^{(t)}(x_k|\mathbf{y}), \tag{3.32}$$

that is, the moment-matching condition is satisfied. Following the Gaussian product lemma, we can derive the updated pair $(\gamma_k^{(t)}, \lambda_k^{(t)})$ as

$$\lambda_k^{(t)} = \frac{1}{v_k^{(t)}} - \frac{1}{v_{e,k}^{(t)}}, \tag{3.33}$$

$$\gamma_k^{(t)} = \frac{\hat{x}_k^{(t)}}{v_k^{(t)}} - \frac{x_{e,k}^{(t)}}{v_{e,k}^{(t)}}. \tag{3.34}$$

---

[7] Notably, the covariance matrix of the cavity marginal distribution $q^{(t)\backslash k}(\mathbf{x})$ is assumed to be diagonal due to the independent Gaussian approximation in the EP framework.

Note that the update of $\lambda_k^{(t)}$ can result in a negative value, which should be discarded since $\lambda_k^{(t)}$ is an inverse-variance term. Therefore, the previous value is retained as $\lambda_k^{(t)} = \lambda_k^{(t-1)}$ and $\gamma_k^{(t)} = \gamma_k^{(t-1)}$ when $\lambda_k^{(t)} < 0$. Furthermore, the update is smoothed by a convex combination of the former value to improve the robustness of the algorithm, that is,

$$\lambda_k^{(t)} = \beta \lambda_k^{(t)} + (1 - \beta)\lambda_k^{(t-1)}, \tag{3.35}$$

$$\gamma_k^{(t)} = \beta \gamma_k^{(t)} + (1 - \beta)\gamma_k^{(t-1)}, \tag{3.36}$$

where $\beta \in [0, 1]$ is the damping factor.

The EP algorithm halts when the maximum number of iterations T is reached or the component-wise variation of the mean and covariance is less than a predefined threshold ϵ. Finally, based on the updated mean vector $\boldsymbol{\mu}^*$, the EP detection $\hat{\mathbf{x}}_{\text{EP}}$ can be made by independently deciding on each element:

$$\hat{x}_{\text{EP},k} = \arg \min_{x_k \in \mathcal{A}} |x_k - \mu_k|^2, \text{ for all } k. \tag{3.37}$$

3.2.5 MCMC-Based Detector

MCMC [26] is a widely used stochastic sampling-based method for approximate inference. Unlike previous examples that resort to analytical approximations of the posterior distribution, MCMC approximates the target distribution by generating a sequence of random samples from the distribution of interest. More often than not, we are interested in computing expectations of a random variable rather than its exact probability distribution. Based on the estimated expectation, we can make predictions for problems such as MIMO detection. To illustrate, let us consider the problem of finding the expectation of a function $f(\mathbf{z})$ based on a probability distribution $p(\mathbf{z})$, which can be expressed as

$$\mathbb{E}_p[f] = \int f(\mathbf{z})p(\mathbf{z}) \, d\mathbf{z}. \tag{3.38}$$

Stochastic sampling, namely, the Monte Carlo statistical technique [26], can be used to approximate this expectation without exactly inferring $p(\mathbf{z})$. The rationale behind this idea is that the expectation can be approximated by using the mean of N samples independently drawn from the distribution $p(\mathbf{z})$, that is,

$$\hat{f}_N = \frac{1}{N} \sum_{n=1}^{N} f(\mathbf{z}^{(n)}), \tag{3.39}$$

$$\hat{f}_N \xrightarrow{\text{a.s.}} \mathbb{E}_p[f] \quad \text{as } N \to \infty, \tag{3.40}$$

where the second line follows from the law of large numbers. This illustration of computing expectations of the distribution provides valuable insights for utilizing MCMC in addressing the MIMO detection problem.

Gibbs Sampling for MIMO Detection

We first introduce Gibbs sampling [128], a specific type of MCMC, where the random samples are generated by iterative sampling from each of its conditional distributions. To sample the distribution $p(\mathbf{z}) = p(z_1, \ldots, z_K)$ for constructing the Markov chain, the Gibbs sampling process replaces the value of variable z_k (which is one of the elements of the vector $\mathbf{z}$) with a value sampled from the conditional probability $p(z_k|\mathbf{z}_{\backslash k})$ in each step. Here, $\mathbf{z}_{\backslash k}$ denotes the vector $\mathbf{z}$ with the kth element excluded.

In the context of MIMO detection, this procedure initializes with a selected symbol vector $\mathbf{x}^{(t=1)}$ and then iteratively updates the symbol vector following the aforementioned style, where the variables of the symbol vector are cycled through in some order. Specifically, the update in the tth iteration can be organized as follows if we cycle through the variables in turn [128]:

$$
\begin{aligned}
x_1^{(t+1)} &\sim p(x_1|x_2^{(t)}, x_3^{(t)}, \ldots, x_K^{(t)}, \mathbf{y}, \mathbf{H}), \\
x_2^{(t+1)} &\sim p(x_2|x_1^{(t+1)}, x_3^{(t)}, \ldots, x_K^{(t)}, \mathbf{y}, \mathbf{H}), \\
&\vdots \\
x_k^{(t+1)} &\sim p(x_k|x_1^{(t+1)}, \ldots, x_{k-1}^{(t+1)}, x_{k+1}^{(t)}, \ldots, x_K^{(t)}, \mathbf{y}, \mathbf{H}), \\
&\vdots \\
x_K^{(t+1)} &\sim p(x_K|x_1^{(t+1)}, x_2^{(t+1)}, \ldots, x_{K-1}^{(t+1)}, \mathbf{y}, \mathbf{H}).
\end{aligned}
\tag{3.41}
$$

When the maximum number of iterations is reached, the estimated symbol vector $\hat{\mathbf{x}}$ is chosen as the sample that has the minimum MLD metric among the generated sample list $\mathcal{X}$ from all iterations:

$$
\hat{\mathbf{x}}_{\text{MCMC}} = \arg\min_{\mathbf{x} \in \mathcal{X}} \|\mathbf{y} - \mathbf{H}\mathbf{x}\|^2.
\tag{3.42}
$$

In MIMO detection, the conditional probability in (3.41) can be further written as

$$
p(x_k = a|\mathbf{x}_{\backslash k}^{(t)}, \mathbf{y}, \mathbf{H}) = \frac{e^{-\frac{1}{2\alpha^2\sigma_w^2}\left\|\mathbf{y}-\mathbf{H}\mathbf{x}_{k|a}^{(t)}\right\|^2}}{\sum_{\tilde{a}\in\mathcal{A}} e^{-\frac{1}{2\alpha^2\sigma_w^2}\left\|\mathbf{y}-\mathbf{H}\mathbf{x}_{k|\tilde{a}}^{(t)}\right\|^2}} \mathbf{1}\{a \in \mathcal{A}\},
\tag{3.43}
$$

where $\mathbf{x}_{k|a}^{(t)} \triangleq [\mathbf{x}_{1:k-1}^{(t+1)}, a, \mathbf{x}_{k+1:K}^{(t)}]^T$, α is a tunable positive parameter to control the convergence of the Markov chain [122], which is also called temperature, and a large α would result in a fast mixing of the Markov chain.

Figure 3.5 presents the BER performance comparison of different approximate inference-based detection methods, where the Gibbs sampling-based MCMC detector utilizes a temperature $\alpha = 2$. From the figure, all approximate inference methods outperform the conventional MMSE detector. In addition, EP achieves an advantage over the AMP-based algorithms when the SNR increases. Moreover, the MCMC detector consistently exhibits performance comparable to or exceeding the EP detector across the SNR range.

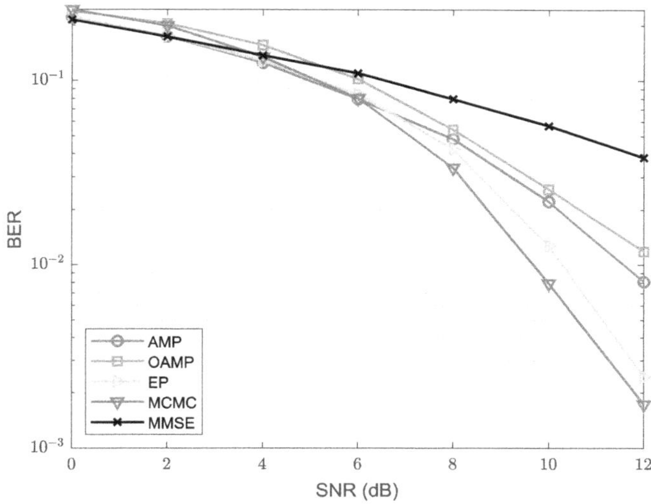

Figure 3.5 The BER performance comparison of different statistical machine learning-based detectors under 8×8 MIMO Rayleigh fading channels with QPSK. The number of iterations in AMP and OAMP is selected as 10, and the counterpart in EP is 5. The Gibbs sampling-based MCMC detectors run 320 iterations before stopping.

Metropolis-Adjusted Langevin Algorithm for MIMO Detection

More recently, a novel gradient descent-aided MCMC method has been proposed [129]. This approach enhances the random walk of the Metropolis–Hastings algorithm, a generalization of Gibbs sampling, by incorporating gradients. The gradient descent provides the Markov chain with the likely directions along which the optimum exists and results in a highly efficient algorithm avoiding the bulk of ineffective search [129]. A similar approach is seen in the stochastic gradient Langevin dynamics method [130], which combines gradient descent with Bayesian sampling. This method adds Gaussian noise to standard gradient descent so that the trajectory of the parameter updates in stochastic gradient descent can converge to the samples distributed according to the posterior distribution. For MIMO detection, this approach enables the Metropolis adjustment to correct the gradient descent through the continuous search space and reach global minima, which are otherwise unattainable due to the inherent discrete property of the QAM constellations.

Along this line of research, a theoretically grounded approach has been proposed in [131], which performs exact sampling in discrete spaces along the gradient descent direction. Assuming a uniform prior distribution, the posterior distribution of the transmitted symbol vector $\mathbf{x}$ in (3.4) can be written as

$$\pi(\mathbf{x}) = \frac{1}{Z} \exp\left(f(\mathbf{x})\right) \prod_{k=1}^{K} \mathbf{1}\{x_k \in \mathcal{A}\}, \tag{3.44}$$

where Z is the normalizing constant, which is intractable to compute in general, and $f(\mathbf{x})$ is the cost function given by

$$f(\mathbf{x}) = -\frac{1}{2\sigma_w^2}\|\mathbf{y} - \mathbf{H}\mathbf{x}\|^2. \tag{3.45}$$

Note that $\pi(\mathbf{x})$ is a probability mass function that can be considered as the restriction of a continuous distribution over $\mathbb{R}^K$ to the discrete space $\mathcal{A}^K$. Hence, gradients of the logarithm of the underlying continuous distribution, i.e., $f(\mathbf{x})$, provide valuable information for sampling from the discrete distribution $\pi(\mathbf{x})$. To leverage such information, we begin with the Langevin dynamics described in Section 1.2.2. Starting from a random sample $\mathbf{x}^{(1)} \in \mathbb{R}^K$, this algorithm generates a proposal $\mathbf{x}'$ at the tth iteration:

$$\mathbf{x}' = \mathbf{x}^{(t)} + \frac{\epsilon}{2}\nabla f(\mathbf{x}^{(t)}) + \sqrt{\epsilon}\mathbf{w}^{(t)}, \tag{3.46}$$

where $\epsilon > 0$ is the step size, $\mathbf{w}^{(t)} \sim \mathcal{N}(\mathbf{0}, \mathbf{I}_K)$ denotes the random perturbation, and ∇f denotes the gradient of $f(\mathbf{x})$ given by

$$\nabla f(\mathbf{x}) = \frac{1}{\sigma_w^2}\mathbf{H}^T(\mathbf{y} - \mathbf{H}\mathbf{x}). \tag{3.47}$$

The proposal generation in (3.46) is equivalent to randomly drawing from a Gaussian distribution

$$\mathcal{N}\left(\mathbf{x}^{(t)} + \frac{\epsilon}{2}\nabla f(\mathbf{x}^{(t)}), \epsilon\mathbf{I}_K\right). \tag{3.48}$$

Taking into account the variable domain $\mathcal{A}^K$, the discrete proposal function using the same rationale as (3.46) is derived as follows:

$$q(\mathbf{x}'|\mathbf{x}^{(t)}) = \frac{\exp\left(-\frac{1}{2\epsilon}\|\mathbf{x}' - \mathbf{x}^{(t)} - \frac{\epsilon}{2}\nabla f(\mathbf{x}^{(t)})\|^2\right)}{Z_A(\mathbf{x}^{(t)})}\prod_{k=1}^{K}\mathbf{1}\{x_k' \in \mathcal{A}\}, \tag{3.49}$$

where the normalizing constant can be expressed as

$$Z_A(\mathbf{x}^{(t)}) = \sum_{\mathbf{x}' \in \mathcal{A}^K}\exp\left(-\frac{1}{2\epsilon}\|\mathbf{x}' - \mathbf{x}^{(t)} - \frac{\epsilon}{2}\nabla f(\mathbf{x}^{(t)})\|^2\right), \tag{3.50}$$

which requires computing the summation over the full space of size M^K and is thus computationally intractable. However, it is important to note that the discrete proposal function can be *elementwise* factorized as [132]

$$q(\mathbf{x}'|\mathbf{x}^{(t)}) = \prod_{k=1}^{K}q_k(x_k'|x_k^{(t)}), \tag{3.51}$$

where $q_k(x_k'|x_k^{(t)})$ is a categorical distribution:

$$q_k(x_k'|x_k^{(t)}) = \sigma\left(\mu(x_k')\right)\mathbf{1}\{x_k' \in \mathcal{A}\}, \tag{3.52}$$

where $\sigma(\cdot)$ denotes the softmax function, and $\mu(\cdot)$ is derived by factorizing the squared ℓ_2 norm term in (3.49) as follows:

$$\mu(x'_k) = \frac{1}{2}\left[\nabla f(\mathbf{x}^{(t)})\right]_k (x'_k - x^{(t)}_k) - \frac{(x'_k - x^{(t)}_k)^2}{2\epsilon}, \tag{3.53}$$

where $\left[\nabla f(\mathbf{x}^{(t)})\right]_k$ denotes the kth element of the gradient vector. The above factorization allows for the update of each element in $\mathbf{x}^{(t)}$ in parallel, i.e., $x'_k \sim q_k(\cdot|x^{(t)}_k)$. The primary computation lies in determining the gradient, which scales only as $\mathcal{O}(K^2)$, enabling an efficient exploration of the search space without incurring excessive costs.

The iteration step in (3.46) is derived from discretizing the Langevin diffusion process, inevitably resulting in discretization errors and bias to the target distribution [133]. The Metropolis adjustment is further incorporated to mitigate these errors [133], guaranteeing the convergence to the target distribution π. Specifically, instead of letting $\mathbf{x}^{(t+1)} = \mathbf{x}'$, the proposal $\mathbf{x}'$ is accepted as the new sample $\mathbf{x}^{(t+1)}$ with a probability computed as

$$A(\mathbf{x}'|\mathbf{x}^{(t)}) = \min\left\{1, \frac{\pi(\mathbf{x}')q(\mathbf{x}^{(t)}|\mathbf{x}')}{\pi(\mathbf{x}^{(t)})q(\mathbf{x}'|\mathbf{x}^{(t)})}\right\}$$

$$= \min\left\{1, \exp\left(f(\mathbf{x}') - f(\mathbf{x}^{(t)})\right)\frac{q(\mathbf{x}^{(t)}|\mathbf{x}')}{q(\mathbf{x}'|\mathbf{x}^{(t)})}\right\}, \tag{3.54}$$

where $q(\mathbf{x}^{(t)}|\mathbf{x}')$ denotes the reverse proposal, computed in a similar manner as in (3.51)–(3.53) for the forward proposal $q(\mathbf{x}'|\mathbf{x}^{(t)})$. Under this criterion, the corresponding Markov chain $\{\mathbf{x}^{(t)}\}$ admits π as its invariant distribution.

The effectiveness of the exact gradient-based MCMC method proposed in [131] is illustrated by convergence diagnostics. A 2×2 MIMO system with SNR $= 8$ dB and QPSK modulation is considered. Figure 3.6 presents the total variation distance[8] between the sampled distributions from gradient-based MCMC approaches and the target posterior distribution with respect to the iteration index t. The sampled distributions are obtained by gathering the samples from 10^5 independent samplers at each iteration t and tallying the occurrences to represent the probability of each state within the search space. The theoretical convergence curve, derived from the convergence theorem of MCMC [134, theorem 4.9], is also shown as a benchmark. From the figure, the empirical curve of the exact gradient-based MCMC method's total variation distance tightly aligns with the theoretical convergence curve, approaching zero as t increases. However, the total variation distance of the inexact method from [129] converges to approximately 0.2, indicating a bias toward the target. These results highlight the effectiveness and superiority of the rigorous probabilistic sampling design. Its practicality and achievement of near-optimal performance in MIMO detection have also been corroborated by extensive empirical studies in [131].

---

[8] The total variation distance between two distributions π_1 and π_2 over $\mathcal{A}^K$ is given by

$$\|\pi_1 - \pi_2\|_{\mathrm{TV}} = \frac{1}{2}\sum_{\mathbf{x}\in\mathcal{A}^K} |\pi_1(\mathbf{x}) - \pi_2(\mathbf{x})|.$$

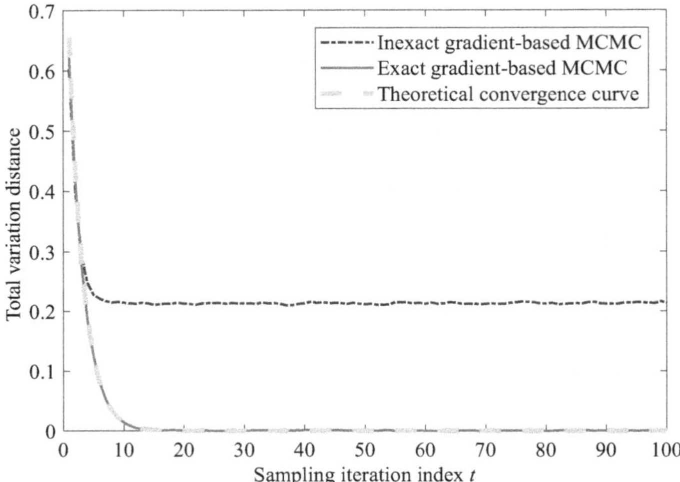

Figure 3.6 Total variation distance with respect to the sampling iteration index in a 2×2 MIMO system with Rayleigh fading channels, QPSK modulation, and SNR $= 8$ dB.

Example 3.2 In this example, we illustrate how to extend the MCMC methods to perform soft-output detection and provide reliability information, i.e., log-likelihood ratios (LLRs), about the transmitted data. This information is usually input to the channel decoder for enhanced decoding performance in modern communication systems. Denote the number of bits in a symbol as Q and the transmitted bit sequence as $\mathbf{b} = [b_1, b_2, \ldots, b_{KQ}]^T$, which is further mapped to the transmitted symbol vector $\mathbf{x} \in \mathcal{A}^K$. The LLR for the kth entry b_k of the transmitted bit sequence $\mathbf{b}$ is defined as

$$L_k = \ln \frac{P\left(b_k = +1 | \mathbf{y}\right)}{P\left(b_k = -1 | \mathbf{y}\right)}. \tag{3.55}$$

To calculate the LLRs, the conditional probabilities $P\left(b_k = +1 | \mathbf{y}\right)$ and $P\left(b_k = -1 | \mathbf{y}\right)$ should be evaluated for $k = 1, 2, \ldots, KQ$. Defining $\mathbf{b}_{\backslash k} \triangleq [b_1, \ldots, b_{k-1}, b_{k+1}, \ldots, b_{KQ}]^T$, the conditional probabilities can be derived as

$$
\begin{aligned}
P\left(b_k = +1 | \mathbf{y}\right) &= \sum_{\mathbf{b}_{\backslash k}} P\left(b_k = +1 | \mathbf{b}_{\backslash k}, \mathbf{y}\right) P\left(\mathbf{b}_{\backslash k} | \mathbf{y}\right), \\
P\left(b_k = -1 | \mathbf{y}\right) &= \sum_{\mathbf{b}_{\backslash k}} P\left(b_k = -1 | \mathbf{b}_{\backslash k}, \mathbf{y}\right) P\left(\mathbf{b}_{\backslash k} | \mathbf{y}\right).
\end{aligned}
\tag{3.56}
$$

These calculations are generally intractable due to the summations over all possible values of $\mathbf{b}_{\backslash k}$, and Monte Carlo summation can be used for approximation. Specifically, considering that the Gibbs sampler is used to generate N_s samples from the posterior distribution $P(\mathbf{b}|\mathbf{y})$ for calculating the LLRs, $P\left(b_k = +1 | \mathbf{y}\right)$ can then be approximated by the empirical average as

$$P\left(b_k = +1|\mathbf{y}\right) \approx \frac{1}{N_s} \sum_{t=1}^{N_s} P(b_k = +1|\mathbf{b}_{\backslash k}^{(t)}, \mathbf{y}), \tag{3.57}$$

where $\mathbf{b}_{\backslash k}^{(t)} \triangleq [b_1^{(t)}, \ldots, b_{k-1}^{(t)}, b_{k+1}^{(t)}, \ldots, b_{KQ}^{(t)}]^T \sim P(\mathbf{b}_{\backslash k}|\mathbf{y})$ is obtained by dropping the kth entry in the sample $\mathbf{b}^{(t)}$ drawn in the tth iteration. Thus, the problem is converted to calculating the full conditional probability $P(b_k = +1|\mathbf{b}_{\backslash k}^{(t)}, \mathbf{y})$, which can be expressed as

$$\begin{aligned} P(b_k = +1|\mathbf{b}_{\backslash k}^{(t)}, \mathbf{y}) &= \frac{P(\mathbf{y}, \mathbf{b}_{\backslash k}^{(t)}, b_k = +1)}{P(\mathbf{y}, \mathbf{b}_{\backslash k}^{(t)})} \\ &= \frac{P(\mathbf{y}|\mathbf{b}_{k+}^{(t)})P(\mathbf{b}_{k+}^{(t)})}{P(\mathbf{y}, \mathbf{b}_{\backslash k}^{(t)})}, \end{aligned} \tag{3.58}$$

where $\mathbf{b}_{k+}^{(t)} = [b_1^{(t)}, \ldots, b_{k-1}^{(t)}, +1, b_{k+1}^{(t)}, \ldots, b_{KQ}^{(t)}]^T$, and we can similarly define $\mathbf{b}_{k-}^{(t)} = [b_1^{(t)}, \ldots, b_{k-1}^{(t)}, -1, b_{k+1}^{(t)}, \ldots, b_{KQ}^{(t)}]^T$. To avoid numerical instability, a log-domain implementation is employed for this calculation by defining

$$\begin{aligned} \gamma_k &= \ln \frac{P(b_k = +1|\mathbf{b}_{\backslash k}^{(t)}, \mathbf{y})}{P(b_k = -1|\mathbf{b}_{\backslash k}^{(t)}, \mathbf{y})} \\ &= \ln \frac{P(\mathbf{y}|\mathbf{b}_{k+}^{(t)})P(\mathbf{b}_{k+}^{(t)})}{P(\mathbf{y}|\mathbf{b}_{k-}^{(t)})P(\mathbf{b}_{k-}^{(t)})} \\ &= \frac{1}{2\sigma_w^2} \left(\|\mathbf{y} - \mathbf{H}\mathbf{x}_{k-}^{(t)}\|^2 - \|\mathbf{y} - \mathbf{H}\mathbf{x}_{k+}^{(t)}\|^2 \right), \end{aligned} \tag{3.59}$$

where $\mathbf{x}_{k+}^{(t)}$ and $\mathbf{x}_{k-}^{(t)}$ are the transmitted symbol vectors mapped from $\mathbf{b}_{k+}^{(t)}$ and $\mathbf{b}_{k-}^{(t)}$, respectively. The final line in (3.59) is derived by considering the likelihood $p(\mathbf{y}|\mathbf{x})$ defined in (3.6) and assuming no a priori information. Considering that $P(b_k = +1|\mathbf{b}_{\backslash k}^{(t)}, \mathbf{y}) = 1 - P(b_k = -1|\mathbf{b}_{\backslash k}^{(t)}, \mathbf{y})$, we can derive

$$P(b_k = +1|\mathbf{b}_{\backslash k}^{(t)}, \mathbf{y}) = \frac{1}{1 + \exp(-\gamma_k)}. \tag{3.60}$$

Hence, after evaluating (3.59) and (3.60) for each sample, $P\left(b_k = +1|\mathbf{y}\right)$ can be approximated by using (3.57). Similarly, we can calculate $P\left(b_k = -1|\mathbf{y}\right)$, and the LLR computation in (3.55) can then be realized. Figure 3.7 shows the performance of the MCMC detector with soft-output detection, exhibiting significant gain over the counterpart that performs hard decisions.

Moreover, the other statistical machine learning detectors mentioned above, including the AMP, OAMP, and EP methods, can also achieve soft-output detection. For detailed information, we recommend referring to [135, 136].

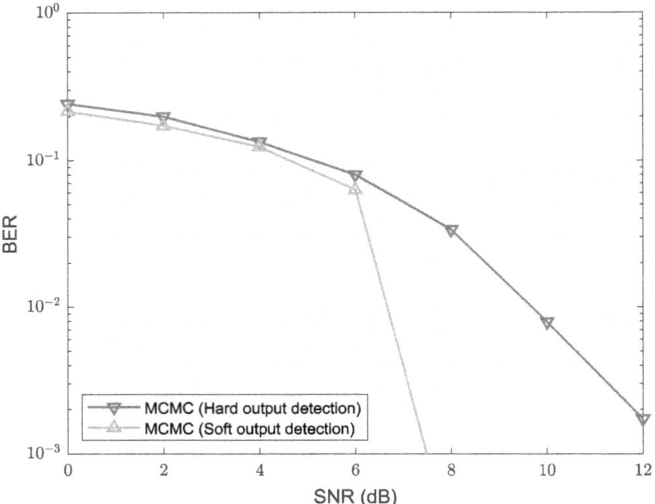

Figure 3.7 The BER performance of MCMC with hard/soft decision outputs. The simulation settings are the same as Figure 3.5, except that low-density parity-check (LDPC) codes with a code rate of 3/4 and block length of 1944 bits are used for channel coding.

3.3 Learning-Based MIMO Detection: Model-Driven Methods

Model-driven deep learning, which integrates domain knowledge into neural network design, has shown great potential in solving traditionally challenging problems and significantly reduced the training cost of data-driven methods. In this section, we introduce model-driven learning-based MIMO detectors and illustrate how conventional iterative detectors can be improved by incorporating learning ingredients.

3.3.1 DetNet for Signal Detection

Recall that the real-valued system model for the MIMO detection problem is given by (3.4). The derivation for DetNet [137] begins with transforming the observation vector $\mathbf{y}$ into the compressed sufficient statistic, i.e.,

$$\mathbf{H}^T\mathbf{y} = \mathbf{H}^T\mathbf{H}\mathbf{x} + \mathbf{H}^T\mathbf{w}, \tag{3.61}$$

where $\mathbf{H}^T\mathbf{y}$ and $\mathbf{H}^T\mathbf{H}\mathbf{x}$ are the two main ingredients used by the architecture to obtain the estimation $\hat{\mathbf{x}}$. Then, DetNet is derived by unfolding a projected gradient descent solution for maximum likelihood optimization, in which the iterations are of the form

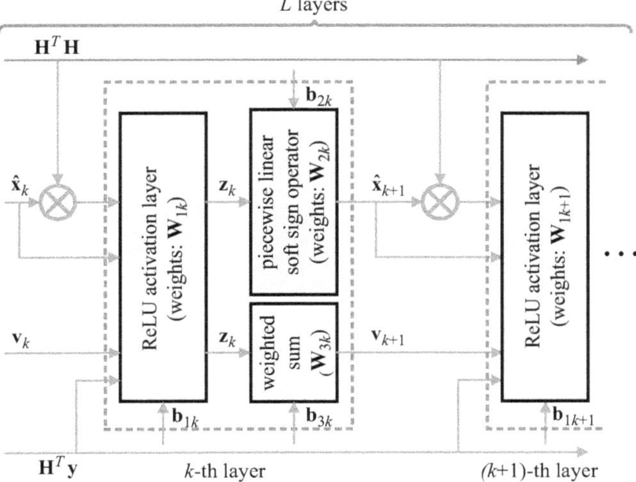

Figure 3.8 The single layer structure of the DetNet.

$$\hat{\mathbf{x}}_{k+1} = \Pi\left[\hat{\mathbf{x}}_k - \delta_k \frac{\partial \frac{1}{2}\|\mathbf{y} - \mathbf{Hx}\|^2}{\partial \mathbf{x}}\bigg|_{\mathbf{x}=\hat{\mathbf{x}}_k}\right]$$

$$= \Pi\left[\hat{\mathbf{x}}_k - \delta_k \mathbf{H}^T\mathbf{y} + \delta_k \mathbf{H}^T\mathbf{H}\hat{\mathbf{x}}_k\right], \tag{3.62}$$

where δ_k is the step size of the gradient descent, and $\hat{\mathbf{x}}_k$ is the estimated symbol vector in the kth iteration. $\Pi[\cdot]$ is the non-linear operator projecting the linear combination of $\hat{\mathbf{x}}_k$, $\mathbf{H}^T\mathbf{y}$, and $\mathbf{H}^T\mathbf{Hx}$ into a higher-dimensional space by using trainable transformation matrices (weights). As shown in Figure 3.8, each iteration is mapped to a layer of the network, which performs as

$$\mathbf{z}_k = \rho\left(\mathbf{W}_{1k}\begin{bmatrix}\mathbf{H}^T\mathbf{y} \\ \hat{\mathbf{x}}_k \\ \mathbf{H}^T\mathbf{H}\hat{\mathbf{x}}_k \\ \mathbf{v}_k\end{bmatrix} + \mathbf{b}_{1k}\right),$$

$$\hat{\mathbf{x}}_{k+1} = \psi_{t_k}\left(\mathbf{W}_{2k}\mathbf{z}_k + \mathbf{b}_{2k}\right),$$

$$\hat{\mathbf{v}}_{k+1} = \mathbf{W}_{3k}\mathbf{z}_k + \mathbf{b}_{3k},$$

$$\hat{\mathbf{x}}_1 = \mathbf{0},$$

$$\hat{\mathbf{v}}_1 = \mathbf{0}, \tag{3.63}$$

where $\rho(a) = \max(0, a)$ denotes a standard ReLU activation function; $\{\mathbf{W}_{1k}, \mathbf{W}_{2k}, \mathbf{W}_{3k}\}_{k=1}^{L}$ and $\{\mathbf{b}_{1k}, \mathbf{b}_{2k}, \mathbf{b}_{3k}\}_{k=1}^{L}$ are the trainable weights and bias, respectively; $\psi_t(\cdot)$ is a piecewise linear soft sign operator that is parameterized by t, that is,

$$\psi_t(x) = -1 + \frac{\rho(x+t)}{|t|} - \frac{\rho(x-t)}{|t|}; \tag{3.64}$$

and $\{t_k\}_{k=1}^L$ for all layers are the trainable parameters in the network. The computation of $\mathbf{z}_k$ in (3.63) can be viewed as an implicit implementation of the gradient descent stage in (3.62). Notably, an explicit gradient descent can also be used in DetNet as follows:

$$\mathbf{q}_k = \hat{\mathbf{x}}_k - \delta_{1k}\mathbf{H}^T\mathbf{y} + \delta_{2k}\mathbf{H}^T\mathbf{H}\hat{\mathbf{x}}_k,$$

$$\mathbf{z}_k = \rho\left(\mathbf{W}_{1k}\begin{bmatrix} \mathbf{q}_k \\ \mathbf{v}_{k-1} \end{bmatrix} + \mathbf{b}_{1k}\right), \tag{3.65}$$

where δ_{1k} and δ_{2k} are the learnable gradient step sizes. Another vital component in the architecture is the sub-layer to learn the non-linear projection operator by approximating the operation with the soft sign activation layer, which leads to the calculation of $\hat{\mathbf{x}}_{k+1}$ in (3.63). Meanwhile, $\hat{\mathbf{v}}_{k+1}$ in (3.63) is used as the unconstrained information passed from one layer to another of the network to improve the ability of the network in capturing complicated features. Overall, a total of L layers are applied in DetNet, and the final estimate is denoted by $\hat{\mathbf{x}}_{L+1}$.

To mitigate the problems of gradient vanishing, the loss function for tuning the parameters $\boldsymbol{\theta} = \{\mathbf{W}_{1k}, \mathbf{b}_{1k}, \mathbf{W}_{2k}, \mathbf{b}_{2k}, \mathbf{W}_{3k}, \mathbf{b}_{3k}, t_k\}_{k=1}^L$ in DetNet is specially designed by taking the output of all layers into account, which can be expressed as

$$\ell\left(\mathbf{x}; \hat{\mathbf{x}}_{\boldsymbol{\theta}}(\mathbf{H}, \mathbf{y})\right) = \sum_{k=1}^L \ln(k)\left\|\mathbf{x} - \hat{\mathbf{x}}_k\right\|^2. \tag{3.66}$$

Last but not least, a residual feature is added in DetNet to further enhance the performance, similar to ResNet [100] in the field of computer vision:

$$\hat{\mathbf{x}}_{k+1} = \alpha\hat{\mathbf{x}}_k + (1 - \alpha)\hat{\mathbf{x}}_{k+1}. \tag{3.67}$$

Equation (3.67) shows that the output of the current layer is the weighted average with the output of the previous layer, and α is the weight coefficient and is chosen as $\alpha = 0.9$ in the implementation.

DetNet is compared with the classical zero-forcing (ZF) detector, the AMP detector [120], and the semidefinite relaxation (SDR) method [138]. The performance of the compared detectors is evaluated in a MIMO channel with input and output sizes of $K = 30$ and $N = 60$, respectively. To test the robustness of DetNet in complex channels, two types of channels are considered. The first is a fixed channel (FC) with a deterministic yet ill-conditioned channel matrix $\mathbf{H}$. The matrix $\mathbf{H}$ is generated such that $\mathbf{H}^T\mathbf{H}$ is a Toeplitz matrix with $[\mathbf{H}^T\mathbf{H}]_{i,j} = 0.55^{|i-j|}$, which is challenging for detection. The second is a varying channel (VC) model, where the channel matrix elements $H_{i,j}$ are randomly chosen from the distribution $\mathcal{N}(0, 1/N)$. DetNet is trained under the corresponding channel model (either the FC or VC model), and the SNR is uniformly chosen from the test range [SNRmin, SNRmax].

Figure 3.9 presents the BER performance of different detectors under the FC model. The AMP detector performs poorly in this scenario because of the high condition number of the channel matrix. On the other hand, DetNet-FC, which is trained

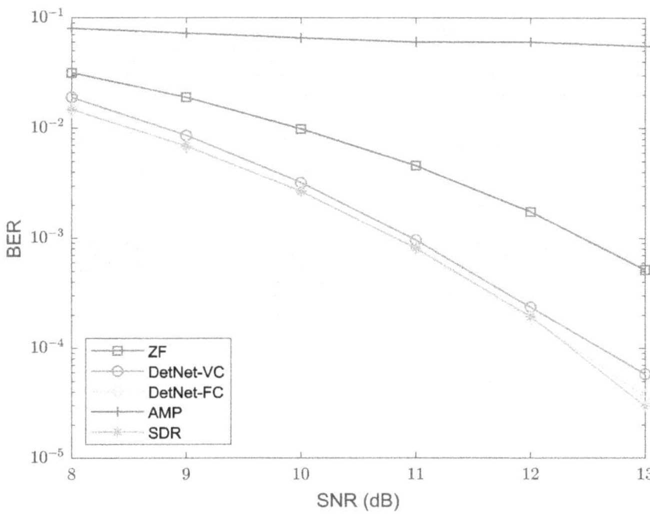

Figure 3.9 The BER performance comparison under the fixed channel with BPSK modulation.

under the same scenario, outperforms AMP significantly and achieves the same performance as the SDR method, which is more computationally expensive. DetNet-FC runs 30 times faster than SDR in a single detection, indicating the efficiency of the learning-based detector. Additionally, DetNet-VC, which is trained using the VC model, also achieves remarkable performance with only 0.2 dB performance loss compared to DetNet-FC, indicating the significant robustness of DetNet against channel mismatch.

The BER performance comparison under the VC model is presented in Figure 3.10. As expected, the AMP- and SDR-based detectors show excellent performance since they are theoretically optimal under the i.i.d. Gaussian channel distribution. However, DetNet-VC performs comparably to AMP and SDR in this scenario. For instance, the gap between DetNet-VC and AMP is only 0.3 dB if we target BER $= 10^{-4}$ in Figure 3.10. Notably, DetNet does not require knowledge of the system SNR, which is essential information for AMP, giving it an advantage in practical situations where SNR values are unknown or hard to be estimated.

DetNet is one of the first works that brings the learning method into MIMO detection. The proposed method incorporates model-based domain knowledge into the neural network design, which is different from the data-driven DNN and is more interpretable. The derived network can improve the accuracy of the iterative detector from which it originates and reach comparable performance to near-optimal solutions under the examined scenarios, but it runs much faster. However, DetNet still contains a large number of parameters to be tuned, requires substantial training data, and performs poorly under high-order modulation or small-sized MIMO channels.

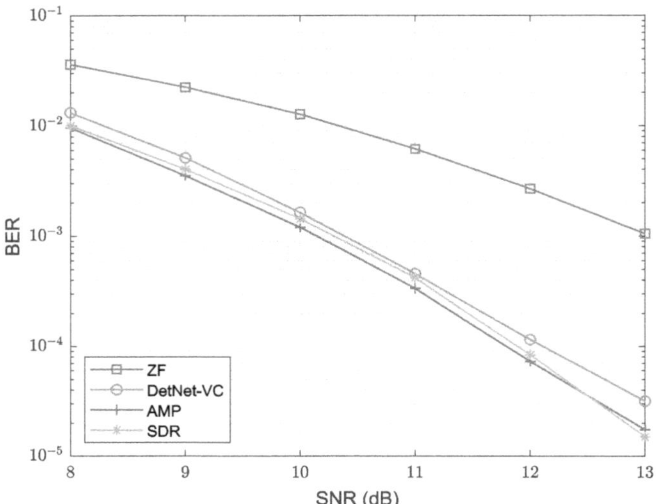

Figure 3.10 The BER performance comparison under the varying channel with BPSK modulation.

3.3.2 OAMP-Net for Signal Detection

OAMP-Net [139] is built by unfolding the OAMP iterations and incorporating only four trainable parameters into each layer of the network. This structure substantially reduces the training cost compared to DetNet. By tuning the parameters to adapt to realistic environments through deep learning techniques, the network can rapidly converge and demonstrate adequate flexibility to harsh channel conditions.

The OAMP algorithm can be used to detect the transmitted symbol vector $\mathbf{x}$ in (3.4) based on the known $\mathbf{y}$, $\mathbf{H}$, and noise variance σ_w^2. According to Section 3.2.3, initialized with $\hat{\mathbf{x}}_1 = \mathbf{0}$, the iterative process proceeds as

$$\mathbf{r}_t = \hat{\mathbf{x}}_t + \mathbf{W}_t \left(\mathbf{y} - \mathbf{H}\hat{\mathbf{x}}_t \right), \tag{3.68}$$

$$\tau_t^2 = \frac{1}{K}\text{tr}(\mathbf{B}_t\mathbf{B}_t^T)v_t^2 + \frac{1}{2K}\text{tr}(\mathbf{W}_t\mathbf{W}_t^T)\sigma_w^2, \tag{3.69}$$

$$\hat{\mathbf{x}}_{t+1} = \mathbb{E}\left[\mathbf{x}|\mathbf{r}_t, \tau_t\right], \tag{3.70}$$

$$v_{t+1}^2 = \frac{\|\mathbf{y} - \mathbf{H}\hat{\mathbf{x}}_{t+1}\|^2 - N\sigma_\omega^2}{\text{tr}\left(\mathbf{H}^T\mathbf{H}\right)}, \tag{3.71}$$

where the de-correlated matrix $\mathbf{W}_t$ takes the LMMSE structure, that is, $\mathbf{W}_t = \frac{K}{\text{tr}(\hat{\mathbf{W}}_t^{\text{LMMSE}}\mathbf{H})}\hat{\mathbf{W}}_t^{\text{LMMSE}}$, where $\hat{\mathbf{W}}_t^{\text{LMMSE}}$ is given by (3.18). Moreover, the matrix $\mathbf{B}_t$ in (3.69) is given by $\mathbf{B}_t = \mathbf{I} - \mathbf{W}_t\mathbf{H}$ with $\text{tr}(\mathbf{B}_t) = 0$ since $\mathbf{W}_t$ is de-correlated, and (3.69) and (3.71) represent the empirical estimators for the error variance τ_t^2 and v_{t+1}^2 in (3.20), respectively. Notably, the divergence-free function η^{df} in the NLE is replaced by a simple MMSE estimator in (3.70) to avoid cumbersome calculations at the expense of performance. The MMSE estimate for each element x_k of the symbol vector $\mathbf{x}$ is given by

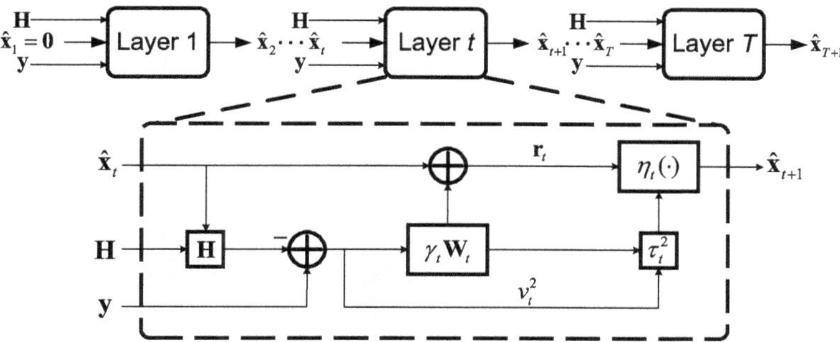

Figure 3.11 The architecture of OAMP-Net.

$$\mathbb{E}[x_k|r_{t,k},\tau_t] = \sum_{a_m\in\mathcal{A}} a_m \frac{\mathcal{N}(x_k = a_m : r_{t,k},\tau_t^2)}{\sum\limits_{a_m\in\mathcal{A}} \mathcal{N}(x_k = a_m : r_{t,k},\tau_t^2)}. \tag{3.72}$$

Equation (3.70) also reveals that $\mathbf{r}_t$ and τ_t^2 are the prior mean and variance in the MMSE estimator that control the convergence and accuracy of the estimated result $\hat{\mathbf{x}}$. Therefore, the step sizes for updating $\mathbf{r}_t$ and τ_t^2 have a significant impact on the final performance. The design of OAMP-Net utilizes the deep learning approach to find the appropriate step sizes and construct a divergence-free NLE, which significantly improves the detection performance without using complicated analytical methods.

Figure 3.11 presents the structure of the T-layer OAMP-Net, which is developed by unfolding the OAMP detector. Each layer of the network contains several layer-dependent trainable parameters. The input of OAMP-Net includes the channel matrix $\mathbf{H}$ and the received signal $\mathbf{y}$. Furthermore, the estimation $\hat{\mathbf{x}}_t$ generated by the $(t-1)$th layer is another input for the tth layer of the network. The final output is the estimated symbol $\hat{\mathbf{x}} = \hat{\mathbf{x}}_{T+1}$ ready to be demapped.

The LE and NLE are the core of each layer, which integrate four scalar trainable variables $\Omega_t = \{\gamma_t, \theta_t, \phi_t, \xi_t\}$. The tth OAMP iteration is thus transformed into[9]

$$\mathbf{r}_t = \hat{\mathbf{x}}_t + \gamma_t \mathbf{W}_t (\mathbf{y} - \mathbf{H}\hat{\mathbf{x}}_t), \tag{3.73}$$

$$\tau_t^2 = \frac{1}{K}\mathrm{tr}(\mathbf{C}_t\mathbf{C}_t^T)v_t^2 + \frac{\theta_t^2}{2K}\mathrm{tr}(\mathbf{W}_t\mathbf{W}_t^T)\sigma_w^2, \tag{3.74}$$

$$\hat{\mathbf{x}}_{t+1} = \eta_t(\mathbf{r}_t, \tau_t^2; \phi_t, \xi_t). \tag{3.75}$$

The LE in (3.73) and the error variance estimator in (3.74) are respectively the revised versions of (3.68) and (3.69) by incorporating the parameters (γ_t, θ_t). Specifically, the parameter γ_t can be viewed as the step size for the update of the mean $\mathbf{r}_t$. The matrix $\mathbf{C}_t = \mathbf{I} - \theta_t\mathbf{W}_t\mathbf{H}$ in (3.74) is modified from $\mathbf{B}_t$ in the OAMP detector with the parameter θ_t to provide an appropriate step size to regulate the update of the variance τ_t^2. The detailed derivation of (3.74) can be found in appendix B of [139].

---

[9] The update of v_t^2 is omitted since it remains unchanged compared to (3.71).

The NLE η_t in (3.75) can be written as

$$\eta_t(\mathbf{r}_t, \tau_t^2; \phi_t, \xi_t) = \phi_t(\mathbb{E}[\mathbf{x}|\mathbf{r}_t, \tau_t] - \xi_t \mathbf{r}_t), \tag{3.76}$$

which is a linear combination of the prior mean, $\mathbf{r}_t$, and the contribution of the MMSE denoiser in (3.70). Actually, the OAMP detector can also utilize (3.76) to construct a divergence-free NLE, but the parameters (ϕ_t, ξ_t) are related to the prior distribution of the transmitted signal and thus difficult to compute. However, in OAMP-Net, the parameters (ϕ_t, ξ_t) in (3.76) can be tuned through deep learning to construct a divergence-free η_t without relying on prior information. Thus, the orthogonality between the two local estimators can be preserved and the stability of the detector improved. Furthermore, the update of the NLE error variance v_t^2 is omitted for OAMP-Net since it takes the same form as (3.71), and a small positive constant $\epsilon = 5 \times 10^{-13}$ is used to substitute v_t^2 when the calculation result of (3.71) is negative, avoiding the numerical instability problem.

Notably, OAMP-Net is reduced to its prototype OAMP when $\gamma_t = \theta_t = \phi_t = 1$ and $\xi_t = 0$. The introduction of these variables originates from the fact that the prerequisites of the conventional AMP-type algorithms, i.e., under a large system with unitarily invariant channel matrix $\mathbf{H}$, are hard to achieve in realistic finite-dimensional systems. For instance, when the system size is small or the channel exhibits strong correlations, the performance of the OAMP detector deteriorates significantly. However, OAMP-Net can reach effective performance improvement under practical scenarios by adaptively learning the appropriate variables for the specific environment. Meanwhile, the total number of trainable parameters in the network is only $4T$, which is independent of the system size. Therefore, the training process of OAMP-Net is time- and resource-efficient, demonstrating an advantage over the data-driven black-box approach and DetNet.

The empirical risk function for OAMP-Net during the training phase is chosen as the MSE between the transmitted vectors (labels) $\mathbf{x}^{(i)}$ and the output prediction $\hat{\mathbf{x}}_{T+1}^{(i)}$ of the network, that is,

$$L_2(\mathbf{\Omega}) = \frac{1}{D} \sum_{i=1}^{D} \left\| \mathbf{x}^{(i)} - \hat{\mathbf{x}}_{T+1}^{(i)}(\mathbf{y}^{(i)}, \mathbf{H}^{(i)}) \right\|^2, \tag{3.77}$$

where $\mathbf{\Omega} = \{\Omega\}_{t=1}^{T}$ is the set of trainable parameters in the network, and D is the batch size.

Figure 3.12 presents the performance comparison of the OAMP, OAMP-Net, LMMSE, sphere decoding (SD), and DetNet detectors under i.i.d. Rayleigh fading channels. The number of layers/iterations in OAMP and OAMP-Net is set as $T = 4$ for a desirable balance between performance and complexity. From the figure, OAMP-Net outperforms other detectors except for SD, which is computationally difficult for large-scale MIMO systems with high-order modulation. Notably, OAMP-Net has an advantage over OAMP for all SNRs and both the QPSK and 64-QAM modulation schemes, indicating that the introduction and tuning of the four trainable parameters can bring about performance gains for finite-dimensional

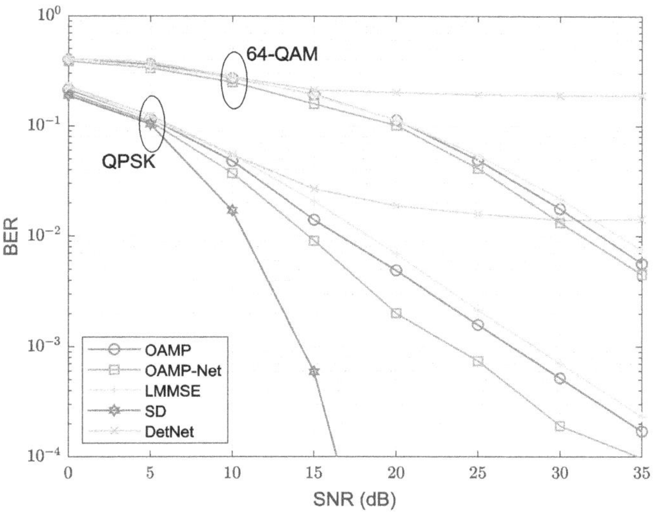

Figure 3.12 The BER performance comparison of OAMP-Net with other competitive detectors under 4×4 MIMO Rayleigh fading channels with different orders of modulation.

MIMO systems. In contrast, DetNet experiences severe performance degradation in these small-sized MIMO systems.

OAMP-Net is a model-driven MIMO detection network developed by unfolding the OAMP algorithm and adding trainable parameters that can be learned. The network inherits the expert knowledge from the baseline algorithm and combines it with the superiority of advanced deep learning techniques, thus demonstrating significant performance gain over conventional MIMO detectors. Furthermore, OAMP-Net is superior to the data-driven learning-based methods in training efficiency since only a few adjustable parameters need to be optimized.

3.3.3 Other Model-Driven Signal Detection Methods

DetNet and OAMP-Net have sparked significant research interest in leveraging learning for MIMO signal detection, particularly model-driven learning and deep unfolding techniques. Recent works have developed advanced model-driven MIMO detectors that are adaptive to environment changes. This subsection provides a brief overview of two notable works in this area.

MMNet [140]: OAMP-Net suffers performance degradation in real-world channels with strong correlations and cannot generalize across a wide range of channel distributions. To solve these limitations, MMNet is proposed by combining the model-driven learning-based detector with the online training mechanism. Specifically, MMNet uses the general iterative shrinkage thresholding (ISTA) algorithm as the prototype and adds a complete trainable matrix in the architecture, which is different from OAMP-Net and provides adequate flexibility and degrees of freedom. Overall, MMNet reaches a desirable tradeoff between network flexibility and complexity. Meanwhile, the online training mechanism for MMNet is based on the temporal and spectral correlation and locality of real channels, that is, adjacent channels have strong similarities in time or frequency. Therefore, the training cost for a specific channel can be shared among adjacent channels, and the trained results for a certain channel can serve as an initial model for the training of adjacent channels, which makes the training more efficient. This accelerates the online training for each channel realization and also avoids frequent retraining. Simulation results show that the proposed method accelerates the online training by more than two orders of magnitude compared to retraining the model from scratch, and MMNet can achieve performance within 2 dB of the optimal maximum likelihood detector and significantly outperforms OAMP-Net in realistic channels.

Deep Hypernetwork-Based MIMO Detection [141]: The hypernetwork provides another way to improve the generalization of the model-driven learning-based detector. The core idea of this method is to use an additional network to generate the parameters for the main network, which is similar to the relationship between a genotype (the hypernetwork) and a phenotype (the main network) [142]. In contrast to the online training scheme, the hypernetwork can adaptively adjust the parameters according to changing scenarios without retraining, which significantly improves deployment efficiency. A hypernetwork-aided MMNet (HyperMMNet) is introduced in [141] to adjust the MIMO detector to different channel realizations and noise levels. In particular, the hypernetwork takes the channel matrix after QR decomposition and the noise variance as inputs, and generates the weights of the MMNet. The number of outputs of the hypernetwork is further reduced through the utilization of a weight-sharing strategy. The hypernetwork learns the varying trends of adjustable parameters within the signal detection network across different scenarios, enabling it to provide suitable parameters for the MMNet. Therefore, aided by the hypernetwork, the model-driven detection network can fully unleash its potential and exhibit remarkable adaptability. Simulation results show that the proposed approach can reach a performance comparable to state-of-the-art schemes without retraining.

3.4   Channel Decoding

Channel coding encompasses a set of encoding and decoding algorithms utilized to detect and correct bit errors by introducing redundancy in the transmitted signals. As pointed out by Claude Shannon's celebrated *A Mathematical Theory of Communication* [143], reliable communications can be achieved with appropriate channel coding given that the data rate does not exceed the channel capacity. In the past several decades, it has attracted much attention from both academia and industry in designing effective channel encoding and decoding algorithms. With the discovery of turbo codes and the rebirth of low-density parity-check (LDPC) codes, channel coding has demonstrated remarkable performance that approaches the capacity limit established in Shannon's landmark paper [143]. In this section, we will introduce some basics of channel coding and conventional channel decoding algorithms. Recent learning-based decoding algorithms, including data-driven and model-driven approaches, will be discussed in Sections 3.5 and 3.6.

3.4.1 Channel Decoding Problem

Channel coding uses redundancy to protect transmitted data against errors that may occur during transmission. It involves the insertion of redundant information at the transmitter, where the channel encoder $E(\cdot)$ maps a length-k sequence of information bits $\mathbf{u}$ to a length-n sequence of coded symbols $\mathbf{c} = E(\mathbf{u})$ called a codeword. The code rate is defined as $R = k/n$. The codewords are transmitted via a noisy channel, and a decoder $D(\cdot)$ is utilized at the receiver to estimate the transmitted message from the noisy received sequence $\mathbf{y}$. Both the BER and block error rate (BLER) are commonly used to measure the reliability.

The ML and MAP decoders provide the optimal decoding performance. The ML decoder seeks to estimate the transmitted codeword $\hat{\mathbf{c}}$ as the one that maximizes the likelihood function $p(\mathbf{y}|\mathbf{c})$, i.e.,

$$\hat{\mathbf{c}}_{\mathrm{ML}} = \arg\max_{\mathbf{c}\in\mathcal{C}} \ p(\mathbf{y}|\mathbf{c}),$$

where $\mathbf{y}$ is the received vector, $\mathcal{C}$ is the set of all possible codewords, and $p(\mathbf{y}|\mathbf{c})$ is the conditional probability of $\mathbf{y}$ given $\mathbf{c}$, which is determined by the channel model, e.g., the additive white Gaussian noise (AWGN) or binary symmetric channel. The MAP decoder, in contrast, seeks to estimate the transmitted codeword $\hat{\mathbf{c}}$ as the one that maximizes the a posteriori probability function $p(\mathbf{c}|\mathbf{y})$, i.e.,

$$\hat{\mathbf{c}}_{\mathrm{MAP}} = \arg\max_{\mathbf{c}\in\mathcal{C}} \ p(\mathbf{c}|\mathbf{y}) = \arg\max_{\mathbf{c}\in\mathcal{C}} \ p(\mathbf{y}|\mathbf{c})p(\mathbf{c}).$$

The MAP decoder is identical to the maximum-likelihood decoder when all codewords have equal probabilities, and it can also handle non-uniform prior distributions. The well-known Bahl–Cocke–Jelinek–Raviv (BCJR) algorithm, which will be discussed in the following subsection, is a bit-wise MAP algorithm designed to minimize the bit-error probability for convolutional and turbo codes.

3.4.2 Convolutional Codes and BCJR Decoding

Discovered by Elias in 1955 [144], convolutional codes have been widely applied in wireless communications due to their convenient encoding and decoding schemes. Unlike block codes, which encode data in fixed-size blocks, convolutional codes are designed to handle continuous streams of data that arrive sequentially.

To encode data using a convolutional code, a sequence of input bits is fed into a linear finite-state machine called a convolutional encoder. The encoder consists of an L-stage shift register, which stores a history of the most recent input bits, and modulo-two adders, which perform bitwise addition on the shift register's contents. As the input data bits are fed into the encoder, they are shifted into the shift register

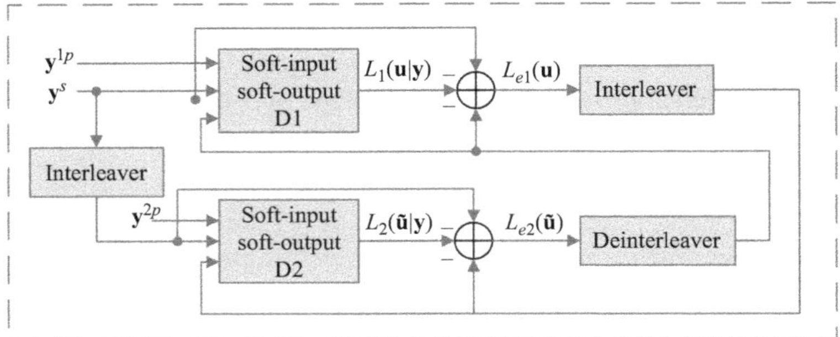

Figure 3.13 The structure of the turbo decoder.

and combined with the values in the adders to produce an output sequence of symbols.

Turbo codes are a type of convolutional code that use two parallel convolutional encoders. They were first introduced in the 1990s and have gained recognition for their exceptional performance, particularly at low SNRs. In a turbo code, the input data is encoded by two parallel recursive systematic convolutional encoders. These two encoders are separated by an interleaver, which is vital in effectively reducing error floors. The codeword is then modulated and transmitted over the channel. At the receiver, two parallel decoders use soft-input soft-output decoding algorithms to decode the two encoded streams. The decoded streams are then interleaved and fed back into the decoder for further iterations, improving the overall decoding performance.

Proposed in 1974 [145], the BCJR algorithm is a famous MAP decoding algorithm for convolutional and turbo codes. As shown in Figure 3.13, the traditional turbo decoder contains two soft-input soft-output decoders (denoted by $D1$ and $D2$) that have the same structure. The BCJR algorithm can be applied for $D1$ and $D2$ via solving a MAP problem, with which the posterior LLRs for information bits can be computed and expressed as

$$
\begin{aligned}
L(u_k \mid \mathbf{y}) &= \ln\left(\frac{P(u_k = 1 \mid \mathbf{y})}{P(u_k = 0 \mid \mathbf{y})}\right) \\
&= \ln\left(\frac{\displaystyle\sum_{(s',s)\in S^1} P(s',s,\mathbf{y})}{\displaystyle\sum_{(s',s)\in S^0} P(s',s,\mathbf{y})}\right),
\end{aligned}
\tag{3.78}
$$

where $s \in S_R$ is the state of the encoder at time k and $s' \in S_R$ is the state of the encoder at time $k - 1$, for the set of all encoder states S_R. Accordingly, $S^1 = \{(s',s): u_k = 1\}$ is the set of ordered pairs (s',s) corresponding to all state transitions $s' \to s$ caused by data input $u_k = 1$, and S^0 refers to data input $u_k = 0$.

With the AWGN channel, the sequence received after time k is only related to the state of the encoder at time k regardless of the previous states. On the basis of Bayes' formula, we have

$$P(s', s, \mathbf{y}) = P(s', s, \mathbf{y}_1^{k-1}, y_k, \mathbf{y}_{k+1}^K)$$
$$= P(\mathbf{y}_{k+1}^K \mid s)P(y_k, s \mid s')P(s', \mathbf{y}_1^{k-1})$$
$$= \beta_k(s)\gamma_k(s', s)\alpha_{k-1}(s'), \tag{3.79}$$

where $\alpha_{k-1}(s') = P(s', \mathbf{y}_1^{k-1})$ and $\beta_k(s) = P(\mathbf{y}_{k+1}^K \mid s)$ can be computed through forward and backward recursions, i.e.,

$$\alpha_k(s) = \sum_{s' \in S_R} \alpha_{k-1}(s')\gamma_k(s', s) \tag{3.80}$$

and

$$\beta_{k-1}(s') = \sum_{s \in S_R} \beta_k(s)\gamma_k(s', s). \tag{3.81}$$

Moreover, $\gamma_k(s', s) = P(y_k, s \mid s')$ is computed as follows:

$$\gamma_k(s', s) = \exp\left\{ \frac{1}{2}\left(x_k^s y_k^s + x_k^{1p} y_k^{1p} \right) + \frac{1}{2}u_k L(u_k) \right\}, \tag{3.82}$$

where $L(u_k)$ is the a priori probability LLR for bit u_k. Consequently, we obtain the extrinsic LLR $L_e(u_k)$ as follows:

$$L_e(u_k) = L(u_k \mid \mathbf{y}) - y_k^s - L(u_k), \tag{3.83}$$

which can be used as the a priori probability LLR input of the subsequent soft-input soft-output decoder D2 after it is interleaved.

In addition, let $\bar{\alpha}_k(s)$, $\bar{\beta}_k(s)$, and $\bar{\gamma}_k(s', s)$ represent the logarithmic values of $\alpha_k(s)$, $\beta_k(s)$, and $\gamma_k(s', s)$, respectively. Given that $\alpha_{k-1}(s')$ and $\beta_k(s)$ are computed in the MAP algorithm, using the Jacobian logarithmic function, equations (3.80) and (3.81) can be approximately written as

$$\bar{\alpha}_k(s) = \max_{s' \in S_R}(\bar{\alpha}_{k-1}(s') + \bar{\gamma}_k(s', s)), \tag{3.84}$$
$$\bar{\beta}_{k-1}(s') = \max_{s \in S_R}(\bar{\beta}_k(s) + \bar{\gamma}_k(s', s)), \tag{3.85}$$

and

$$L(u_k \mid \mathbf{y}) = \max_{(s', s) \in S^1}(\bar{\alpha}_{k-1}(s') + \bar{\gamma}_k(s', s) + \bar{\beta}_k(s))$$
$$- \max_{(s', s) \in S^0}(\bar{\alpha}_{k-1}(s') + \bar{\gamma}_k(s', s) + \bar{\beta}_k(s)), \tag{3.86}$$

namely the max-log-MAP algorithm.

3.4.3 Linear Block Codes and BP Decoding

The linear block code, including Bose–Chaudhuri–Hocquenghem (BCH), polar, and LDPC codes, is another famous family of error-correcting codes in addition to convolutional codes. Consider an (n, k) linear block code, containing k information bits and $n - k$ parity check bits. These base codewords are often arranged in the

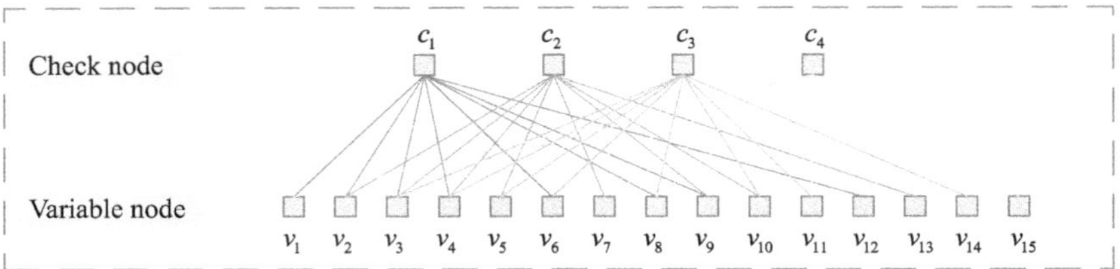

Figure 3.14 The structure of the Tanner graph for BCH(15,11).

rows of a matrix $\mathbf{G}$, known as the generator matrix. Given the information $\mathbf{u}$, the codeword $\mathbf{c}$ is generated as $\mathbf{c} = \mathbf{u}\mathbf{G}$.

In addition to the generator matrix, the parity-check matrix $\mathbf{H}$ of a linear block code is a matrix that describes the linear relationships that a codeword must satisfy. Specifically, a codeword $\mathbf{c}$ is valid if and only if the matrix–vector product $\mathbf{H}\mathbf{c} = \mathbf{0}$.

From $\mathbf{H}$, we can construct a bipartite graph T, known as a Tanner graph. Specifically, the vertex set of T consists of the set n variable nodes (VNs) and r check nodes (CNs), where $r = n - k$. An edge between the ith check node and the jth variable node exists in T if and only if $\mathbf{H}_{i,j} = 1$. For instance, the structure of the Tanner graph for the BCH(15, 11) code is illustrated in Figure 3.14, where the variable nodes are denoted by v, and the check nodes are denoted by c.

Belief propagation (BP) is a message-passing algorithm used for solving the inference problem in factor graphs and is widely adopted in LDPC decoding. Given the Tanner graph, BP decoding messages are calculated on the variable and check nodes and sent over the edges of the graph. The main steps of the BP algorithm involve initializing messages on the edges, updating messages on the variable nodes based on messages received from check nodes, and updating messages on the check nodes based on messages received from variable nodes. BP decoding continues until convergence is achieved or a predetermined maximum number of iterations is reached.

The main steps of the decoding process are as follows. First, the messages transmitted over edges of the Tanner graph are initialized, i.e., for each edge in the Tanner graph, set

$$\Gamma_{v \to c}(C_v) = l_v,$$
$$\Lambda_{c \to v}(C_v) = 0. \tag{3.87}$$

The messages defined by

$$\Lambda_{c \to v}(C_v) = 2 \tanh^{-1}\left\{ \prod_{v' \neq v} \tanh\left[\frac{1}{2}\Gamma_{v' \to c}(C_v) \right] \right\} \tag{3.88}$$

are then sent from the check nodes to the variable nodes, where $\tanh(x) = (e^x - e^{-x})/(e^x + e^{-x})$ is the hyperbolic tangent function. After that, the messages sent from the variable nodes to the check nodes are defined by

$$\Gamma_{v \to c}(C_v) = l_v + \sum_{c' \neq c} \Lambda_{c' \to v}(C_v). \tag{3.89}$$

In addition, the reliability value is evaluated from

$$\Gamma_v(C_v) = l_v + \sum_{c'} \Lambda_{c' \to v}(C_v). \tag{3.90}$$

Finally, the reliability value calculated by means of (3.90) is used to obtain an estimated value $\hat{x}_v$ of the codeword bit C_v, according to the following rule:

$$\hat{x}_v = \begin{cases} 1 & \text{if } \Gamma_v(C_v) \geq 0 \\ 0 & \text{if } \Gamma_v(C_v) < 0. \end{cases} \tag{3.91}$$

The syndrome of the estimated codeword $\hat{\mathbf{x}}$ through the matrix $\mathbf{H}$ is computed. If the syndrome is null, i.e., the parity check is successful, decoding stops and gives the estimated codeword as the output of the algorithm. Otherwise, the algorithm reiterates, going back to (3.88) with the updated values of $\Gamma_{v \to c}(C_v)$ until a predetermined maximum number of iterations is reached.

BP decoding can be applied to decode LDPC codes and achieves near-Shannon channel capacity when the block length is relatively large. However, the performance of the BP algorithm is not competitive for high-density parity-check (HDPC) codes, such as BCH and Reed–Muller (RM) codes, owing to the large number of short cycles in the graph.

In addition to LDPC codes, polar codes [146] represent a major breakthrough in linear block codes. They achieve the capacity of binary-input discrete memoryless channels via a channel polarization technique, which transforms a set of identical channels into polarized channels – some nearly error-free and others nearly useless. Decoding of polar codes is typically performed using successive cancellation, which exploits the polarization structure by decoding bits sequentially based on previous decisions. Due to its exceptional performance and manageable complexity, the polar code has been adopted in the standards for 5G mobile communication technology.

3.5 Learning-Based Channel Decoding: Data-Driven Methods

Typically, conventional decoding algorithms, such as BCJR and BP, require many iterations to achieve near-capacity performance, which can be computationally expensive. Furthermore, these algorithms depend on prior knowledge of the channel, and their performance can be adversely affected when the channel distribution changes in practice.

An advantage of deep learning-based decoders over traditional decoding algorithms is that they do not rely on predefined models of channels. Instead, these decoders learn the mapping between received and transmitted signals. Deep learning-based decoders can be categorized into two classes: data-driven and model-driven. Data-driven deep-learning decoders require no prior knowledge of the encoding process and treat the encoding process and channel as a black box. Conversely, in model-driven deep learning schemes, the decoder architectures are built on the traditional decoding algorithms such as BP and BCJR. This model-driven learning approach improves the efficiency of training and the scalability to high dimensions. In this section, we focus on data-driven learning-based decoders.

In [16], a data-driven decoding approach was proposed for polar codes and random codes by optimizing a DNN. At the transmitter, the k information bits are encoded into a codeword of length N. The coded bits are then modulated and transmitted over a noisy channel. At the receiver, the noisy codeword is received and decoded by the DNN decoder. In contrast to conventional iterative decoding methods, which require multiple iterations through the decoding algorithm, the DNN decoder is able to produce an estimate by passing through each layer only once.

The loss function used for training is the ℓ_2 loss and the binary cross-entropy, defined respectively as

$$
\begin{aligned}
\ell_2(b_i, \hat{b}_i) &= \frac{1}{k} \sum_i \left(b_i - \hat{b}_i \right)^2, \\
\ell_{\text{BCE}}(b_i, \hat{b}_i) &= -\frac{1}{k} \sum_i \left[b_i \ln\left(\hat{b}_i \right) + (1 - b_i) \ln\left(1 - \hat{b}_i \right) \right],
\end{aligned}
\tag{3.92}
$$

where $b_i \in \{0, 1\}$ is the ith information bit and $\hat{b}_i \in [0, 1]$ the DNN soft estimate.

The DNN decoder is composed of three hidden layers, containing 128, 64, and 32 nodes, respectively. To implement the data-driven approach, binary phase shift keying (BPSK) modulation and the AWGN channel are adopted [16]. The DNN decoder also contains an additional layer for the BPSK modulation and a noise layer that generates new noise realizations on the fly, ensuring that the input to the DNN decoder is never the same. The decoder's output neurons are enforced to assume values between zero and one using the sigmoid function at the output layer. Overall, the additional layers in the DNN decoder architecture enable the network to learn the mapping from the received symbols to the transmitted bits in the presence of channel noise. The performance of the DNN decoder is evaluated by measuring the BER for both random codes and polar codes with a codeword length of $N = 16$ and a code rate $r = 0.5$. For the random codes, codewords are randomly selected from the codeword space with Hamming distances larger than two. Polar codes are constructed using a generator matrix of block size $N = 2^n$, which is defined as

$$
\mathbf{G}_N = \mathbf{F}^{\otimes n}, \quad \mathbf{F} = \begin{bmatrix} 1 & 0 \\ 1 & 1 \end{bmatrix},
\tag{3.93}
$$

where $\mathbf{F}^{\otimes n}$ denotes the nth Kronecker power of $\mathbf{F}$. The codewords are then computed by multiplying $\mathbf{u}$, containing k information bits and $N - k$ frozen bits, with the generator matrix, i.e., $\mathbf{x} = \mathbf{u}\mathbf{G}_N$.

Figure 3.15 presents the BER performance of the DNN decoder for different numbers of training epochs, ranging from $M_{\text{ep}} = 2^{10}$ to 2^{18}. As the number of training epochs increases, the gap between the performance of the DNN decoder

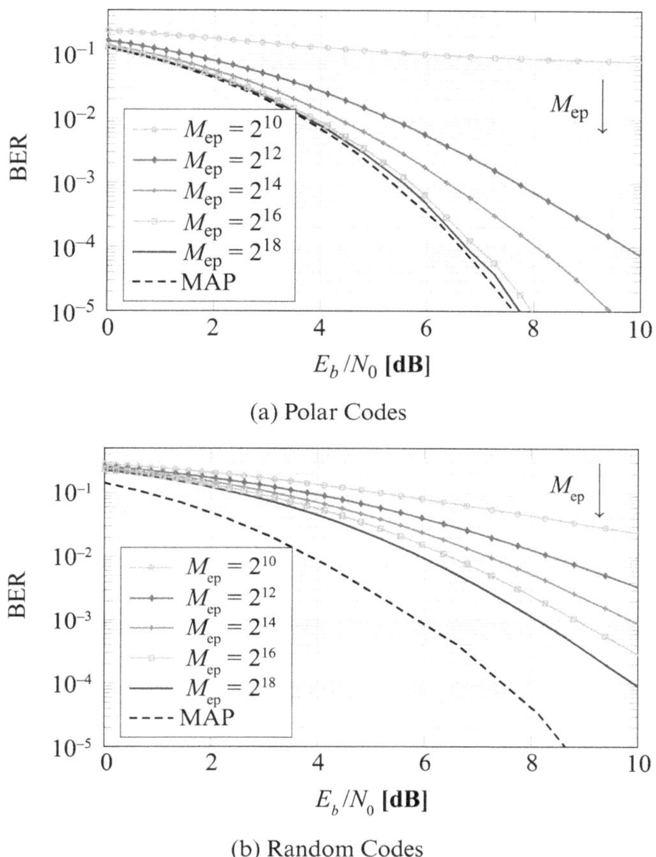

(a) Polar Codes

(b) Random Codes

Figure 3.15 Influence of the number of epochs M_{ep} on the BER of a 128-64-32 neural network for 16 bit-length codes with code rate $r = 0.5$ [16].

and that of the MAP decoder becomes smaller. However, for random codes, the gap remains substantial even with $M_{\mathrm{ep}} = 2^{18}$ epochs, while for polar codes, the DNN decoder performance is already close to that of the MAP decoder with the same number of epochs. This suggests that the deep learning decoder is better suited for learning the decoding algorithms of structured codes, such as polar codes, rather than random codes.

To address the scalability issue of the data-driven deep learning decoder, a partitioning approach is proposed in [147]. This approach involves concatenating multiple neural decoders, where each neural decoder is responsible for decoding a subset of the codeword. To propagate the results among these decoders, a conventional BP algorithm is utilized. By dividing the codeword into multiple subsets, this approach is able to handle longer block lengths than using a single neural decoder without partitioning. Simulations in [147] demonstrate that the partitioning approach achieves comparable performance to conventional decoding algorithms such as BP while reducing the latency for decoding. These results suggest that the deep learning decoder, when combined with partitioning, has the potential to scale to longer block lengths without compromising performance.

In addition to DNN-based designs, recurrent neural networks (RNNs) have been used for decoding convolutional codes and turbo codes in [148]. Inspired by the recurrent structure of the convolutional encoder, a two-layer bidirectional gated recurrent unit is trained as a neural decoder for convolutional codes. In terms of training the model, the received codewords from the channel are used as input to the RNN network, and the predicted information bits are the output, as previously described in the DNN-based design. The RNN decoder can also be extended for turbo codes, which combine two convolutional codes with an interleaver. Similar to a conventional turbo decoder, two RNN decoders are trained to work in an iterative manner, where the outputs of one decoder are interleaved and fed to the other decoder at the next iteration. Simulations show that RNN decoder performance on the AWGN channel is comparable to optimal decoders such as Viterbi and BCJR decoding algorithms.

3.6   Learning-Based Channel Decoding: Model-Driven Methods

The curse of dimensionality is a significant challenge for data-driven deep-learning decoders. Specifically, the code block size in a communication system needs to be long enough to ensure a sufficient coding gain. However, as the size of possible codewords grows exponentially with the code block size, the portion of unseen codewords during training will significantly increase accordingly. Therefore, most data-driven deep-learning works on decoding can only address short block codes.

To ensure the scalability of the deep learning decoder, its architecture is often designed to match that of conventional decoding algorithms, known as model-driven deep learning decoders. In particular, each iteration of the conventional decoding algorithm is represented as one layer in the neural network. As a result, model-driven decoders can be viewed as executing the conventional decoding algorithm with a fixed number of iterations while offering potential benefits such as reduced computational complexity and improved decoding performance.

3.6.1   Model-Driven Deep Learning Decoders with Unfolding BP

A model-driven deep learning decoder is proposed for linear codes in [9], where the DNN architecture is built by unfolding the BP algorithm, discussed in Section 3.4.3 with L full iterations. The key idea is to introduce trainable parameters on the edges of the factor graph so that the standard BP decoding algorithm can be made adaptive to the training data. The model-driven deep learning architecture is depicted in Figure 3.16, with all nodes in the hidden layers corresponding to messages transmitted over edges in the Tanner graph. The input layer is of size N, representing the LLRs from the symbol-detection module, and the output layer also has length N, representing the final decoded codeword. Between the input and output layers, there

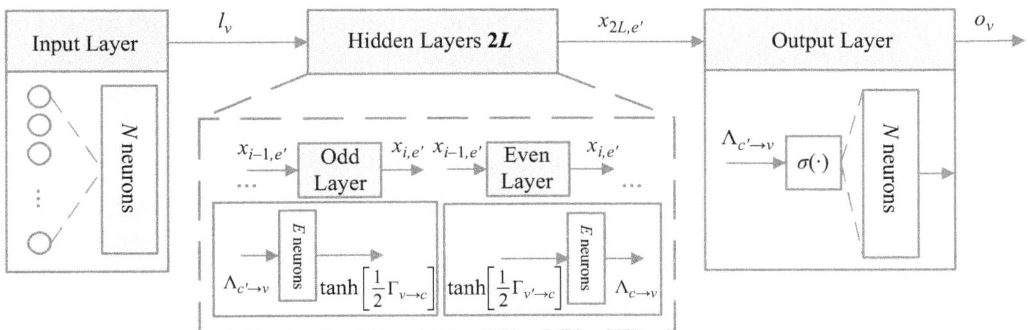

Figure 3.16 Architecture of the model-driven DNN decoder.

are $2L$ hidden layers, with the odd and even layers corresponding to the variable and check nodes, respectively. The weights assigned to the edges in the Tanner graph are trained using the stochastic gradient descent (SGD) method, and the output message of the processing element corresponding to the variable node v and the check node c, i.e., $e = (v, c)$, is denoted by $x_{i,e}$. For odd i, we have

$$x_{i,e} = \tanh\left(\frac{1}{2}\left(w_{i,v}l_v + \sum_{e'=(v,c'),c'\neq c} w_{i,e,e'}x_{i-1,e'}\right)\right),\tag{3.94}$$

and for even i,

$$x_{i,e} = 2\tanh^{-1}\left(\prod_{e'=(v',c),v'\neq v} x_{i-1,e'}\right);\tag{3.95}$$

the final vth output of the network is given by

$$o_v = \sigma\left(w_{2L+1,v}l_v + \sum_{e'=(v,c')} w_{2L+1,v,e'}x_{2L,e'}\right),\tag{3.96}$$

where $\sigma(x) = (1 + e^{-x})^{-1}$ is a sigmoid function. Cross-entropy is applied as the loss function

$$\ell(o,y) = -\frac{1}{N}\sum_{v=1}^{N}y_v\ln(o_v) + (1 - y_v)\ln(1 - o_v).\tag{3.97}$$

The sparse and structural architecture of the model-driven deep learning decoder enables the network to be trained solely with the zero codeword, whereas the data-driven deep learning decoder needs to be trained with all possible types of codewords. The objective is to optimize the parameters $w_{i,v}$, $w_{i,e,e'}$, and $w_{i,v,e'}$ to minimize the discrepancy between the N-dimensional output codeword of the decoder and the zero codeword.

The training data used to train the model-driven deep learning decoder was generated using an AWGN channel with SNR values ranging from 1 to 6 dB. The trained model consists of 10 hidden layers, corresponding to 5 full iterations of the

BP algorithm. The model parameters were optimized using SGD with mini-batches consisting of 20 codewords for each SNR value.

The model-driven deep learning decoder demonstrated better performance compared to the BP algorithm for both BCH(63,45) and BCH(127,106) codes. For BCH(127,106), an improvement of up to 0.75 dB in the high-SNR region is obtained. Moreover, the loss function can be extended to include the outputs at each odd i layer, i.e.,

$$\ell(o, y) = -\frac{1}{N} \sum_{i=1,3,\ldots}^{2L-1} \sum_{v=1}^{N} y_v \ln(o_{v,i}) + (1 - y_v) \ln(1 - o_{v,i}), \tag{3.98}$$

and after training the decoders with the loss function as represented in (3.98), an improvement of up to 0.9 dB is obtained over the plain BP algorithm for BCH(63,45). Moreover, the detection accuracy that the BP decoder achieves with 50 iterations can be attained by the deep neural decoder with 5 iterations.

3.6.2 Model-Driven Deep Learning Decoders with Unfolding BCJR

Another model-driven learning decoder is TurboNet, introduced in [149], and its architecture is depicted in Figure 3.17. TurboNet is designed by unrolling the conventional BCJR decoding algorithm, as described in Section 3.4.2, to obtain an "unfolded" structure with M iterations, where each iteration is represented by a DNN decoding unit. The mth DNN decoding unit consists of two identical interleavers, two subnets with the same structure (weights are not shared), and one deinterleaver, where $m = 0, 1, \ldots, M - 1$.

The two subnets are based on the neural max-log-MAP algorithm. Each subnet contains one input layer, $K + 1$ hidden layers, and one output layer. The input layer consists of $3K$ neurons, and the input of the neurons constitutes the triplet $I_k =$

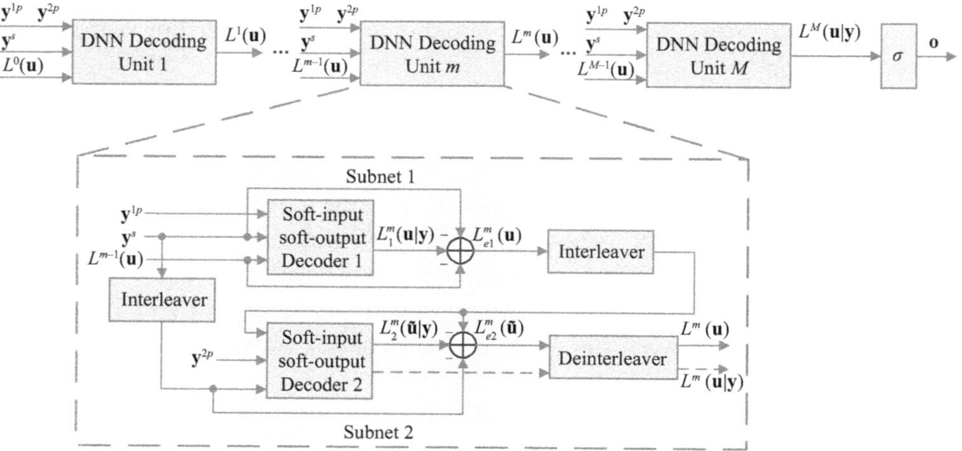

Figure 3.17 Architecture of TurboNet.

$\{y_k^s, y_k^{1p}, L^{m-1}(u_k)\}$ for $k = 1, 2, \ldots, K$. The first hidden layer has $16K$ neurons, and the output of the neurons is denoted by $N^1 = \{\bar{\gamma}_k(s', s): (s', s) \in S, k = 1, 2, \ldots, K\}$. There are $2 \times |S_R| = 16$ neurons in each of the following $K - 1$ hidden layers, and the output of all neurons in the zth hidden layer constitutes the set $N^z = N^z_{odd} \cup N^z_{even}$, where $N^z_{odd} = \{\bar{\alpha}_k(s): k = z - 1, s \in S_R\}$ is the set of neuron outputs for all odd positions in the zth hidden layer, and $N^z_{even} = \{\bar{\beta}_{k-1}(s'): k = K - z + 2, s' \in S_R\}$ is the set of neuron outputs for all even positions in the zth hidden layer, for $z = 2, 3, \ldots, K$. Hidden layer $K + 1$, which is the last hidden layer, contains K neurons, and the output is denoted by $N^{K+1} = \{L^m_1(u_k \mid \mathbf{y}) : k = 1, 2, \ldots, K\}$. The output layer consists of K neurons, and the output of all neurons in this layer is denoted by $N^{Out} = \{L^m_{e1}(u_k): k = 1, 2, \ldots, K\}$.

Given that the output of the Mth DNN decoding unit is $L^M(u_k \mid \mathbf{y})$, the sigmoid function $\sigma(x)$ is used so that the final network output $o_k = \sigma(L^M(u_k \mid \mathbf{y}))$ is in the range $[0, 1]$. A loss function defined as (3.99) is used to measure the difference between TurboNet and the traditional log-MAP algorithm:

$$\ell_{\text{turbo}} = \frac{1}{K} \sum_{k=1}^{K} \left(L^M(u_k \mid \mathbf{y}) - L^T_{log-MAP}(u_k \mid \mathbf{y}) \right)^2, \tag{3.99}$$

where $L^M(u_k \mid \mathbf{y})$ represents the a posteriori LLR obtained by TurboNet, and $L^T_{log-MAP}(u_k \mid \mathbf{y})$ represents the a posteriori LLR calculated by the traditional log-MAP algorithm with T iterations.

TurboNet is trained with the SGD algorithm. The objective is to minimize the loss function of the network by optimizing the trainable parameters in hidden layer 1, hidden layer $K + 1$, and the output layer. The final decoding results can be obtained by a hard decision shown below:

$$\hat{u}_k = \begin{cases} 1 & o_k \geq 0.5, \\ 0 & o_k < 0.5. \end{cases} \tag{3.100}$$

Moreover, an improved TurboNet structure can be obtained, called TurboNet+, which only retains the extrinsic LLR weights defined in the output layer as the trainable parameters and shows better convergence rate and BER performance.

In the following, model-driven turbo-decoding algorithms are compared with traditional turbo-decoding algorithms in terms of BER and computational complexity. TurboNet, which has three DNN decoding units corresponding to three full iterations, is trained for turbo codes on randomly generated training data on an AWGN channel. The loss function in (3.99) is adopted and $T = 6$ is used to obtain the target LLR $L^T_{log-MAP}(u_k \mid \mathbf{y})$.

Figure 3.18 compares the BER performance of the model-driven turbo decoder and the traditional max-log-MAP algorithm under code lengths $K = 120$ with code rate $R = 1/2$, where the MAP algorithm is plotted as the benchmark. The training SNR with 16-QAM modulation mode is 6 dB. We can observe that the TurboNet decoder achieves better BER performance compared with the traditional max-log-MAP algorithm with the same number of iterations.

Table 3.1 Complexity analysis for turbo decoders.

Decoder	Iterations	Parameters	Time (seconds)
max-log-MAP	3	–	3.02×10^{-3}
log-MAP	6	–	5.93×10^{-2}
TurboNet	3	4.5×10^4	3.22×10^{-3}
TurboNet+	3	1.8×10^3	3.04×10^{-3}

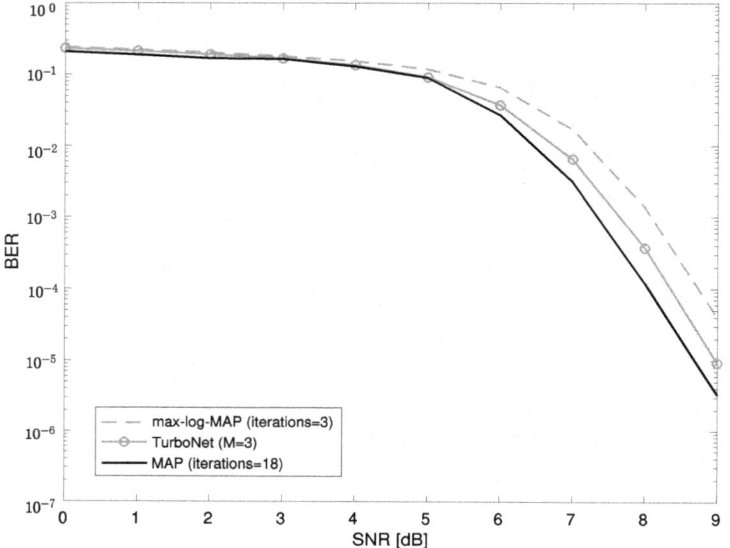

Figure 3.18 BER performance curves of the TurboNet.

Table 3.1 presents a comparison between the decoding complexities of model-driven turbo decoders and conventional algorithms with $K = 100$ and $R = 1/3$. It is observed that TurboNet and TurboNet+ attain significant performance gains compared to the max-log-MAP algorithm and approximate the upper bound that the MAP algorithm provides. Moreover, TurboNet and TurboNet+ introduce negligible computational complexity compared to the max-log-MAP algorithm, thereby maintaining the computational cost advantage.

3.6.3 Other Model-Driven Decoding Methods

Recently, there has been a surge in efforts to enhance model-driven decoders by adopting more advanced architectures that go beyond merely unfolding conventional decoding algorithms, such as BP and BCJR. In this regard, two notable works can be summarized as follows.

RNN-Based BP Decoder

In [150], RNNs are utilized to improve model-driven decoders. The network unfolds the BP decoding algorithm and employs the same trellis representation as [9], but the weights on the edges in the Tanner graph are learned in a data-driven manner. In addition, these weights are tied, i.e., set to be equal in each iteration. This tying transfers the feed-forward architecture into the RNN architecture and allows for a significant reduction in the number of parameters to be trained. Simulation results demonstrate that this method can achieve up to 1.5 dB improvement over the standard BP algorithm in high-SNR regions.

Neural-Enhanced BP for LDPC Code

A new approach, termed neural enhanced belief propagation (NEBP), has been introduced in [151] to improve the performance of factor graph-based decoders by extending graph neural networks (GNNs) to factor graphs. The NEBP algorithm works by feeding BP messages into the GNN after each BP iteration, which runs for two iterations and updates the BP messages. This recursive procedure is repeated for a few iterations, and the decoding results can be obtained by calculating the marginals from the refined BP messages. The novelty of this approach lies in the use of data-driven messages passing through the GNN, which are learned from the data, thereby improving the robustness of the decoder. Simulation results show that the proposed NEBP achieves superior performance to the conventional LDPC decoder for realistic bursty noise distributions.

3.7 Exercises

Exercise 3.1   Section 3.1 introduces learning-based signal detection methods for OFDM systems. Please complete the following tasks based on reading the text and the reference [15].

(a) Considering an OFDM system with 64 subcarriers and 64-QAM modulation and assuming that eight identical FC-DNNs are used to recover the transmitted bits, please calculate the size of the input and output layers of each FC-DNN model as shown in Figure 3.2. Furthermore, do you think it is sufficient to simulate only one DNN instead of all parallel DNNs for BER evaluation? Explain your idea briefly.

(b) Reproduce the simulation results in Figure 3.3. Detailed simulation setups can be found in [15], and reference code can be found at https://github.com/le-liang/wcmlbook/tree/main/ch3/Exercise_3.1.

(c) Repeat task (b) when the modulation scheme changes from QPSK to 64-QAM and other settings remain unchanged. Compare the results with Figure 3.3 and discuss your findings.

(d) Train a single FC-DNN to predict all transmitted bits in a data symbol vector (e.g., 128 bits for the OFDM system with 64 subcarriers and QPSK modulation), and compare the BER performance with that derived by using 8 identical DNNs with a smaller output size.

Exercise 3.2 Please derive the mathematical formulation for the linear MMSE detector in (3.8). *Hint:* Consider the detection problem as designing the weight matrix $\mathbf{W}$ for $\hat{\mathbf{x}} = \mathbf{W}\mathbf{y}$. The MMSE weight matrix is derived by the following criterion:

$$\mathbf{W}_{\mathrm{MMSE}} = \arg\min_{\mathbf{W}} \mathbb{E}[\|\mathbf{W}\mathbf{y} - \mathbf{x}\|^2]. \tag{3.101}$$

Exercise 3.3 Extend the EP algorithm introduced in Section 3.2.4 to soft-output detection, and derive the corresponding approximation of the LLRs in (3.55).

Exercise 3.4 The EP-based inference of approximating the posterior distribution in (3.6) assumes that the CSI is perfectly known. However, in practical systems, exact CSI is not available for a variety of reasons, such as channel estimation errors. In MIMO systems, consider a length-T transmission block with T_{p} symbols serving as pilot sequences and the remaining $T_{\mathrm{d}} = T - T_{\mathrm{p}}$ symbols used for data transmission. The CSI can be estimated from the pilot sequences using LMMSE estimation:

$$\hat{\mathbf{h}} = \mathbf{R}_{\mathbf{hh}}\mathbf{A}_{\mathrm{p}}^{H}\left(\mathbf{A}_{\mathrm{p}}\mathbf{R}_{\mathbf{hh}}\mathbf{A}_{\mathrm{p}}^{H} + \sigma_w^2\mathbf{I}_{T_{\mathrm{p}}N_r}\right)^{-1}\mathbf{y}_{\mathrm{p}}, \tag{3.102}$$

where $\mathbf{R}_{\mathbf{hh}} \in \mathbb{C}^{N_r N_t \times N_r N_t}$ denotes the channel correlation matrix, $\mathbf{A}_{\mathrm{p}} = \mathbf{X}_{\mathrm{p}} \otimes \mathbf{I}_{N_r} \in \mathbb{C}^{T_{\mathrm{p}}N_r \times N_r N_t}$ is the pilot matrix, and $\mathbf{y}_{\mathrm{p}} = \mathrm{vec}(\mathbf{Y}_{\mathrm{p}}) \in \mathbb{C}^{T_{\mathrm{p}}N_r \times 1}$ is the vectorization of the received pilot components. The imperfect CSI $\hat{\mathbf{h}} \in \mathbb{C}^{N_r N_t \times 1}$ results in severe performance degradation [152].

(a) A promising solution for mitigating the side effects of the imperfect CSI is to treat the channel estimation errors, i.e., $\Delta\mathbf{h} = \mathbf{h} - \hat{\mathbf{h}}$, as part of the equivalent noise vector, and to modify the EP-based approximation of transmitted symbols $\mathbf{X}_{\mathrm{d}}$ accordingly. Derive the corresponding equivalent MIMO model.
(b) Derive the covariance of $\Delta\mathbf{h}$, i.e., $\mathbf{R}_{\Delta\mathbf{h}} = \mathbb{E}\{\Delta\mathbf{h}\Delta\mathbf{h}^{H}\}$, for correction of the equivalent noise covariance.

Exercise 3.5 Derive the conditional probability in (3.43). *Hint:* Use Bayes' theorem in (3.6), noting that the transmitted signals are equiprobable and we assume an AWGN channel.

Exercise 3.6 Given the clues for the computation of $P(b_k = +1|\mathbf{b}_{\backslash k}^{(t)}, \mathbf{y})$ from equations (3.58)–(3.60), derive the mathematical expression for $P(b_k = -1|\mathbf{b}_{\backslash k}^{(t)}, \mathbf{y})$.

Exercise 3.7 Section 3.2.5 introduces the basic MCMC-based MIMO detection algorithm using Gibbs sampling. Recently, the combination of MCMC and gradient descent, namely gradient-based MCMC, has shown potential in solving non-convex optimization problems, including the integer least-squares problem of MIMO detection. In gradient-based MCMC, the Metropolis–Hastings algorithm is generally more favored than Gibbs sampling since its inherent vector-level operations facilitate gradient computations. We recommend the reader investigates [26, section 11.2] and [129] to get a better understanding of the Metropolis–Hastings algorithm and gradient-based MCMC, respectively, and accordingly complete the following tasks. Reference code can be found at https://github.com/le-liang/wcmlbook/tree/main/ch3/Exercise_3.7.

(a) Compare the Metropolis–Hastings and Gibbs sampling algorithms, and list their respective pros and cons concerning designing MIMO detectors.

(b) The gradient-based MCMC algorithm introduced in [129] performs a Metropolis–Hastings random walk in the (preconditioned) gradient descent direction of the continuous-relaxed least-squares surface to accelerate the search for the optimum. Implement this algorithm for MIMO detection, and compare the BER performance with the other methods shown in Figure 3.5.

Exercise 3.8   By combining equations (3.52) and (3.54), the discrete Metropolis-adjusted Langevin algorithm (DMALA) is obtained, which is a sampling algorithm for discrete distributions utilizing gradient information. This algorithm can be further simplified when the variable $\mathbf{x}$ (i.e., the symbol vector) is binary-valued. Please derive the simplified algorithm.

Exercise 3.9   Sections 3.2.3 and 3.3.2 introduce the OAMP and OAMP-Net algorithms for signal detection, respectively. Please use Python, with the Numpy package and deep learning frameworks (e.g., Tensorflow or Pytorch) to complete the following programming tasks. Reference code can be found at https://github.com/le-liang/wcmlbook/tree/main/ch3/Exercise_3.9.

(a) Implement the OAMP algorithm to solve the MIMO signal-detection problem in an 8×8 MIMO system with QPSK modulation and Rayleigh fading channels. Plot the BER performance with SNRs assuming values of 0 to 25 dB with an increment of 5 dB.

(b) Implement the OAMP-Net algorithm with (γ_t, θ_t) in (3.73) and (3.74) as learnable parameters (fix (ϕ_t, ξ_t) in (3.76) as $\phi_t = 1$ and $\xi_t = 0$). Optimize the parameters in a supervised manner using the loss function defined in (3.77). Plot the BER performance in the same settings as the above task and compare the results.

(c) Implement the OAMP-Net algorithm with $(\gamma_t, \theta_t, \phi_t, \xi_t)$ as learnable parameters. Optimize the parameters in a supervised manner using the loss function defined in (3.77) and use the same training datasets as in task b). Plot the BER performance with the same system settings as the above tasks and compare the results.

Exercise 3.10   In Section 3.3.2, OAMP-Net is built by unfolding the OAMP iterations for signal detection. Likewise, the EP iterations can be unfolded as a model-based deep learning-enhanced detector. Specifically, The values of the damping factor β are crucial for the stability and convergence speed of the EP solution. Therefore, the deep unfolded scheme can be utilized in EP iterations, where the damping factor at each layer is set as the trainable parameter, termed EPNet [142]. Use Python, with the Numpy package and deep learning frameworks (e.g., Tensorflow or Pytorch) to complete the following programming task. Reference code can be found at https://github.com/le-liang/wcmlbook/tree/main/ch3/Exercise_3.9.

(a) Implement the EP algorithm to solve the MIMO signal detection problem in an 8×8 MIMO system with QPSK modulation and Rayleigh fading channels. Plot the BER performance with SNRs assuming values of 0 to 25 dB with an increment of 5 dB.
(b) Implement the EPNet algorithm with damping factor β as a learnable parameter. Optimize the parameters in a supervised manner using the loss function defined in (3.77), and plot the BER performance with the same system settings as the above tasks and compare the results.

Exercise 3.11   Discuss the advantages and disadvantages of the deterministic and stochastic approximate inference-based MIMO signal detection methods introduced in Section 3.2.

Exercise 3.12   The parity check matrix of a (6,2) LDPC code is given by

$$\mathbf{H} = \begin{bmatrix} 1 & 1 & 0 & 1 & 0 & 0 \\ 0 & 1 & 1 & 0 & 1 & 0 \\ 1 & 0 & 0 & 0 & 1 & 1 \\ 0 & 0 & 1 & 1 & 0 & 1 \end{bmatrix}. \tag{3.103}$$

(a) Please list all the parity check equations.
(b) Please draw the resulting Tanner graph according to the listed equations.

Exercise 3.13   Briefly describe the advantages and disadvantages of using deep learning-based designs instead of traditional algorithms in decoding problems.

Exercise 3.14   Consider the following parity-check matrix for (7,4) Hamming code:

$$\mathbf{H} = \begin{bmatrix} 1 & 1 & 1 & 0 & 1 & 0 & 0 \\ 1 & 1 & 0 & 1 & 0 & 1 & 0 \\ 1 & 0 & 1 & 1 & 0 & 0 & 1 \end{bmatrix}. \tag{3.104}$$

Please draw the Tanner graph and depict different types of nodes (i.e., check nodes and variable nodes) in the graph according to the given parity-check matrix.

4   End-to-End Learning of Wireless Communication Systems

In this chapter, we explore advanced learning-based communication systems, where deep learning models are employed in both transmitters and receivers. These models are jointly trained through an end-to-end paradigm, which transforms the entire communication system into a data-driven framework. This novel end-to-end paradigm eliminates various modules commonly seen in traditional communication transmitters and receivers, including the channel encoder and decoder, modulator, channel estimator, and signal detector. The learned end-to-end communication architecture draws inspiration from the autoencoder network, where the transmitter functions as the encoder network and the receiver serves as the decoder network. Nevertheless, the challenges of training learned end-to-end communication systems are significant due to the complexities of wireless channels.

Within this chapter, we begin with the presentation of an illustrative example, a learning-based system operating over an additive white Gaussian noise (AWGN) channel with a short block size. Subsequently, we shift our focus to recent research endeavors dedicated to making this end-to-end learning paradigm applicable in real-world wireless communication scenarios. These efforts encompass addressing the complexities inherent in wireless channels and expanding the block size. Furthermore, we explore extensions of learned end-to-end communication systems to various semantic communication tasks. We introduce fundamental concepts in semantic communication and give illustrative applications and algorithms within this burgeoning field.

4.1 From Module-Based Communication to Learned End-to-End Communication

4.1.1 Introduction of Learned End-to-End Communication

Conventional communication systems adhere to a compartmental design, where both the transmitter and receiver are structured as sequential arrays of signal-processing blocks. Each of these blocks is dedicated to specific tasks, such as source coding, modulation, and detection. This discrete approach provides the advantage of fine-tuning individual components independently. However, it also raises the question of whether the entire system is operating at its optimum, as these blocks are not jointly optimized. Furthermore, many of the underlying signal-processing algorithms assume ideal conditions, e.g., stationary, linear, and Gaussian behaviors. However, these assumptions often do not hold in real-world scenarios.

In contrast, a novel paradigm has emerged to address these challenges: the learned end-to-end communication system. As demonstrated in Chapter 3, learning methods can be employed on the receiver side to enhance the performance of individual processing blocks, including channel decoding and signal detection. The learned end-to-end communication system takes this idea a step further by modeling both the transmitter and receiver using deep neural networks (DNNs). This groundbreaking approach departs from the rigid discrete structure and instead concentrates on optimizing the entire communication system, attempting to achieve optimal end-to-end performance.

In essence, the architecture of a learned end-to-end communication system can be understood as an autoencoder system, a widely employed approach for learning latent representations. In this analogy, the transmitter and receiver align with the roles of the autoencoder and autodecoder, respectively. The transmitter uses an encoder network to transform the transmitted symbols into encoded data denoted by $\mathbf{x}$, which is subsequently sent through the communication channel. Conversely, the receiver learns to recover the transmitted symbols based on the signal received from the channel, denoted by $\mathbf{y}$. In this way, the traditional communication modules at the transmitter are replaced with the encoder network. Similarly, the modules at the receiver are also replaced with the decoder network. By doing so, the block structure in conventional communication systems is no longer required.[1] By optimizing the encoding and decoding networks jointly based on gradient descent algorithms, the communication system can minimize an end-to-end loss so that the performance is directly optimized. In Example 4.1, we demonstrate how to mobilize this approach to design a learned end-to-end communication system in an AWGN channel.

Example 4.1   An example of a learned end-to-end communication system is shown in Figure 4.1. The detailed architecture of the autoencoder is provided in Table 4.1. The input $\mathbf{s}$ is encoded as a one-hot vector $\mathbf{1_s} \in \mathbb{R}^M$, with the sth element set to one. This vector $\mathbf{1_s}$ is mapped to the transmitted signal $\mathbf{x} \in \mathbb{R}^K$ through a feed-forward neural network (NN) with multiple dense layers, followed by a normalization layer that ensures compliance with physical constraints for $\mathbf{x}$. An AWGN channel is under consideration, represented as

$$\mathbf{y} = \mathbf{x} + \mathbf{n}, \tag{4.1}$$

---

[1] In fact, the autoencoder architecture does not require the complete replacement of the transmitter and receiver. It can selectively replace specific blocks, provided that the other blocks remain fixed and differentiable.

Table 4.1 Layout of the autoencoder.

Layer	Output dimension
Input	M
Dense+ReLU	M
Dense+linear	K
Normalization	K
Noise	K
Dense+ReLU	M
Dense+softmax	M

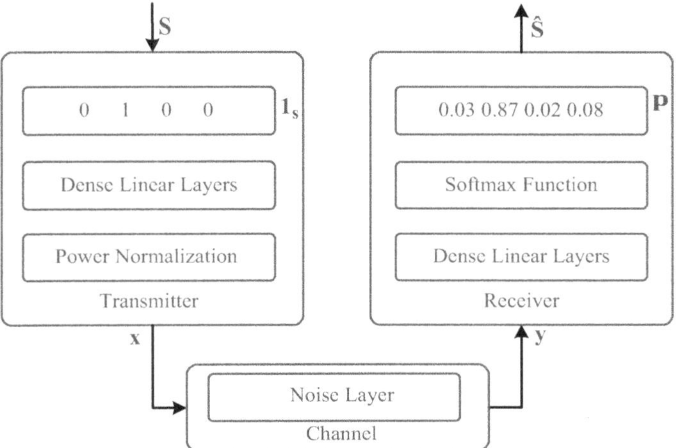

Figure 4.1 A communication system over an AWGN channel represented as an autoencoder.

where $\mathbf{n} \sim \mathcal{CN}(0,\sigma^2)$ represents the noise. At the receiver, the received signal $\mathbf{y}$ is mapped to a probability distribution $\mathbf{p} = [p_1, \ldots, p_M] \in (0,1)^M$ over all possible messages through a feedforward NN with softmax activation function, introduced in Section 1.2.2, where p_i represents the likelihood that the ith element is equal to one. According to $\mathbf{p}$, the most probable message is selected as the output for $\hat{s}$.

The autoencoder is trained in an end-to-end fashion using stochastic gradient descent (SGD). Since the channel is differentiable, back-propagation can be employed for training both the receiver and transmitter. The loss function for each training pair is defined as the categorical cross-entropy loss between $\mathbf{1_s}$ and $\mathbf{p}$,

$$\ell(\mathbf{1_s}, \mathbf{p}) = \sum_{i=0}^{M-1} 1_s^i \log(p_i), \qquad (4.2)$$

where 1_s^i denotes the ith element of $\mathbf{1_s}$.

Figure 4.2 presents a comparison of the block error rate (BLER) between a conventional communication system and a trained autoencoder with energy constraint

$\|\mathbf{x}\|_2^2 = K$. E_b represents the energy per bit, and N_0 represents the power spectral density of noise. The training process of the autoencoder is conducted at $E_b/N_0 = 7$ dB, employing the Adam optimizer with a learning rate of 0.001. The conventional communication system employs binary phase-shift keying (BPSK) modulation with a Hamming (7,4) code, utilizing either binary hard-decision decoding or maximum likelihood decoding (MLD) . Both systems operate at rate R = 4/7. Additionally, the BLER of uncoded BPSK (4,4) is provided for comparison. Simulation results demonstrate that the autoencoder, trained without any domain-specific knowledge, has effectively learned an encoder and decoder function. It performs comparably to Hamming (7,4) coding with MLD.

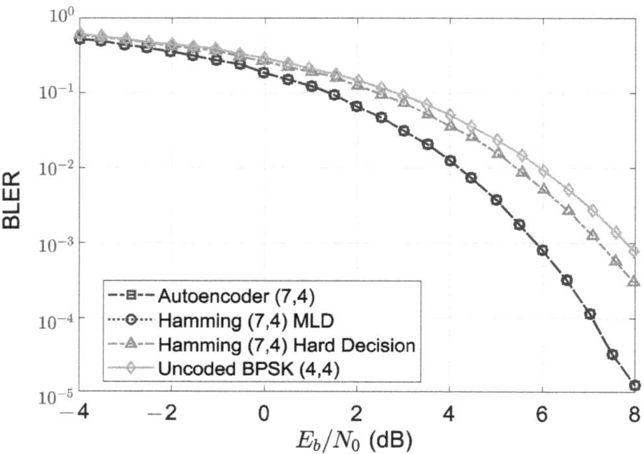

Figure 4.2 BLER versus E_b/N_0 for the autoencoder and several baseline communication schemes [10].

4.2   Extension to General Wireless Channels

Section 4.1 illustrates how a learned end-to-end communication system can serve as an alternative architecture to the conventional module-based design. Nevertheless, there are two significant challenges that currently impede the practical application of the end-to-end architecture in real-world communication systems:

- **Complexity of wireless channel effects**: In real communication systems, intricate factors such as hardware imperfections and varying channel conditions introduce complexities far beyond the scope of the idealized AWGN channel assumption.
- **The curse of dimensionality:** Practical communication systems often require relatively long block sizes to achieve significant coding gains. However, as the block size increases, the number of candidate messages grows exponentially. This leads to a substantial portion of these messages being unseen during training, which causes difficulties in learning a robust and well-performing model.

In this section, we explore recent research endeavors to address these challenges. We classify these end-to-end systems into two distinct categories: those that are learned with and without channel models. This classification is based on whether a perfect channel model is assumed to be available.

4.2.1 Learning with Channel Models

Let us begin by assuming that the wireless channel model is readily available, facilitating the direct application of back-propagation through the entire communication system. In the following, we will explore three specific examples: the Rayleigh fading channel, the multipath fading channel, and the multiple-input multiple-output (MIMO) channel.

Rayleigh Fading Channel

Mathematically, in Rayleigh fading channels the received signal $\mathbf{y}$ can be expressed as

$$\mathbf{y} = h\mathbf{x} + \mathbf{n}, \tag{4.3}$$

where $\mathbf{x}$ denotes the transmitted signal, $h \sim \mathcal{CN}(0, 1)$ denotes the fading as a complex Gaussian random variable, and $\mathbf{n} \sim \mathcal{CN}(0, \sigma^2)$ represents the noise with variance σ^2.

Recall that in Section 4.1, fully connected layers are employed in the transmitter and receiver for encoding and decoding, respectively, to build a learned end-to-end communication system based on the autoencoder. However, the message block length is severely limited due to the constraints associated with these fully connected layers. Inspired by convolutional codes, where the encoding process can be represented by a convolutional transform, hierarchical one-dimensional convolutional layers are used in [153] for both the transmitter and the receiver, as shown in Figure 4.3. In a fully

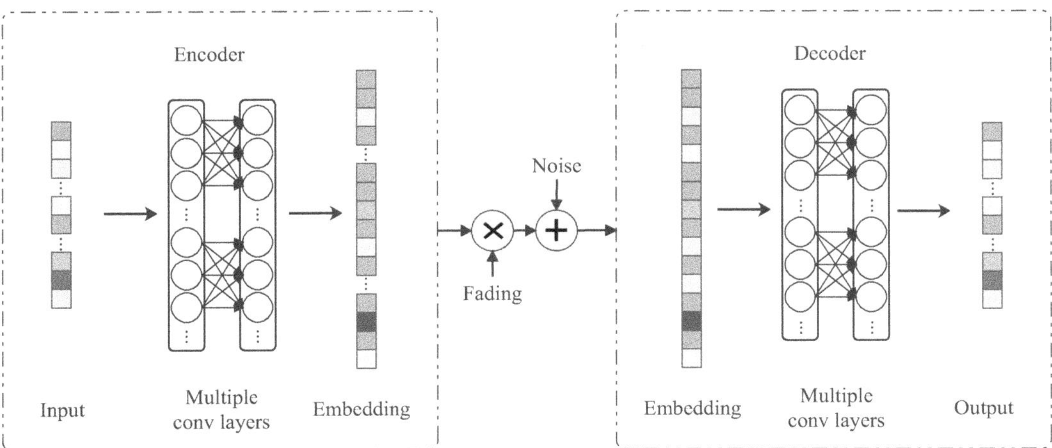

Figure 4.3 A communication system over a Rayleigh fading channel based on convolutional layers.

Table 4.2 Model parameters of CNN.

Type of layer	Kernel size/Annotation	Output size
Transmitter		
Input	Input layer	$K \times 1$
Conv+ReLU	5	$K \times 256$
Conv+ReLU	3	$K \times 128$
Conv+ReLU	3	$K \times 64$
Conv	3	$K \times 2$
Normalization	Power normalization	$K \times 2$
Receiver		
Conv+ReLU	5	$K \times 256$
Conv+ReLU	5	$K \times 128$
Conv+ReLU	5	$K \times 128$
Conv+ReLU	5	$K \times 128$
Conv+ReLU	5	$K \times 64$
Conv+ReLU	5	$K \times 64$
Conv+ReLU	5	$K \times 64$
Conv+sigmoid	3	$K \times 1$

connected layer, each neuron is connected to all neurons in the previous layer. In contrast, in a convolutional layer, each neuron is only connected to a few nearby neurons in the previous layer, which is called the receptive field of this neuron, and the same set of weights is shared by all neurons in a layer, as discussed in Section 1.2.2 in Chapter 1. When both the transmitter and the receiver are represented by convolutional neural networks (CNNs), introduced in Section 1.2.2, the codes learned by a CNN can be recovered more easily at the receiver than conventional handcrafted codes. We adopt the model architecture in [153], and the parameters of the CNN are shown in Table 4.2.

The autoencoder learns to map N information bits, $\mathbf{s} \in \{0, 1\}^N$, into a fixed-length embedding of length K, $\mathbf{x} \in \mathbb{C}^K$ and sends the embedding to the channel, while the autodecoder learns to recover the original information according to the received signal $\mathbf{y}$ from the channel. The distance between the original information bits, $\mathbf{s}$, and the recovered information bits, $\hat{\mathbf{s}} \in [0, 1]^N$, will be calculated. The binary cross-entropy loss function is used to measure the distance, which can be expressed as

$$\ell(\mathbf{s}, \hat{\mathbf{s}}) = \sum_{n=0}^{N-1} (s_n \log(\hat{s}_n) + (1 - s_n) \log(1 - \hat{s}_n)), \qquad (4.4)$$

where s_n and $\hat{s}_n$ represent the nth elements of $\mathbf{s}$ and $\hat{\mathbf{s}}$, respectively. An additional equalization layer is employed for the Rayleigh fading channel. The input of the autodecoder is the equalized signal $\hat{\mathbf{y}} = \frac{\mathbf{y}}{h}$, where h is obtained from the pilot information.

The end-to-end approach is compared with a conventional communication baseline, where quadrature amplitude modulation (QAM) is used as the modulation, and recursive systematic convolutional (RSC) coding of code rate 1/2 is used as channel coding. In each block, 64 information bits will be transmitted; thus the input size of the end-to-end approach is 64. In experiments, the end-to-end approach achieves similar performance to traditional methods while showing more robustness to different types of channel distortions in terms of the bit error rate (BER).

Frequency-Selective Fading Channel

To deal with frequency-selective fading, orthogonal frequency division multiplexing (OFDM) divides the wideband signal into many narrowband subcarriers, each exposed to flat-fading. In [154], the learned end-to-end communication system is extended to an OFDM scheme to enable reliable transmission over challenging multipath channels. As in Figure 4.4, the source message $\mathbf{s} \in \mathbb{M}^N$ is mapped to n complex-valued symbols $\mathbf{X} \in \mathbb{C}^{N \times n}$ by the autoencoder at the transmitter. Instead of directly transmitting the encoder's output $\mathbf{X}$, an inverse discrete Fourier transform (IDFT) with size N is applied to it. As each transmission of the autoencoder output still requires n channel uses, n complex-valued OFDM symbols $\mathbf{x} \in \mathbb{C}^{N \times n}$ are generated. To counter inter-symbol interference, a cyclic prefix (CP) of length N_{cp} is added to the transmitted signal $\mathbf{x}$. Thus, the complex-valued symbols $\mathbf{x_{cp}} \in \mathbb{C}^{(N+N_{cp}) \times n}$ are subsequently transmitted over the multipath channels with the added CP. As discussed in Section 2.1.1, the baseband complex channel impulse response can be expressed as

$$h(\tau, t) = \sum_i \alpha_i e^{j\phi_i(t)} \delta(\tau - \tau_i) \tag{4.5}$$

and

$$\phi_i(t) = \phi_i - 2\pi f_c \tau_i + 2\pi f_{D,i} t, \tag{4.6}$$

where α_i, τ_i, $f_{D,i}$, and ϕ_i are the amplitude, delay, Doppler shift, and random phase offset, respectively, associated with the ith propagation path. At the receiver, after removing the CP, the received signal $\mathbf{y}$ is transformed to $\mathbf{Y} \in \mathbb{C}^{N \times n}$ by performing a discrete Fourier transform (DFT). Finally, $\hat{\mathbf{s}}$ is reconstructed from $\mathbf{Y}$ through the decoder. Note that dense layers are employed for both the transmitter and receiver.

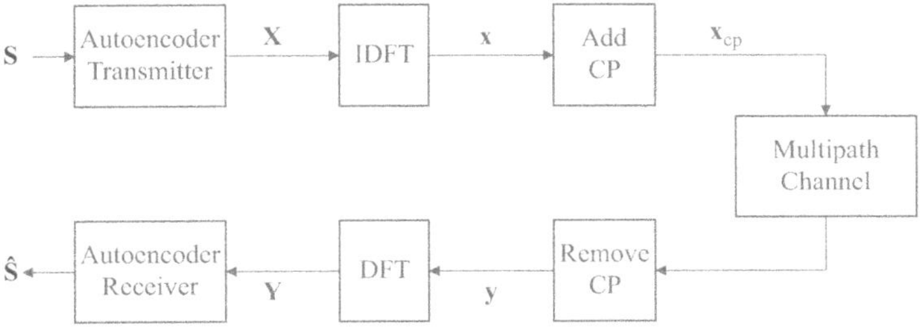

Figure 4.4 Illustration of the learned OFDM end-to-end communication system.

The entire network is trained using the end-to-end cross-entropy loss function as in (4.4).

The channel state information (CSI) plays a crucial role in the autoencoder with fading channels and is typically acquired through channel estimation. Additional pilots can be added to the message for channel estimation and equalization. The radio transformer network equalizer operates on a per-subcarrier basis on sequences of $n + 1$ symbols, i.e., a message plus one fixed pilot tone [155]. Meanwhile, similar to conventional systems, the per-subcarrier minimum mean square error (MMSE) channel estimation and equalization can be used in the autoencoder [154]. This method only requires element-wise operations, making it computationally efficient for end-to-end communication systems training. For the received symbol $y_i \in \mathbb{C}$ of the ith subcarrier, the channel estimate is given by

$$\hat{h}_i = \frac{y_i \cdot p^*}{|p|^2 + \sigma^2}, \tag{4.7}$$

where $\hat{h}_i$ represents the estimated channel, p represents the pilot symbol, and σ^2 represents the variance of noise.

The experiments in [154] illustrate that the autoencoder outperforms the quadrature phase-shift keying (QPSK) baseline by around 2 dB across the entire tested signal-to-noise ratio (SNR) range while achieving the same spectral efficiency. It is worth noting that the baseline could be extended to include higher-order modulation schemes combined with channel coding for additional enhancements. It is further demonstrated that even when there is no dedicated pilot and equalization, the system performs as well as the baseline. This suggests that the autoencoder does not require explicit pilots or explicit equalization, offering space for further spectral efficiency improvement.

Example 4.2 Consider a learned end-to-end OFDM system based on Table 4.3.

a) Compute the number of subcarriers N.
b) Determine the dimensions of the source message $\mathbf{s}$, $\mathbf{x}$, and the receiver input with an explicit pilot.

Solution

a) Given that Bandwidth = Subcarrier spacing $\times$ N, the number of subcarriers N can be calculated as

Table 4.3 Parameters for OFDM scheme.

Parameters	Value
Bandwidth	960 kHz
Number of OFDM symbols	7
Subcarrier spacing	15 kHz
Carrier frequency	3.5 GHz

$$N = \frac{\text{Bandwidth}}{\text{Subcarrier spacing}} = \frac{960 \text{ kHz}}{15 \text{ kHz}} = 64.$$

b) Since the number of subcarriers N equals 64, the dimension of s is $\mathbb{M}^{64}$. Additionally, according to Table 4.3, the number of OFDM symbols is $n = 7$, so the dimension of $\mathbf{x}$ is $\mathbb{C}^{64 \times 7}$. Finally, considering the receiver input dimension (per subcarrier) as $n + 1$, the dimension of the receiver input is $\mathbb{C}^{64 \times 8}$.

MIMO Channel

The learned end-to-end communication system can also be extended to MIMO channels. In [156], a pilot-free end-to-end paradigm with block size K is developed for flat-fading MIMO channels, where the wireless channels are modeled as a stochastic convolutional layer. As illustrated in Figure 4.5, the learned end-to-end communication paradigm follows the structure of the autoencoder, where the transmitter network and the receiver network correspond to the encoder and the decoder, respectively. The source data, $\mathbf{s}$, is encoded by the encoder network at the transmitter into an embedded vector $\mathbf{x}$. The embedded vector $\mathbf{x}$ is then transmitted through the wireless channel, and the receiver obtains the channel output, $\mathbf{y}$. For transmission of the vector $\mathbf{x}$, a system with multiple transmitter and receiver antennas over a narrowband channel is considered. The channel matrix $\mathbf{H}$ is modeled following the standard complex Gaussian distribution, i.e., each entry of the matrix being independent and identically distributed (i.i.d.) as $\mathcal{CN}(0, 1)$. The channel output can be expressed as $\mathbf{y} = \mathbf{Hx} + \mathbf{n}$, where $\mathbf{n}$ is the noise vector and $\mathbf{y}$ is the received signal vector. For correlated channels, the channel matrix can be expressed as $\mathbf{H_c} = \sqrt{\mathbf{R_t}}\mathbf{H}\sqrt{\mathbf{R_r}}$, where $\mathbf{R_t}$ and $\mathbf{R_r}$ represent the transmitter and receiver channel correlation matrices, respectively. The correlation matrix $\mathbf{R_t}$ can be expressed as

$$\mathbf{R_t} = \begin{bmatrix} 1 & \rho_t & \rho_t^4 & \cdots \\ \rho_t & 1 & \rho_t & \cdots \\ \vdots & & & \ddots \\ \rho_t^{(N_t-1)^2} & & \cdots & 1 \end{bmatrix}, \tag{4.8}$$

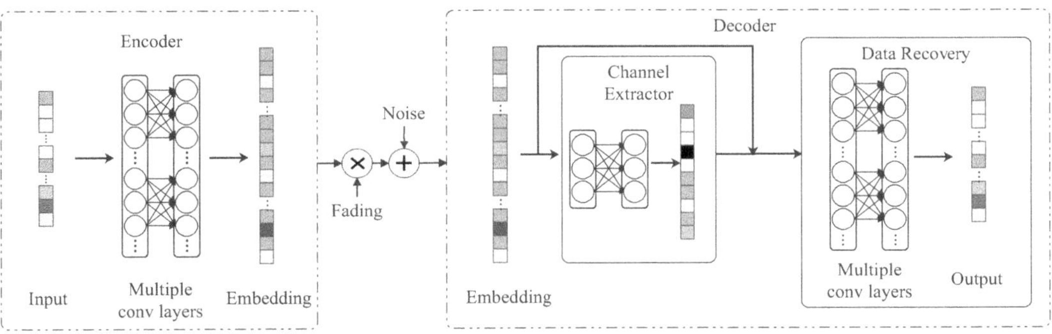

Figure 4.5 Structures of a learned end-to-end communication system for flat-fading MIMO channels.

where ρ_t denotes the correlation coefficients in the transmitter, and N_t denotes the numbers of antennas in the transmitter. Similarly, the correlation matrix at the receiver, $\mathbf{R_r}$, is structured like $\mathbf{R_t}$ but parameterized by the receiver's correlation coefficient ρ_r and the number of receiver antennas N_r.

At the receiver, the decoder network learns to recover the original data, $\hat{\mathbf{s}}$, based on the received signal $\mathbf{y}$. Although the learned end-to-end approach is pilot-free, it is still necessary to infer the current wireless channel, as the end-to-end reconstruction loss heavily depends on the current channel. Without the pilots, the channel information can only be obtained from the received data $\mathbf{y}$. As shown in Figure 4.5, two networks are employed at the receiver for channel information extraction and data reconstruction, respectively. To leverage the received data for channel information extraction, fully connected networks (FCNs) are employed to obtain global channel features, $\mathbf{z} \in \mathbb{R}^{l_z}$. These features are then integrated with the received data, $\mathbf{y} \in \mathbb{R}^{[K \times N_r]}$, through a bilinear operation, commonly used in computer vision [157]. With the bilinear operation, the received signal, $\mathbf{y}$, is augmented at each position by the extracted channel feature, $\mathbf{z}$. In particular, the received signal is first reshaped into a vector, and channel feature $\mathbf{z}$ is then multiplied by each item of the vector. As a result, an augmented signal $\mathbf{Y} \in \mathbb{R}^{K \cdot N_r \times l_z}$ is obtained via $\mathbf{Y} = \text{vec}(\mathbf{y}) \otimes \mathbf{z}$, where $\otimes$ represents the Kronecker product operation. The augmented signal $\mathbf{Y}$ is further reshaped into a matrix of size $[K, l_z \times N_r]$. Therefore, the bilinear operation increases the number of feature maps of the received signal by l_z times. Following the bilinear operation, the combined features are then utilized in the data recovery module to reconstruct the original data, $\hat{\mathbf{s}}$. The binary cross-entropy loss function is utilized to measure the distance between $\mathbf{s}$ and $\hat{\mathbf{s}}$ as in (4.4).

As illustrated in Figure 4.6, the pilot-free learned end-to-end system is compared with zero-forcing (ZF) and MMSE signal detection approaches. The block size K is set as 256. Both approaches utilize MMSE for the channel estimation and require

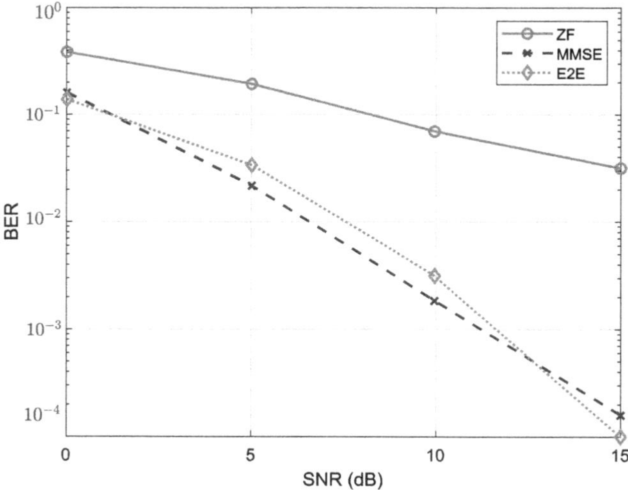

Figure 4.6 BER versus SNR on MIMO channel.

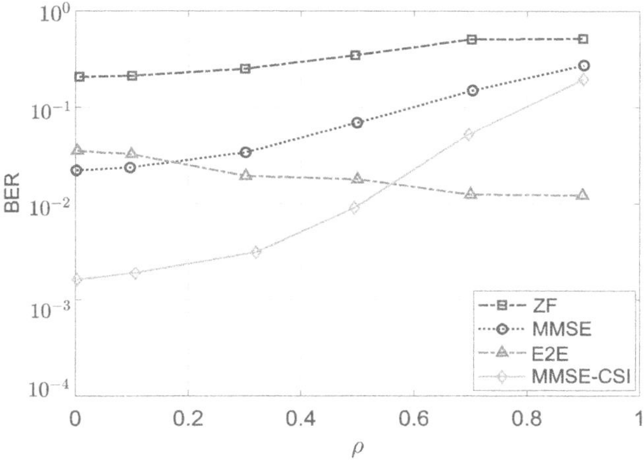

Figure 4.7 BER versus channel correlation ρ.

additional pilot overhead that consumes 1/8 of the block size. QPSK and convolutional code with rate 1/2 are used for modulation and channel coding, respectively. The learned end-to-end system, which does not require pilots, outperforms the ZF detection approach by a large margin. Across the whole tested SNRs, it performs measurably close to the MMSE detection method and even exceeds MMSE when the SNR is above 12.5 dB. This clearly demonstrates its significant potential in reducing pilot overhead without compromising communication performance.

The performance of the learned end-to-end method with different correlation coefficients is shown in Figure 4.7, where $\rho = \rho_t = \rho_r$. The learned end-to-end approach and the baselines are evaluated with SNRs equal to 5 dB. With the learned end-to-end approach, the correlation of the channel can be leveraged automatically to improve the performance, while the correlation of the channel degrades the BER performance of the baseline methods. As shown in Figure 4.7, the MMSE approach with perfect CSI (labeled as "MMSE-CSI") outperforms the learned end-to-end system when $\rho_t = \rho_r = 0$. But when $\rho_t = \rho_r = 0.5$, the pilot-free learned end-to-end system can achieve similar results to the MMSE approach assuming perfect CSI. In addition, when $\rho_t = \rho_r = 0.9$, the baseline approaches can hardly handle the channel correlation even with perfect CSI, while the pilot-free learned end-to-end approach remains effective.

When deployed in a real environment, the channel distribution may differ from that in the training phase, which may degrade the performance. If the distributional difference is significant, the model may need to be fine-tuned or retrained with the new data. But we still wish the system to be robust to small variances in the channels. In response, the robustness of the learned end-to-end system to discrepancies in training and testing channel distributions, particularly in terms of the power delay profile and channel correlation coefficients, is evaluated in [156]. The results demonstrate that the learned end-to-end system achieves remarkable robustness against such discrepancies, thereby showing significant potential in real-world deployment.

4.2.2 Learning without Channel Models

While the assumption of known and differentiable channel models has enabled learned end-to-end systems to exhibit competitive performance compared to traditional communication systems, certain real-world complexities defy accurate representation in a differentiable manner. The simplified channel models, if applied, may sometimes misguide the trained system due to the discrepancy between these models and the actual physical channels. For this reason, training learned end-to-end communication systems without a specific channel model is desirable. However, if the channel transfer function, $\mathbf{y} = f_h(\mathbf{x})$, is not available, the back-propagation of the gradients from the receiver network to the transmitter network is blocked, preventing the overall training of the end-to-end system.

To address this issue, two approaches have been investigated. The first involves the development of a channel-agnostic end-to-end communication system, as explored in [158]. In this approach, the distributions of channel outputs are learned through a conditional generative adversarial network (GAN) [159], eliminating the need for explicit knowledge of the channel transfer function. The second approach employs a reinforcement learning-based (RL-based) framework, as seen in [11], to optimize the learned end-to-end communication system without relying on the channel transfer function. The channel and the receiver are treated as part of the environment during the training of the transmitter, creating a more adaptable and robust system in scenarios where explicit channel information is unavailable or hard to model accurately.

GAN-Based Approach

As introduced in Chapter 1, the GAN model comprises two DNNs, namely the generator denoted by G and the discriminator denoted by D. The generator takes a noise vector $\mathbf{z}$ as input and produces a generated sample denoted by $G(\mathbf{z})$. In comparison, the discriminator's role is to determine whether a given sample input is sampled from the real dataset following the target distribution p_{data} or a generated sample, $G(\mathbf{z})$. The conditional GAN integrates additional information $\mathbf{m}$ into both the generator G and the discriminator D, as depicted in Figure 4.8. With the additional information, the optimization objective functions for the generator and the discriminator can be defined as

$$\mathcal{L}_G = \min_{\theta_G} \mathbb{E}_{\mathbf{z} \sim p_z}[\log(1 - D(G(\mathbf{z}, \mathbf{m}), \mathbf{m}))], \tag{4.9}$$

$$\mathcal{L}_D = \max_{\theta_D} \mathbb{E}_{\mathbf{x} \sim p_{data}}[\log(D(\mathbf{x}, \mathbf{m}))] + \mathbb{E}_{\mathbf{z} \sim p_z}[\log(1 - D(G(\mathbf{z}, \mathbf{m}), \mathbf{m}))], \tag{4.10}$$

respectively, where p_z denotes the input noise distribution, and θ_G and θ_D are the trainable parameters of the generator and the discriminator, respectively.

In [158], a conditional GAN was used to learn channel effects and to act as a bridge for the gradients to pass between transmitters and receivers. Using the conditional GAN, the output distribution of the channel can be approximated in a data-driven

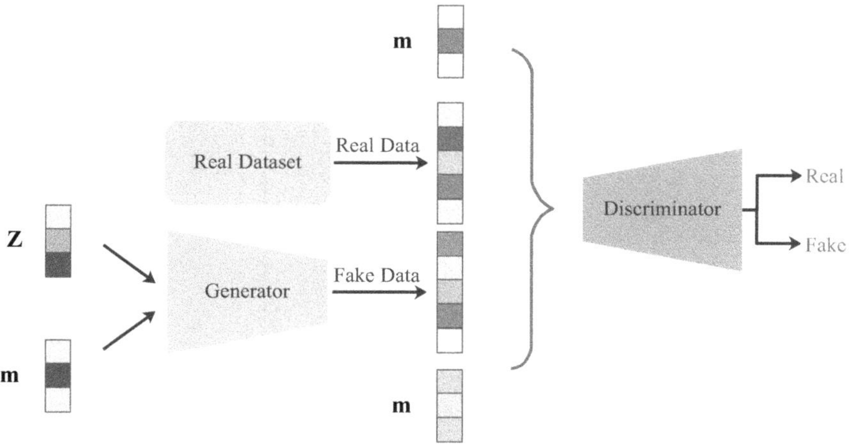

Figure 4.8 Structure of the conditional GAN.

Algorithm 4.1 Channel GAN Training Algorithm

1: **for** number of training iterations **do**
2:      *% Update the Generator*
3:      Sample minibatch of transmit data **s** and channel **h**.
4:      Obtain the minibatch of condition information **m** from the output of the transmitter DNN and the received pilot signal from the channel **h**.
5:      Sample minibatch of samples **z**.
6:      Update the generator by ascending the stochastic gradient of the loss function (4.9).
7:      *% Update the Discriminator*
8:      Sample minibatch of transmit data **s** and channel **h**.
9:      Obtain the minibatch of condition information **m** from the output of the transmitter DNN and the received pilot signal from the channel **h**.
10:     Sample minibatch of real data by collecting the output of the channel.
11:     Sample minibatch of samples **z**.
12:     Update the discriminator by descending the stochastic gradient of the loss function (4.10).

---

manner. Since the channel output, $\mathbf{y}$, for a given input, $\mathbf{x}$, is determined by the conditional distribution, $p(\mathbf{y}|\mathbf{x})$, a conditional GAN can be employed for learning the output distribution of a channel by taking $\mathbf{x}$ as the condition information. The generator will try to produce samples similar to the output of the real channel, while the goal of the discriminator is to distinguish between the data coming from the real channel output and the data coming from the generator. In practice, pilot symbols are usually transmitted for the receiver to estimate the channel, which will be used for subsequent symbol detection. In the proposed method, the received signal, $\mathbf{y_p}$, corresponding to the pilot symbols is further added as a part of the conditional

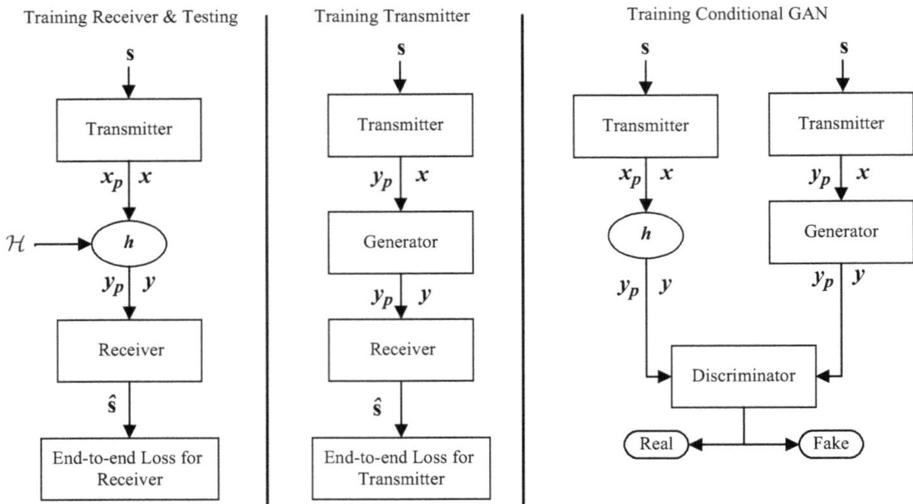

Figure 4.9 Training and testing of the end-to-end system.

information so that the output samples follow the distribution of $\mathbf{y}$ given the input $\mathbf{x}$ and the received pilot data, $\mathbf{y_p}$. If $\mathbf{y_p}$ is excluded from the condition information, the conditional GAN will learn to generate different channel conditions. With $\mathbf{y_p}$ as the conditional information, the output distribution of the conditional GAN will mimic the effect of the current channel rather than exhibiting diverse distributions.

The autoencoder at the transmitter side learns to map N information bits, $\mathbf{s} \in \{0, 1\}^N$, into a fixed length embedding of length K, $\mathbf{x} \in \mathbb{R}^K$, and sends the embedding to the channel, while the autodecoder at the receiver side learns to recover the original information according to the received signal $\mathbf{y}$. The distance between the original information bits, $\mathbf{s}$, and the recovered information bits, $\hat{\mathbf{s}} \in [0, 1]^N$, measured by the binary cross-entropy, will be calculated based on (4.4).

The training and testing of the proposed learned end-to-end communication system are illustrated in Figure 4.9. To obtain a training dataset, the information bits, $\mathbf{s}$, are randomly generated, and the instantaneous CSI is sampled randomly from the channel set. Due to different objectives in the modules, the transmitter, the receiver, and the channel generator can be trained iteratively based on the training data. When training one component, the parameters of the others remain fixed. In the testing stage, the end-to-end performance is evaluated on the learned transmitter and receiver with real channels.

- **Training the receiver:** At the receiver, a DNN model is trained for recovering the original data $\mathbf{s}$, where the input is the original data corresponding to the original data, $\mathbf{y}$, while the output is the estimation $\hat{\mathbf{s}}$. By comparing the $\mathbf{s}$ and $\hat{\mathbf{s}}$, the loss function can be calculated on the basis of (4.4). The receiver can be trained easily since the loss function is computed at the receiver and thus the gradients of the loss function can be easily obtained. For the time-varying channels, by directly putting the received signal, $\mathbf{y}$, and the received pilot data, $\mathbf{y_p}$, together as the input, the receiver can automatically infer the channel condition and perform the symbol detection without explicitly estimating the channel, as was investigated in Example 3.1 of Chapter 3.

- **Training the transmitter:** With the channel generator as a surrogate channel, the training of the transmitter will be similar to that of the receiver. During the training phase, the transmitter, the generator, and the receiver can be viewed as a whole DNN. The output of the transmitter is the value of the last hidden layer in the transmitter DNN. The end-to-end cross-entropy loss function is computed at the receiver as in (4.4), and the gradients are propagated back to the transmitter through the conditional GAN. The parameters of the transmitter will be updated based on SGD while those of the conditional GAN and the receiver remain fixed. The transmitter can learn the constellation of the embedding, $\mathbf{x}$, so that the received signal can be efficiently detected at the receiver.
- **Training the conditional GAN:** The training procedure of the conditional GAN is illustrated in Algorithm 4.1 in detail. In each iteration, the generator and the discriminator are trained iteratively. The parameters of one model will be fixed when training the other. With the learned transmitter, the real data can be obtained by passing the encoded signal from the transmitter through the real channel, while the fake data is obtained by passing the encoded data through the channel generator. The parameters of the generator and the discriminator are updated according to the loss function in equations (4.9) and (4.10), respectively.

Example 4.3   We consider two types of DNN models to model the channel effects in a data-driven way based on the channel GAN. One is the FCN and the other is the CNN. The FCN is used for small block sizes, and the CNN is used for large block sizes to tackle the curse of dimensionality. The parameters of the FCN and CNN are shown in Table 4.4 and Table 4.5, respectively. Next, the learned end-to-end communication system is compared with a conventional communication system.

We first use the FCN for a block size of four bits over an AWGN channel. The end-to-end BER performance on the AWGN channel is shown in Figure 4.10. At each time step, four information bits are transmitted, and the length of the transmitter output is set to be seven. From the figure, the BER of the learning-based approach is similar to that of Hamming (7,4) code with MLD and the model-aware

Table 4.4 Model parameters of FCN.

Parameters	Values
Transmitter: Neurons in each hidden layers	32, 32
Learning rate	0.001
Receiver: Neurons in each hidden layers	32, 32
Learning rate	0.001
Generator: Neurons in each hidden layers	128, 128, 128
Discriminator: Neurons in each hidden layers	32, 32, 32
Learning rate	0.0001

Table 4.5 Model parameters of CNN.

Type of layer	Kernel size/Annotation	Output size
Transmitter		
Input	Input layer	$K \times 1$
Conv+ReLU	5	$K \times 256$
Conv+ReLU	3	$K \times 128$
Conv+ReLU	3	$K \times 64$
Conv	3	$K \times 2$
Normalization	Power normalization	$K \times 2$
Receiver		
Conv+ReLU	5	$K \times 256$
Conv+ReLU	5	$K \times 128$
Conv+ReLU	5	$K \times 128$
Conv+ReLU	5	$K \times 128$
Conv+ReLU	5	$K \times 64$
Conv+ReLU	5	$K \times 64$
Conv+ReLU	5	$K \times 64$
Conv+sigmoid	3	$K \times 1$
Generator		
Conv+ReLU	5	$K \times 256$
Conv+ReLU	3	$K \times 128$
Conv+ReLU	3	$K \times 64$
Conv	3	$K \times 2$
Discriminator		
Conv+ReLU	5	$K \times 256$
Conv+ReLU	3	$K \times 128$
Conv+ReLU	3	$K \times 64$
Conv+ReLU	3	$K \times 16$
FCN+ReLU	100	100
FCN+sigmoid	1	1

end-to-end autoencoder system trained under the AWGN channel, which proves the effectiveness of using the conditional GAN as a surrogate channel.

To train models with a larger block size, CNNs are then used to mitigate the curse of dimensionality. We train the CNN under the AWGN channel where the noise is added to the hidden layer directly. The network is trained at 3 dB and tested with different SNRs. Figure 4.11 shows the BER curves of the proposed end-to-end method with the length of the transmitted information sequence being 128 bits. From the figure, it can be seen that the performance of the proposed method is similar to RSC code in the low SNRs and significantly outperforms RSC in the high SNRs. Compared with the model-aware end-to-end autoencoder system, the performance degradation with increased block size is negligible.

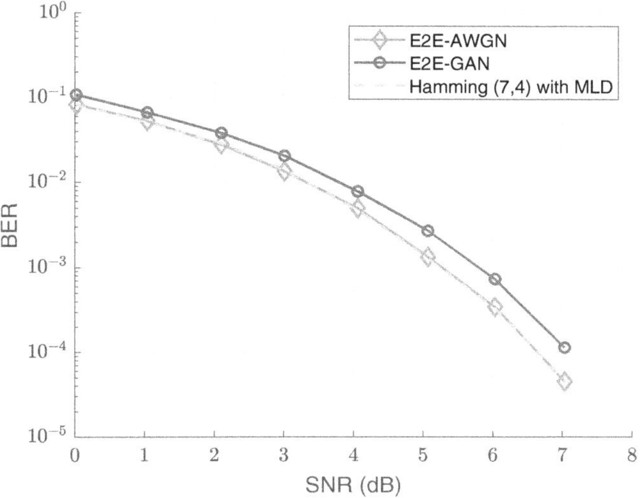

Figure 4.10 BER with a block size of four bits under an AWGN channel.

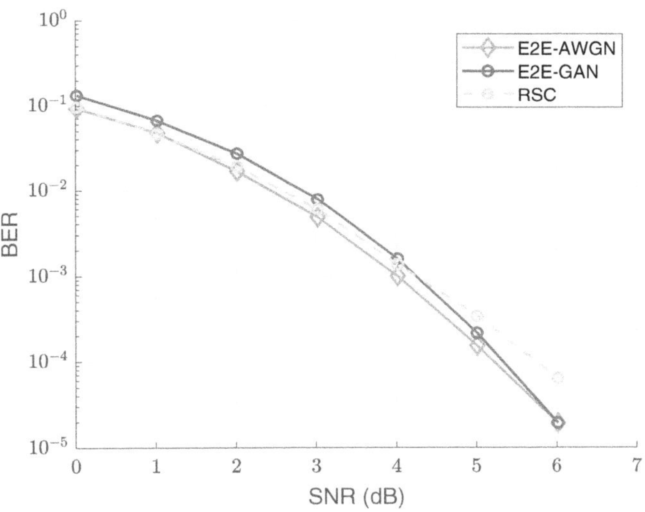

Figure 4.11 BER with a block size of 128 bits under an AWGN channel.

RL-Based Approach

An RL-based learned end-to-end communication system without a specific channel model was developed in [11]. In this framework, training the receiver is formulated as a supervised learning task, while training the transmitter is approached using an RL strategy. In the following, we delve into the details of the training processes for both the transmitter and the receiver.

Formally, the message $s \in \mathcal{M} = \{1, ..., M\}$ is encoded into channel symbols $\mathbf{x}$ at the transmitter. This mapping is represented as $f_{\boldsymbol{\theta}_T}^{(T)}: \mathrm{M} \to \mathbb{R}^{2N}$, where N represents the number of channel uses, and $\boldsymbol{\theta}_T$ denotes the parameters associated with the transmitter. At the receiver, a corresponding mapping $f_{\boldsymbol{\theta}_R}^{(R)}: \mathbb{R}^{2N} \to \left\{ \mathbf{p} \in \mathbb{R}_+^M \mid \sum_{i=1}^M p_i = 1 \right\}$ is employed, where $\boldsymbol{\theta}_R$ represents the parameters of the receiver, and $\mathbf{p}$ denotes a probability distribution over all candidate messages. Consequently, the loss can be formally expressed as

$$\mathcal{L}(\boldsymbol{\theta}_T, \boldsymbol{\theta}_R) \triangleq \mathbb{E}_s \left\{ \int l\left(f_{\boldsymbol{\theta}_R}^{(R)}(\mathbf{y}), s \right) p\left(\mathbf{y} \mid f_{\boldsymbol{\theta}_T}^{(T)}(s) \right) d\mathbf{y} \right\}, \tag{4.11}$$

where $\mathbb{E}_s$ denotes the expectation taken across the message space s, and $l(\mathbf{p}, s) = -\log(p_s)$ denotes the categorical cross-entropy. For gradient descent training algorithms, the gradient of the loss can be decomposed into two components: The first component is the gradient of the loss with respect to θ_R, denoted by $\nabla_{\theta_R}\mathcal{L}$; the second component is the gradient of the loss with respect to θ_T, denoted by $\nabla_{\theta_T}\mathcal{L}$.

Regarding the receiver training, since the unknown channel is not involved, $\nabla_{\theta_R}\mathcal{L}$ can be analytically calculated and represented by

$$\nabla_{\boldsymbol{\theta}_R}\mathcal{L} = \mathbb{E}_{s,\mathbf{y}} \left\{ \nabla_{\boldsymbol{\theta}_R} l\left(f_{\boldsymbol{\theta}_R}^{(R)}(\mathbf{y}), s \right) \right\}. \tag{4.12}$$

For each training batch, the stochastic gradient in (4.12) can be expressed as

$$\underline{\nabla_{\boldsymbol{\theta}_R}\mathcal{L}} \triangleq \frac{1}{N} \sum_{i=1}^N \nabla_{\boldsymbol{\theta}_R} l\left(f_{\boldsymbol{\theta}_R}(\mathbf{y}^{(i)}), s^{(i)} \right), \tag{4.13}$$

where N represents the batch size (the number of samples utilized to approximate the loss value), $s^{(i)}$ refers to the ith training sample, and $\mathbf{y}^{(i)}$ represents the corresponding received signal. As the calculation of equation (4.13) solely necessitates sampling from the channel output, receiver training can be executed without knowledge of the precise channel model, $p(\mathbf{y}|\mathbf{x})$.

Regarding the transmitter training, the gradient of $\mathcal{L}$ with respect to $\boldsymbol{\theta}_T$ is

$$\nabla_{\boldsymbol{\theta}_T}\mathcal{L} = \mathbb{E}_s \left\{ \int l\left(f_{\boldsymbol{\theta}_R}^{(R)}(\mathbf{y}), s \right) \nabla_{\boldsymbol{\theta}_T} f_{\boldsymbol{\theta}_T}^{(T)}(s) \cdot \nabla_{\mathbf{x}} p(\mathbf{y}|\mathbf{x}) \mid_{\mathbf{x}=f_{\boldsymbol{\theta}_T}^{(T)}(s)} d\mathbf{y} \right\}, \tag{4.14}$$

where $\nabla_{\boldsymbol{\theta}_T} f_{\boldsymbol{\theta}_T}^{(T)}(s)$ represents the Jacobian matrix of the transmitter output, and $\nabla_{\mathbf{x}} p(\mathbf{y}|\mathbf{x}) \mid_{\mathbf{x}=f_{\boldsymbol{\theta}_T}^{(T)}(s)}$ is the gradient of the channel $p(\mathbf{y}|\mathbf{x})$. As $p(\mathbf{y}|\mathbf{x})$ is unknown and may not be differentiable, its gradient cannot be calculated and might even be undefined. A solution is to train the transmitter through an RL framework. In this framework, the transmitter is treated as an agent, while both the channel and the receiver are considered part of the environment. At each time step, the source data m serves as the state observed by the transmitter, and the channel input $\mathbf{x}$ is the action taken by the transmitter. To enable exploration, the channel input $\mathbf{x}$ is relaxed to follow a distribution $\hat{\pi}_{\bar{\mathbf{x}},\sigma}$, parameterized by $\bar{\mathbf{x}}$ with a standard deviation σ, which constitutes the stochastic RL policy. Under this policy, the loss (4.11) can be rewritten as

$$\widehat{\mathcal{L}}(\boldsymbol{\theta}_T, \boldsymbol{\theta}_R) \triangleq \mathbb{E}_s \left\{ \int \hat{\pi}_{f_{\boldsymbol{\theta}_T}^{(T)}(s), \sigma}(\mathbf{x}) \cdot \int l\left(f_{\boldsymbol{\theta}_R}^{(R)}(\mathbf{y}), s\right) p\left(\mathbf{y}|\mathbf{x}\right) d\mathbf{y} d\mathbf{x} \right\} \tag{4.15}$$

with gradient

$$\nabla_{\boldsymbol{\theta}_T} \widehat{\mathcal{L}} = \mathbb{E}_{s,\mathbf{x},\mathbf{y}} \left\{ l\left(f_{\boldsymbol{\theta}_R}^{(R)}(\mathbf{y}), s\right) \nabla_{\boldsymbol{\theta}_T} f_{\boldsymbol{\theta}_T}^{(T)}(s) \cdot \nabla_{\bar{\mathbf{x}}} \log\left(\hat{\pi}_{\bar{\mathbf{x}},\sigma}(\mathbf{x})\right) \Big|_{\bar{\mathbf{x}} = f_{\boldsymbol{\theta}_T}^{(T)}(s)} \right\}. \tag{4.16}$$

The differentiability of $p(\mathbf{y}|\mathbf{x})$ is not required when computing $\nabla_{\boldsymbol{\theta}_T} \widehat{\mathcal{L}}$, which can be simply estimated by sampling from the channel distribution,

$$\underline{\nabla_{\boldsymbol{\theta}_T} \widehat{\mathcal{L}}} \triangleq \frac{1}{N} \sum_{i=1}^{N} l\left(f_{\boldsymbol{\theta}_R}^{(R)}(\mathbf{y}^{(i)}), s^{(i)}\right) \nabla_{\boldsymbol{\theta}_T} f_{\boldsymbol{\theta}_T}^{(T)}(s^{(i)})$$
$$\cdot \nabla_{\bar{\mathbf{x}}} \log\left(\hat{\pi}_{\bar{\mathbf{x}},\sigma}(\mathbf{x}^{(i)})\right) \Big|_{\bar{\mathbf{x}} = f_{\boldsymbol{\theta}_T}^{(T)}(s^{(i)})}, \tag{4.17}$$

where $l\left(f_{\boldsymbol{\theta}_R}^{(R)}(\mathbf{y}^{(i)}), s^{(i)}\right)$ can be regarded as the reward provided by the environment, based on the chosen action and the current state.

Furthermore, we consider an alternative interpretation and an understanding of the relation between $\nabla_{\boldsymbol{\theta}_T} \widehat{\mathcal{L}}$ and $\nabla_{\boldsymbol{\theta}_T} \mathcal{L}$ in (4.14) and (4.16). Let us study the difference between $\nabla_{\boldsymbol{\theta}_T} \widehat{\mathcal{L}}$ and $\nabla_{\boldsymbol{\theta}_T} \mathcal{L}$,

$$\nabla_{\boldsymbol{\theta}_T} \mathcal{L} - \nabla_{\boldsymbol{\theta}_T} \widehat{\mathcal{L}} = \mathbb{E}_s \Bigg\{ \int l\left(f_{\boldsymbol{\theta}_R}^{(R)}(\mathbf{y}), s\right) \nabla_{\boldsymbol{\theta}_T} f_{\boldsymbol{\theta}_T}^{(T)}(s) \Bigg[\nabla_{\mathbf{x}} p\left(\mathbf{y}|\mathbf{x}\right) \Big|_{\mathbf{x} = f_{\boldsymbol{\theta}_T}^{(T)}(s)}$$
$$- \mathbb{E}_{\mathbf{x}} \left\{ p\left(\mathbf{y}|\mathbf{x}\right) \nabla_{\bar{\mathbf{x}}} \log\left(\hat{\pi}_{\bar{\mathbf{x}},\sigma}(\mathbf{x})\right) \Big|_{\bar{\mathbf{x}} = f_{\boldsymbol{\theta}_T}^{(T)}(s)} \right\} \Bigg] d\mathbf{y} \Bigg\}. \tag{4.18}$$

It is clear that the difference between the two gradients approaches zero when the true channel gradient $\nabla_{\mathbf{x}} P(\mathbf{y}|\mathbf{x}) \Big|_{\mathbf{x} = f_{\boldsymbol{\theta}_T}^{(T)}(s)}$ is well approximated by

$\mathbb{E}_{\mathbf{x}} \left\{ p\left(\mathbf{y}|\mathbf{x}\right) \nabla_{\bar{\mathbf{x}}} \log\left(\hat{\pi}_{\bar{\mathbf{x}},\sigma}(\mathbf{x})\right) \Big|_{\bar{\mathbf{x}} = f_{\boldsymbol{\theta}_T}^{(T)}(s)} \right\}$. Hence, the relaxation of the transmitter output can be viewed as an approach for estimating and approximating the elusive channel gradient.

With the gradient for the transmitter and receiver, a training algorithm is proposed that involves alternating between training the receiver using the true gradient as (4.12) and training the transmitter using gradient approximation as (4.16). During the receiver training (or transmitter training), the parameters of the transmitter (or receiver) are held constant. Consequently, the training algorithm consists of two distinct phases: one for the receiver and another for the transmitter. This training procedure continues until a pre-defined stopping criterion is met, which can be reaching a fixed number of iterations or observing the training loss getting flat.

4.3   End-to-End Channel Coding

Besides the efforts to extend the capabilities of learned end-to-end communication systems to adapt to general wireless channels, there has also been a parallel focus on increasing the block size of these systems to obtain robust learning-based channel codes.

The journey toward efficient codes with computationally manageable decoders has been characterized by sporadic progress and has often relied on human ingenuity. Since Shannon's groundbreaking work in 1948 [143], it took several decades of research to arrive at the current state-of-the-art codes [146]. These channel codes, such as turbo codes [160], low-density parity-check (LDPC) codes [161], and polar codes [146], approach Shannon capacity on AWGN channels. Nevertheless, traditional coding methods exhibit several limitations:

- **Designed for standard channels**: Traditionally, channel coding algorithms are designed by optimizing specific mathematical properties, such as the minimum code distance [162]. These optimizations are typically tailored for standard channels, typically AWGN channels, where the signal is subjected to i.i.d. Gaussian noise. However, in practical scenarios, when the channel deviates from the AWGN model, ad-hoc heuristic techniques are often necessary to adapt to the non-Gaussian nature of the noise.
- **Infinite block lengths**: Channel codes are typically developed with finite block lengths. While these capacity-achieving codes are theoretically proven optimal as the block length approaches infinity, practical scenarios often involve short or moderate block lengths, where there still exists room for improvement.

The learned end-to-end communication architecture offers a promising avenue for designing novel neural channel codes that can potentially address the challenges mentioned above. By employing neural networks to represent both channel encoders and decoders, these learned codes can readily adapt to realistic and complex channels, as well as accommodating different block lengths, with minimal adjustments to the training data.

Nevertheless, a significant challenge in neural-based channel coding arises due to the dimensionality issue stemming from the vast code space involved. As the code block size increases, the number of potential codewords grows exponentially, leading to a substantial portion of codewords being unseen during training. In practice, many learned end-to-end systems are designed with a very short block size, such as Example 4.1. However, it is imperative for the code block size to be of sufficient length to ensure substantial coding gains. Consequently, the design of neural channel codes with longer block sizes has garnered considerable attention in recent research, which mimics traditional channel coding without considering the modulation.

In the following, we introduce recent advances in learning neural channel codes with larger block sizes in the framework of the learned end-to-end communication architecture.

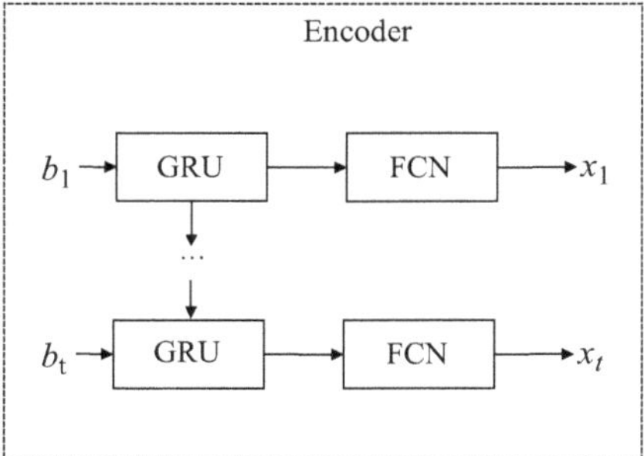
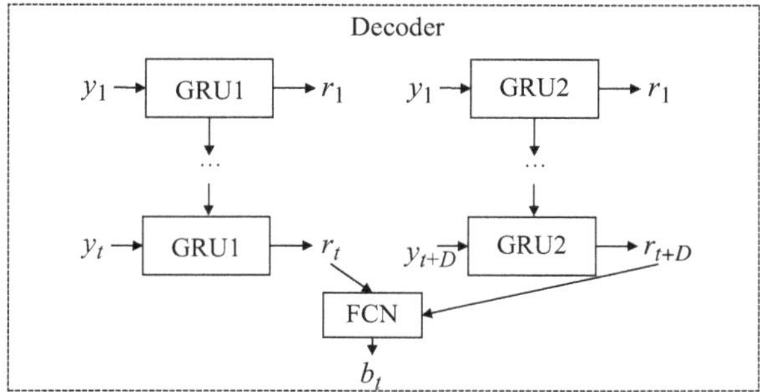

Figure 4.12 Architecture of the encoder and decoder in the learned end-to-end channel codes with RNNs.

4.3.1 Learning End-to-End Channel Codes with RNNs

To increase the scalability of the learned end-to-end communication system, researchers have explored using more advanced deep learning models to replace the fully connected DNNs. Recurrent neural networks (RNNs) and CNNs have been chosen due to their natural connections to classic convolutional codes. In [163], the Bidirectional RNN (Bi-RNN) [2] is employed for both encoder and decoder with an additional low latency constraint. The neural encoder is constructed as an RNN containing two layers of gated recurrent units (GRUs), followed by a fully connected layer. With an RNN-based encoder and decoder, we can effectively train neural codes with a length of roughly 100. These learned neural codes outperform canonical convolutional codes and generalize well to non-Gaussian channels.

The RNN-based encoder and decoder architecture can be effectively employed to design codes with latency constraints. In this context, the decoder's structural delay, denoted by D, represents the number of bits the decoder can look ahead to perform decoding. To enhance latency, the RNN decoder takes a different approach by utilizing two GRUs instead of Bi-RNN structures, as illustrated in Figure 4.12. These two distinct GRUs operate in different time slots: One GRU processes information up to the current time slot, while the other GRU extends further for a specified duration of D steps. The outputs of these two GRUs are subsequently synthesized using a fully connected layer. In this manner, the decoder ensures the efficient utilization of all information bits that comply with the specified delay constraint, achieving this solely through a forward pass. In addition, each GRU exclusively handles one step ahead, resulting in a decoding computational complexity of $O(1)$. By contrast, Viterbi and BCJR low-latency decoders require traversing the trellis and backtracking to the intended position, involving both a forward step and backward movement spanning delay-constraint steps. This results in a computational complexity of $O(D)$ for decoding each bit.

---

[2] The Bi-RNN combines both a forward and a backward RNN, allowing it to infer the current state by evaluating information from both past and future contexts.

4.3.2 Learning End-to-End Channel Codes with TurboAE

In addition to utilizing existing deep learning architectures, researchers have also devised novel deep learning models that aim to increase the block size by integrating some modules of classic channel codes, such as turbo and LDPC codes.

In turbo codes, an interleaver plays a crucial role in generating the encoded symbol sequence from a parallel arrangement of convolutional codes. Formally, the interleaver, denoted by $x^\pi = \pi(x)$, and the de-interleaver, denoted by $x = \pi^{-1}(x^\pi)$, rearrange and restore the input sequence x using a pseudo-random interleaving array that is shared between the encoder and decoder. This process is illustrated on the left side of Figure 4.13.

In [164], TurboAE is introduced, featuring interleaved encoding and iterative decoding, which imparts long-term memory into the code. Specifically, the system uses a code rate of 1/3 for the interleaved encoder f_θ. This encoder comprises three trainable encoding blocks: $f_{i,\theta}(\cdot)$ for $i \in \{1,2,3\}$. Each of these blocks is responsible for encoding $b_i = f_{i,\theta}(u)$ for $i \in \{1,2\}$, while $b_3 = f_{3,\theta}(\pi(u))$ is encoded by the third block. The decoding process for interleaved codes involves an iterative approach, considering both the interleaved and de-interleaved orders, as depicted in Figure 4.14.

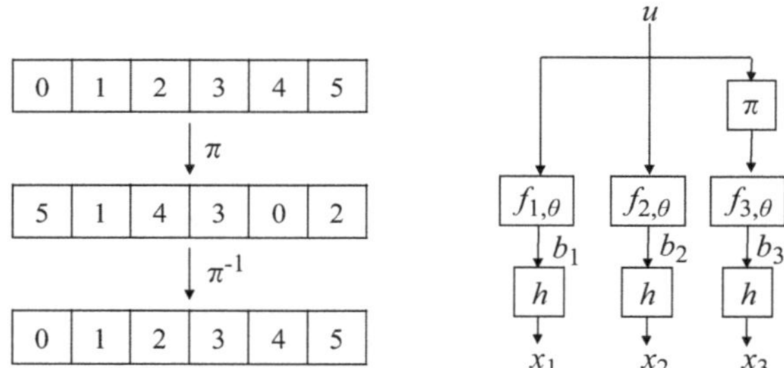

Figure 4.13 Visualization of interleaver and de-interleaver (left); TurboAE encoder on code rate 1/3 (right).

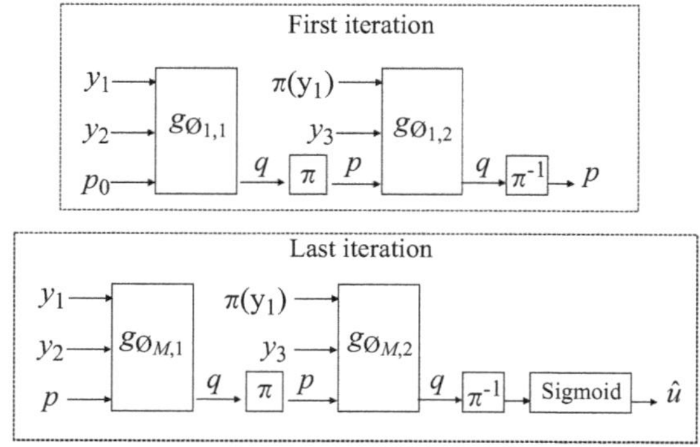

Figure 4.14 TurboAE iterative decoder for code rate 1/3.

The received codewords are generated from both the original message u and the interleaved message $\pi(u)$. As a result, the decoder operates for multiple iterations, and in each iteration, two decoders are employed: $g_{\phi_{i,1}}$ and $g_{\phi_{i,2}}$.

Let y_1, y_2, y_3 denote noisy versions of x_1, x_2, x_3, respectively. The first decoder, $g_{\phi_{i,1}}$, processes the received signals y_1 and y_2 along with the de-interleaved prior p of size (K, F), where F denotes the feature size for each coded bit, and K represents the block length. It then generates a posterior q with the shape (K, F). Meanwhile, the second decoder, $g_{\phi_{i,2}}$, handles the interleaved signal $\pi(y_1)$, y_3, and the interleaved prior to producing a new posterior q. The posterior q from the previous iteration becomes the prior p for the subsequent iteration. The initial iteration employs a prior of 0, and in the final iteration, the posterior is of shape $(K, 1)$, decoded using the sigmoid function introduced in Section 1.2.2 to yield $\hat{u}$.

In TurboAE, the encoder and decoder are parameterized with a one-dimensional CNN. Since the encoder, channel, and decoder are all differentiable, TurboAE can undergo end-to-end training using a gradient descent algorithm. To meet the power constraint requirement, a normalization layer is applied to constrain the power of the code, represented as

$$x_i = \frac{b_i - \mu(b)}{\sigma(b)}, \tag{4.19}$$

where $\mu(b) = \frac{1}{K} \sum_{i=1}^{K} b_i$ and $\sigma(b) = \sqrt{\frac{1}{K} \sum_{i=1}^{K} (b_i - \mu(b))^2}$ represent the mean and standard deviation estimates of the entire block, respectively. $\mu(b)$ and $\sigma(b)$ are estimated from the entire batch during the training phase, while these values are pre-computed using multiple batches in the testing phase. This normalization layer can also be interpreted as batch normalization without affine projection and plays a critical role in stabilizing the training of the encoder.

4.4 Semantic Communication and Semantic Information Theory

4.4.1 From Bit Transmission to Semantic Communication

In [143], Shannon and Weaver categorized communication systems into three distinct levels as depicted in Figure 4.15:

- **Syntactic communication**: This level emphasizes accurate transmission of symbols to maintain their integrity.
- **Semantic communication:** This level focuses on the accurate transmission of the meaning of symbols.
- **Pragmatic communication:** This level addresses how the meaning of symbols influences the system performance.

While the first level of communication, aimed at accurate transmission of symbols, has been extensively explored and successfully implemented in conventional communication systems, there has more recently been a surging interest in semantic communication, driven by the stringent demands of new applications.

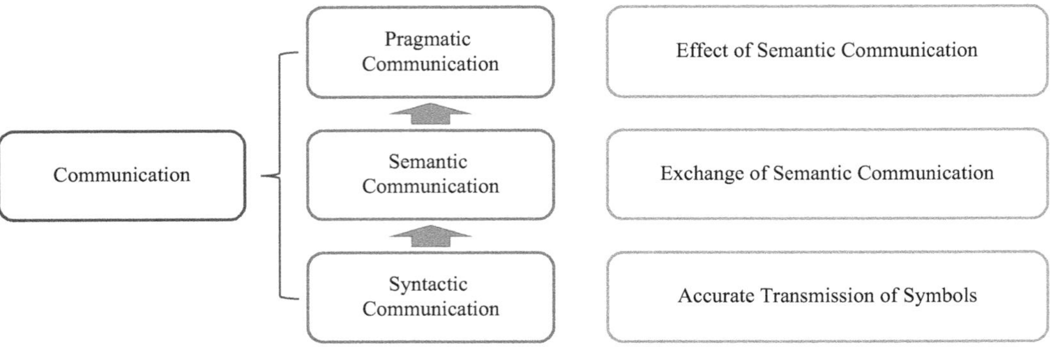

Figure 4.15 Three levels of communication.

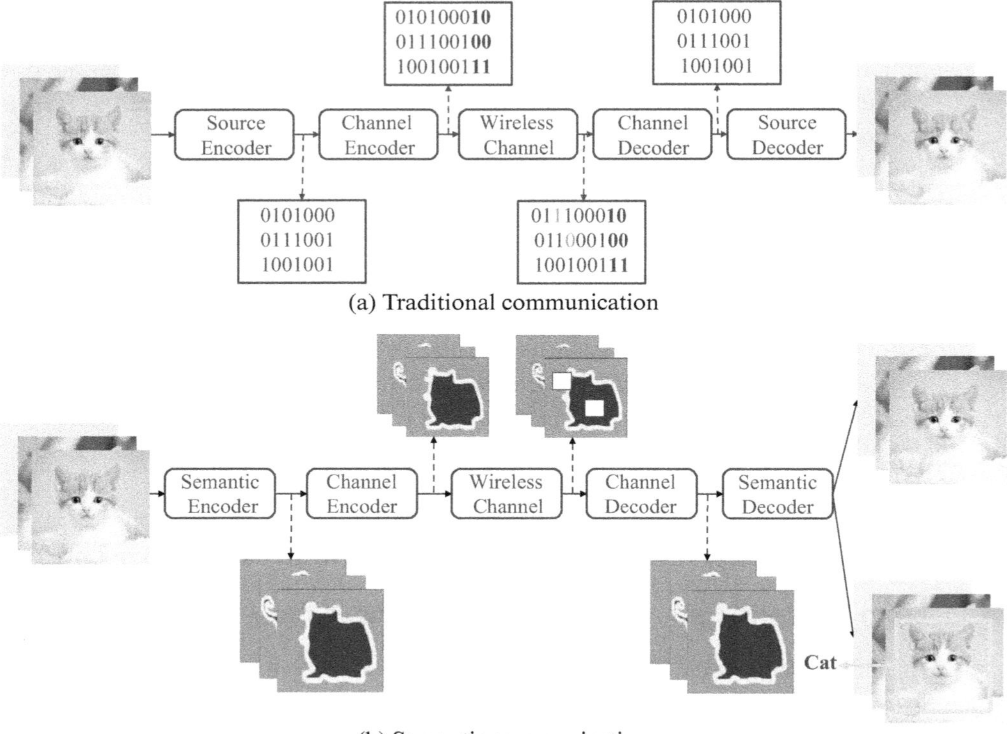

(a) Traditional communication

(b) Semantic communication

Figure 4.16 An illustration of the semantic communication system.

In contrast to conventional communication systems, semantic communication serves as a more efficient approach by transmitting only the information beneficial for a specific task at the receiver [165]. An example case is illustrated in Figure 4.16, where the technology of semantic communication is employed to transmit images for the task of object recognition. As distinct from traditional communication systems, which transmit entire images as sequences of binary bits, semantic communication systems extract the semantic information of images through semantic encoding.

Only this semantic information will be transmitted through the wireless channel, thus reducing the required communication resources significantly. At the receiver, the received data undergoes a semantic decoding process to reconstruct the semantic information of the original images, which is then utilized for object recognition.

Recall from Section 4.2 that the development of learned end-to-end communication systems, empowered by deep learning technologies, has enabled joint optimization of transmitters and receivers. These systems integrate all components in conventional communication systems. On this basis, semantic communication systems can be developed, where the encoding and decoding networks will be enhanced through a more sophisticated design to extract, process, and leverage the semantic information.

4.4.2 Semantic Communication Theory

Before delving into the algorithms and applications of learning-based semantic communication systems, let us begin by examining semantic communication theory. Although the field is still in its infancy, exploring these emerging concepts offers valuable insights, particularly for designing the architecture and loss function for deep learning-based semantic communication systems.

Shannon's information theory was a landmark achievement in providing a rigorous mathematical framework for communication, primarily based on probabilistic models. In this framework, the concept of information is defined as the content, which can be utilized to reduce uncertainty, and the analysis primarily revolves around mutual information within the domain of entropy.

Building upon Shannon's groundbreaking work, researchers have endeavored to extend this theory to include semantic and pragmatic levels of communication. In the following, "*entropy*" serves as an illustrative example to draw a comparison between conventional information theory and semantic information theory.

Entropy in Shannon information theory gives the measurement of information by the uncertainty of the source. Suppose that the set of source symbols can be denoted by $X = \{x_1, x_2, \cdots, x_n\}$ with corresponding probabilities $\{p(x_i)\}_{i=1}^{n}$, where x_i and $p(x_i)$ represent the ith candidate symbol and its corresponding probability, respectively. The source entropy $H(X)$ is defined as

$$H(X) = -\sum_{i=1}^{n} p(x_i) \log_2 p(x_i). \tag{4.20}$$

In semantic information theory, the first definition of semantic entropy was introduced in a specific language system [166] as

$$H(s, e) = -\log c(s, e), \tag{4.21}$$

where $c(s, e)$ represents the degree of confirmation of the sentence s on the evidence e, denoted by

$$c(s, e) = \frac{m(s, e)}{m(e)}, \tag{4.22}$$

where $m(e)$ and $m(s, e)$ represent the logical probability of e and the logical probability of s on e, respectively. Here, logical probability is a metric used to quantify the degree of inductive support or partial entailment between propositions [166]. Analogous to the source entropy in conventional information theory, in the framework of semantic information theory, sentences with more semantic information tend to have higher semantic entropy. However, there still exists a paradox about this definition of semantic entropy, i.e., contradictory sentences are also assigned high semantic entropy.

When it comes to measuring semantic entropy in the context of intelligent tasks, researchers have proposed different definitions tailored to specific tasks. For translation tasks, a definition of semantic entropy was introduced in [167], which takes into account the translational distributions of words. It can be expressed as

$$
\begin{aligned}
H(w) &= H(T|w) + N(w) \\
&= -\sum_{t \in T} p(t|w) \log p(t|w) + p(\text{NULL}|w) \log F(w),
\end{aligned}
\tag{4.23}
$$

where $H(w)$ represents the semantic entropy of the word w, T is the set of target words, $H(T|w)$ is the entropy of the set of target words given the word w, $N(w)$ denotes the contribution of null links of w, and $F(w)$ is the frequency of occurrence of the word w. Considering that one word can be translated into different words, $H(T|w)$ represents the translation diversity of the word w.

For classification tasks, the membership degree in axiomatic fuzzy set theory was introduced in [168] to define the semantic entropy. In a nutshell, the semantic entropy of the concept ς on the dataset $\mathcal{X}$ containing m different classes is expressed by

$$
H(\varsigma) = \sum_{j=1}^{m} H_{C_j}(\varsigma),
\tag{4.24}
$$

where C_j is the jth class, and $H_{C_j}(\varsigma)$ denotes the semantic entropy of class C_j on ς, which can be expressed as

$$
H_{C_j}(\varsigma) = -D_j(\varsigma) \log_2 D_j(\varsigma),
\tag{4.25}
$$

where the matching degree, $D_j(\varsigma)$, is represented by

$$
D_j(\varsigma) = \frac{\sum_{x \in \mathcal{X}_{C_j}} \mu_\varsigma(x)}{\sum_{x \in \mathcal{X}} \mu_\varsigma(x)},
\tag{4.26}
$$

where $\mathcal{X}_{C_j}$ represents the dataset of the jth class, and $\mu_\varsigma(x)$ is the membership degree of x on ς.

Although it is possible to define semantic entropy for specific intelligent tasks, there remains a lack of a universally applied definition. Researchers have proposed some definitions that can be applied in various intelligent tasks. However, these definitions often come with limitations or rely on strong premises and assumptions that are challenging to fulfill in practical scenarios. This lack of a unified definition of semantic entropy poses serious challenges in achieving a unified understanding and measurement of semantic information across different tasks and domains. Nevertheless, recent years having witnessed the rapid development of semantic information theory, and many exciting new advances are being made.

4.5 Deep Learning-Based Semantic Communication

While a uniform and well-established framework for semantic information theory remains elusive, the incorporation of deep learning into communication systems has opened up an avenue for designing and optimizing semantic communication systems via an end-to-end paradigm. In this section, we introduce applications of deep learning-based semantic communication systems for different data modalities.

4.5.1 Semantic Communications for Text

In recent years, significant advancements in natural language processing (NLP) fields, notably driven by the transformer model [33], have inspired researchers to adapt innovations in the NLP domain to the design of semantic communication systems for text. In a recent work [169], the transformer architecture has been extended for use in text-oriented semantic communication systems.

Specifically, the input sentence can be represented as $\mathbf{s} = [\omega_1, \omega_2, \ldots, \omega_L]$, in which ω_l stands for the lth word in the sentence. The transmitter, composed of the semantic encoder and channel encoder, encodes the input sentence $\mathbf{s}$ into the transmitted vector $\mathbf{x}$, represented as

$$\mathbf{x} = C_\alpha \left(S_\beta \left(\mathbf{s} \right) \right), \tag{4.27}$$

where $S_\beta \left(\cdot \right)$ and $C_\alpha \left(\cdot \right)$ represent the semantic encoder parameterized by α and the channel encoder parameterized by β, respectively. At the receiver, the received signal $\mathbf{y}$ over the Rayleigh fading channel can be expressed as

$$\mathbf{y} = h\mathbf{x} + \mathbf{n}, \tag{4.28}$$

where $h \sim \mathcal{CN}(0, 1)$ denotes the fading as a complex Gaussian random variable, and $\mathbf{n} \sim \mathcal{CN}(\mathbf{0}, \sigma^2 \mathbf{I})$ represents the noise. The received signal $\mathbf{y}$ is then fed into the channel decoder and semantic decoder to generate the recovered sentence $\hat{\mathbf{s}}$, represented as

$$\hat{\mathbf{s}} = S_\chi^{-1} \left(C_\delta^{-1} \left(\mathbf{y} \right) \right), \tag{4.29}$$

where C_δ^{-1} and S_χ^{-1} represent the channel decoder parameterized by δ and semantic decoder parameterized by χ, respectively. Note that the semantic encoder and decoder are based on the transformer architecture, while the channel encoder and decoder are designed using FCNs.

In [169], a specially designed loss function, consisting of two components, is employed to train the semantic communication system. First, the cross-entropy loss function, denoted by $\mathcal{L}_{\text{CE}}$, quantifies the disparity between the original and recovered symbols. The cross-entropy loss function is represented as

$$\mathcal{L}_{\mathrm{CE}}(\mathbf{s}, \hat{\mathbf{s}}) = -\sum_{i} q(w_i) \log(p(w_i)) + (1 - q(w_i)) \log(1 - p(w_i)), \tag{4.30}$$

where $\mathbf{s}$ and $\hat{\mathbf{s}}$ denote the transmitted and the recovered sentence, respectively, $q(w_i)$ is the real probability that the word w_i appears in sentence $\mathbf{s}$, and $p(w_i)$ is the predicted probability of its appearance in the estimated sentence $\hat{\mathbf{s}}$.

Second, a mutual information loss function, denoted by $\mathcal{L}_{\mathrm{MI}}$, is designed to gauge the mutual information between the input and output of the wireless channel. Denoting the input and output of the wireless channel by $\mathbf{x}$ and $\mathbf{y}$, respectively, the mutual information between $\mathbf{x}$ and $\mathbf{y}$ is closely related to the channel capacity. Hence $\mathcal{L}_{\mathrm{MI}}$ is included as part of the total loss function to give the channel encoder and decoder guidance to combat the channel impairment and noise, and to preserve the semantic information of the data. $\mathcal{L}_{\mathrm{MI}}$ is defined as

$$\mathcal{L}_{\mathrm{MI}}(\mathbf{x}, \mathbf{y}) = \mathbb{E}_{p(\mathbf{x}, \mathbf{y})}\left[f_T\right] - \log\left(\mathbb{E}_{p(\mathbf{x})p(\mathbf{y})}\left[e^{f_T}\right]\right), \tag{4.31}$$

where $p(\mathbf{x})$ and $p(\mathbf{y})$ are the probability of sending $\mathbf{x}$ and receiving $\mathbf{y}$, $p(\mathbf{x}, \mathbf{y})$ is the joint probability, and f_T represents a parameterized neural network for mutual information estimation. In fact, $\mathcal{L}_{\mathrm{MI}}$ provides a lower bound for the mutual information of $\mathbf{x}$ and $\mathbf{y}$, i.e.,

$$\mathcal{L}_{\mathrm{MI}}(\mathbf{x}, \mathbf{y}) \leq I(\mathbf{x}, \mathbf{y}), \tag{4.32}$$

where $I(\mathbf{x}, \mathbf{y})$ denotes the mutual information between $\mathbf{x}$ and $\mathbf{y}$, which can be expressed as

$$\begin{aligned} I(\mathbf{x}; \mathbf{y}) &= \int xyp(x, y) \log \frac{p(x, y)}{p(x)p(y)} dxdy \\ &= \mathbb{E}_{p(x, y)}\left[\log \frac{p(x, y)}{p(y)p(x)}\right]. \end{aligned} \tag{4.33}$$

Therefore through (4.32), the mutual information can be well estimated by optimizing the neural network f_T. Formally, the overall loss function can be calculated as

$$\mathcal{L}_{\mathrm{total}} = \mathcal{L}_{\mathrm{CE}}(\mathbf{s}, \hat{\mathbf{s}}) + \lambda \mathcal{L}_{\mathrm{MI}}(\mathbf{x}, \mathbf{y}), \tag{4.34}$$

where λ is used for balancing the weight of the two parts in the total loss function.

In contrast to traditional communication systems, which primarily prioritize the transmission reliability of bit-level information, semantic communication systems prioritize the accurate transmission of semantic information. Therefore, these systems require distinct evaluation metrics tailored to their specific modalities. For text-oriented semantic communication systems, there are two frequently used performance metrics: the bilingual evaluation understudy (BLEU) score and sentence similarity.

For the original sentence $\mathbf{s}$ and the reconstructed sentence $\hat{\mathbf{s}}$, the BLEU score is a commonly used metric to quantify the bit difference between two sentences [170]. However, it does not depict the similarity of two sentences at the semantic level. Compared with the BLEU score, sentence similarity is a more reasonable metric as it translates sentences into vectors through the transformer model and compares the semantic similarity of two sentences using these vectors, i.e.,

$$\text{match}(\hat{\mathbf{s}}, \mathbf{s}) = \frac{\mathbf{B}_{\Phi}(\mathbf{s}) \cdot \mathbf{B}_{\Phi}(\hat{\mathbf{s}})^T}{\|\mathbf{B}_{\Phi}(\mathbf{s})\| \, \|\mathbf{B}_{\Phi}(\hat{\mathbf{s}})\|}, \tag{4.35}$$

where $\text{match}(\hat{\mathbf{s}}, \mathbf{s})$ represents the sentence similarity between $\hat{\mathbf{s}}$ and $\mathbf{s}$, and $\mathbf{B}_{\Phi}$ stands for bidirectional encoder representations from transformers (BERT), a state-of-the-art transformer model commonly used in NLP [171]. Given its extensive pre-training on billions of sentences, BERT excels at grasping sentence meanings and effectively maps semantically similar sentences to closely positioned vectors in vector space.

In experiments, for 64-QAM modulation, a decrease in SNR leads to a corresponding decrease in sentence similarity. This decline occurs because conventional approaches face challenges in preserving the semantic information of symbols in low-SNR regimes. Specifically, when the SNR is relatively low, a significant number of words in the sentence are susceptible to transmission errors, leading to a severe degradation in sentence similarity. In contrast, the semantic communication system successfully preserves the semantics of the sentences even in a low-SNR regime, thus remaining high sentence similarity at both low and high SNRs.

4.5.2 Semantic Communications for Speech

In [172], a semantic communication system is developed for speech transmission, named DeepSC-S . DeepSC-S can be divided into a transmitter part and a receiver part. The transmitter consists of an attention-based semantic encoder and a channel encoder based on two-dimensional CNNs. The receiver contains a channel decoder and an attention-based semantic decoder. The entire network is trained in an end-to-end manner to recover the speech sequence at the receiver.

At the transmitter, the input speech sequence of DeepSC-S, denoted by $\mathbf{s}$, is initially reshaped into $\mathbf{m}$ without any feature learning or extraction and is then processed by an attention-based encoder, referred to as the *semantic encoder*. The *semantic encoder* directly extracts the semantic information of speech from $\mathbf{m}$ and outputs the corresponding features $\mathbf{b}$. To extract semantic information, DeepSC-S utilizes an attention mechanism based on a squeeze-and-excitation (SE) network, specifically the SE-ResNet module shown in Figure 4.17a. The SE-ResNet module begins with a split layer consisting of multiple blocks that are subsequently combined, which takes original speech sequences $\mathbf{m}$ as input. The resulting spliced feature undergoes dimension reduction through a transition layer, yielding $\mathbf{p}$. Following this, the SE layer is employed to compress $\mathbf{p}$ further, thereby generating attention factors for each feature. This results in an output denoted by $\mathbf{z}$, where $\mathbf{z}$ encapsulates the semantic significance of individual features. The variation in semantic importance across different features is accentuated by element-wise multiplication between $\mathbf{p}$ and $\mathbf{z}$ along the feature dimension. By doing so, the weight of each feature in $\mathbf{m}$ is reassigned, i.e., the features containing more beneficial speech information are paid more attention

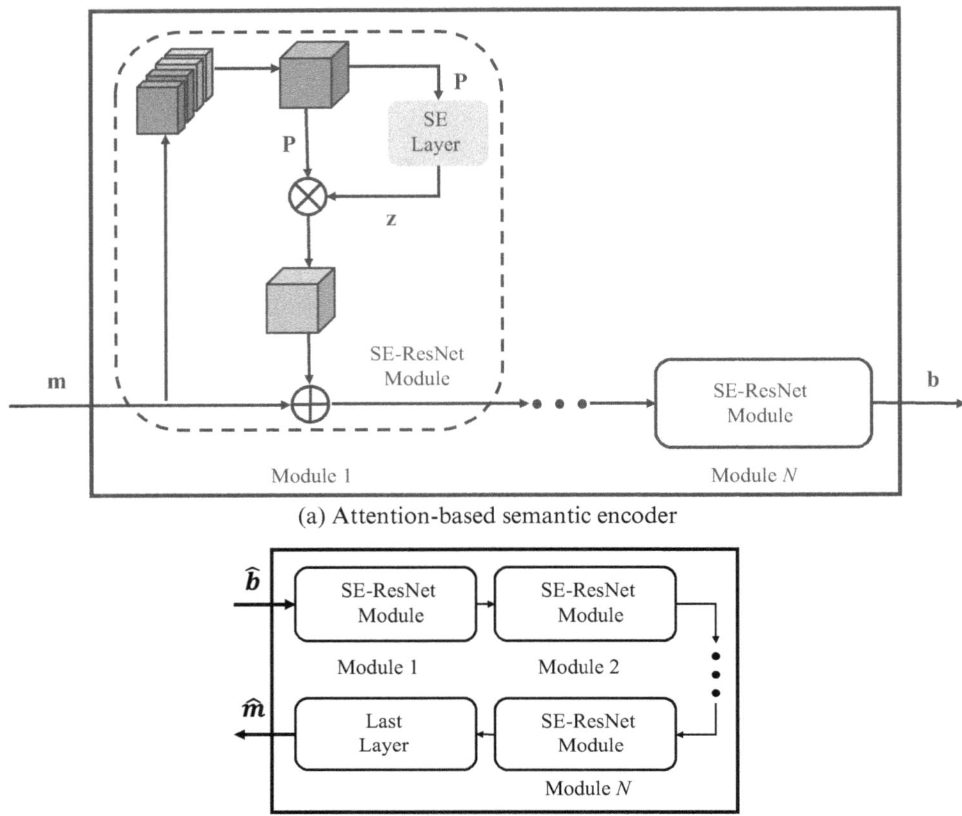

(a) Attention-based semantic encoder

(b) Attention-based semantic decoder

Figure 4.17 Semantic encoder and decoder in DeepSC-S.

to. Finally, the residual network is adopted, and the output is obtained by adding **m** into the output of the SE-ResNet module.

Afterward, the *channel encoder*, implemented as a sequence of two-dimensional CNN blocks, converts **b** into **U**. To transmit **U** through a physical channel, it is reshaped into transmitted symbols, **x**, via a reshape layer. The wireless channel takes the transmitted symbols, **x**, as input and produces **y** at the receiver. The received signal, **y**, is reshaped into **Û** before being fed into the *channel decoder*, which is also a sequence of two-dimensional CNN blocks. The output of the *channel decoder* is **b̂**. As illustrated in Figure 4.17b, the attention-based decoder, known as the *semantic decoder*, consists of several SE-ResNet modules and a last layer including a two-dimensional CNN module with a filter. It converts **b̂** into **m̂**, which is further reshaped into the recovered speech sequence **ŝ**.

As for the training of a DeepSC-S system, the squared loss is utilized to measure the distance between **s** and **ŝ** with the same length W. Hence the empirical risk is given by

$$\mathcal{L} = \frac{1}{N} \sum_{i=1}^{N} \frac{1}{W} \|\mathbf{s}_i - \hat{\mathbf{s}}_i\|^2, \tag{4.36}$$

where N is the number of samples in a batch. $\mathbf{s}_i$ and $\hat{\mathbf{s}}_i$ are the ith original input sentence and its corresponding reconstruction.

The performance of the DeepSC-S algorithm is compared with that of the conventional communication systems over a slow Rayleigh fading channel in [172]. The baseline is a traditional communication scheme with separate source and channel coding techniques, while the semi-traditional benchmark combines feature learning with traditional channel encoding. It is clear that the semi-traditional system consistently achieves a higher signal-to-distortion ratio (SDR) score than the traditional system. Additionally, DeepSC-S obtains a higher SDR score than the other two systems under Rayleigh fading channels, proving its effectiveness.

4.5.3 Semantic Communications for Images

In [173], a semantic communication system for joint source and channel coding (JSCC) of images is introduced. As shown in Figure 4.18, the system employs a combination of multi-layer CNNs and nonlinear activation layers to extract the semantics from an image $\mathbf{s}$ of size n and transform it into complex-valued channel input samples of size k, forming the latent representation. At the receiver, the system recovers the image $\hat{\mathbf{s}}$ by decoding the received signals from the channel. This entire system, including the encoder and decoder, is jointly optimized through end-to-end training. The optimization aims to find the optimal encoding function $\boldsymbol{\theta}$ and decoding function $\boldsymbol{\phi}$ by minimizing a distortion measurement $d(\mathbf{s},\hat{\mathbf{s}})$, represented by

$$\left(\boldsymbol{\theta}^*, \boldsymbol{\phi}^*\right) = \arg\min_{\boldsymbol{\theta},\boldsymbol{\phi}} \mathbb{E}_{p(\mathbf{s},\hat{\mathbf{s}})}[d(\mathbf{s},\hat{\mathbf{s}})], \tag{4.37}$$

where $p(\mathbf{s},\hat{\mathbf{s}})$ represents the joint probability of $\mathbf{s}$ and $\hat{\mathbf{s}}$.

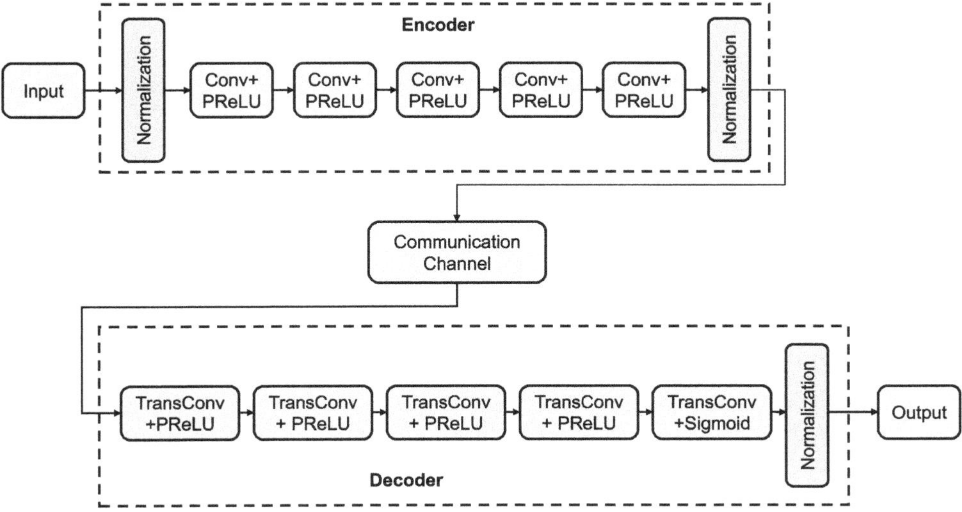

Figure 4.18 Illustration of a learned end-to-end semantic communication system for image transmission.

The squared loss is used in [173] to measure the pixel-by-pixel gap between $\mathbf{s}$ and $\hat{\mathbf{s}}$. Hence the empirical risk is given by

$$\mathcal{L}(\boldsymbol{\theta}, \boldsymbol{\phi}) = \frac{1}{N} \sum_{i=1}^{N} \frac{1}{n} \|\mathbf{s}_i - \hat{\mathbf{s}}_i\|^2, \tag{4.38}$$

where N is the number of samples in a batch, and $\mathbf{s}_i$ and $\hat{\mathbf{s}}_i$ are the ith original input image and its corresponding reconstruction. However, this loss only describes the difference between $\mathbf{s}$ and $\hat{\mathbf{s}}$ at the pixel level or bit level and does not quantitatively describe the difference between the two from the semantic view. Therefore, the average loss in [174] measures the difference between two images in VGG[3] feature space, i.e.,

$$\mathcal{L}(\boldsymbol{\theta}, \boldsymbol{\phi}) = \frac{1}{N} \sum_{i=1}^{N} \frac{1}{n} \|\Psi(\mathbf{s}_i) - \Psi(\hat{\mathbf{s}}_i)\|^2, \tag{4.39}$$

where $\Psi(\cdot)$ represents the pre-trained VGG network.

Furthermore, peak SNR (PSNR) is chosen as the performance metric, defined as

$$\text{PSNR} = 10 \log_{10} \left(\frac{\text{MAX}_I^2}{\text{MSE}} \right), \tag{4.40}$$

where MAX_I is defined as the maximum pixel value of an image, e.g., 255 for the range of [0, 255], and the mean squared error (MSE) between $\mathbf{s} \in \mathbb{R}^{m \times n \times 3}$ and its reconstruction $\hat{\mathbf{s}}$ is calculated as

$$\text{MSE} = \frac{1}{m \times n \times 3} \sum_{i=0}^{m-1} \sum_{j=0}^{n-1} \sum_{k=0}^{2} \left[\mathbf{s}(i,j,k) - \hat{\mathbf{s}}(i,j,k) \right]^2. \tag{4.41}$$

Figure 4.19 illustrates the PSNR performance of the JSCC algorithm over a slow Rayleigh fading channel at various SNRs with respect to the bandwidth compression ratio, defined as k/n. The channel transfer function is $y = hx + n$, where $h \sim \mathcal{CN}(0, 1)$, $n \sim \mathcal{CN}(0, \sigma^2)$, and y denotes the received symbol from the channel. For the baseline scheme, the original image is compressed using JPEG and then transmitted over the channel at the channel capacity C. As shown in the figure, the proposed deep JSCC scheme outperforms the JPEG baseline scheme in all simulation settings. In the low-SNR regime, the JPEG baseline exhibits significant performance degradation due to the "cliff effect," i.e., a system with a separate design may suffer from severe channel fading. On the contrary, the proposed scheme performs well even in low-SNR regimes, showing its ability to mitigate the "cliff effect." For moderate or high SNRs, JPEG exhibits severe performance degradation when the maximum achievable transmission rate, $R_c = Ck/n$, falls below the minimum rate for image transmission, $R_{\min}$. In cases with limited channel bandwidth (i.e., when k/n falls

---

[3] VGG (visual geometry group) is a convolutional neural network architecture known for its depth and effectiveness in image classification tasks. It achieved state-of-the-art performance on various computer vision challenges and played a pivotal role in the development of deep learning for image analysis.

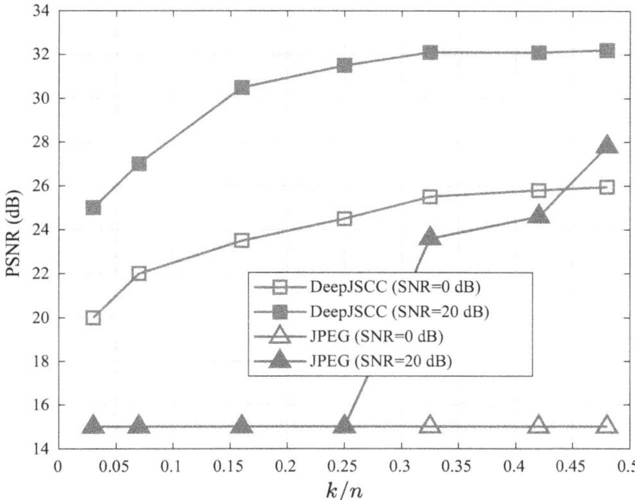

Figure 4.19 Performance of the deep JSCC scheme compared with JPEG [173].

below 0.2), the proposed deep JSCC method significantly outperforms the JPEG method, even assuming reliable transmission at channel capacity. Even as the channel bandwidth becomes less constrained, especially for $k/n > 0.3$, the deep JSCC scheme maintains its superior performance, highlighting its competitiveness compared with the JPEG method.

Example 4.4 As illustrated in Figure 4.19, JPEG outperforms the deep JSCC algorithm in high-SNR scenarios when $k/n = 1/2$. Explain the reason for this phenomenon.

Solution

This phenomenon can be explained by the fact that in high-SNR scenarios, traditional separated source channel coding, which is optimized for such conditions, can fully exploit the available channel capacity. Traditional methods like JPEG are designed to provide high-quality reconstructions by leveraging the high SNR to retain more detailed information. In contrast, deep JSCC, while robust in low-SNR conditions due to its end-to-end learned nature, may not be as efficient as traditional methods in exploiting the channel capacity at high SNR, leading to its relatively lower performance in this regime.

4.5.4 Semantic Communications for Cooperative Perception

Light detection and ranging (LiDAR) is a critical technology for autonomous driving, providing high-resolution point clouds for detection and ensuring accurate perception in various lighting and weather conditions. Hence, similar to other modalities such as text, speech, images, and videos, the LiDAR point clouds also contain a large amount of semantic information, which can be utilized for environmental perception.

Since the perception field of the autonomous vehicle itself is always limited by obstacles such as buildings and trees, a novel detection scheme, known as cooperative perception, is investigated for a broader perception field. For cooperative perception, different connected automated vehicles (CAVs) can exchange and fuse their sensory information through vehicle-to-vehicle communication technology, enabling the provision of multiple viewpoints for the same object to complement one another. A semantic communication framework for cooperative perception has been introduced in a recent work [175]; this employs an importance map to extract significant semantic information at the transmitter and fuses these intermediate features through an attention-based mechanism at the receiver.

The framework of the cooperative perception process with intermediate fusion is illustrated in Figure 4.20. The cooperative perception system with intermediate fusion comprises four main modules: feature extraction, feature sharing, feature fusion, and detection result generation.

Feature extraction. The feature tensor F_i, containing the semantic information of the ith sample, is extracted from the raw data X_i (LiDAR point clouds) of the CAV through a backbone network, such as PointPillars [176]. In the PointPillars model, the raw data would be scattered to form a two-dimensional pseudo-image and subsequently passed to the neural network for further processing. Similarly, the

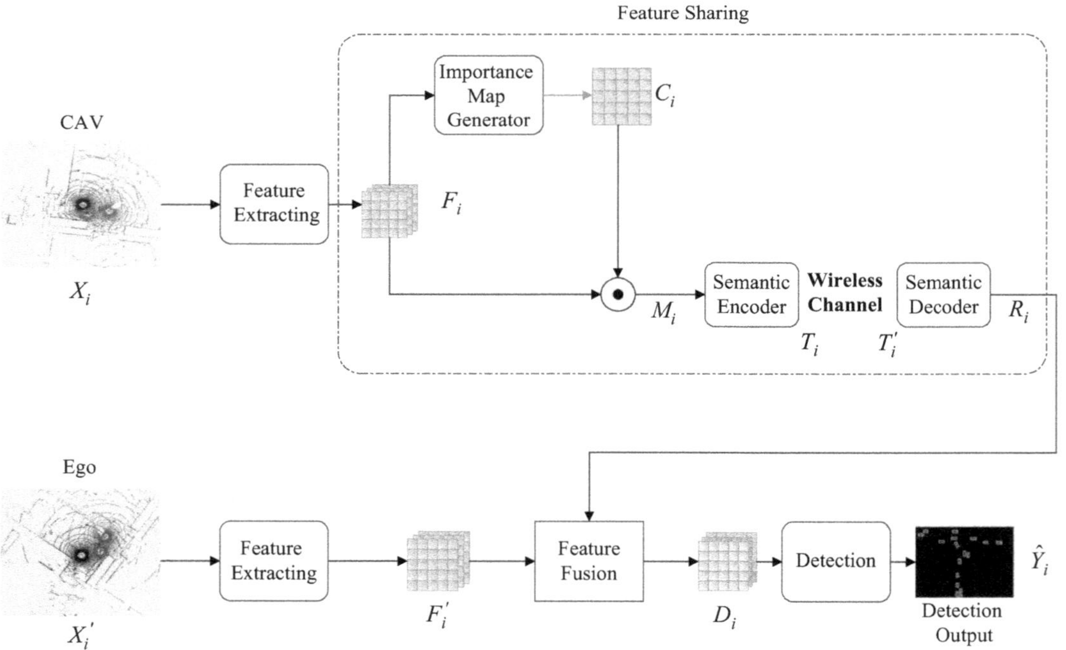

Figure 4.20 Structure of the cooperative perception model based on the importance map.

feature tensor of the ego vehicle[4] F_i' is also extracted from the raw data X_i' of the ego vehicle through the same backbone.

Feature sharing. In this module, the ego vehicle receives the feature tensor F_i from the CAV after the feature extraction. The received intermediate feature is then passed through the remaining networks of the ego vehicle. For more efficient transmission, the system endeavors to identify the significant parts of the feature tensor by leveraging the importance map. Specifically, the feature tensor F_i is fed into an importance map generator based on a neural network $P(\cdot)$ to obtain an importance map C_i, after which F_i will be element-wise multiplied by the generated importance map C_i to generate transmitted data M_i, represented by

$$C_i = P(F_i) \tag{4.42}$$

and

$$M_i = F_i \odot C_i. \tag{4.43}$$

The non-zero element ratio of the importance map is defined as the compression rate (CR), which indicates the data size of the transmitted information. It is found that the CR can achieve a balance between transmission overhead and perception performance when set on the order of 10^{-2}. Consequently, the feature tensor becomes sparse after the element-wise multiplication, and only the non-zero elements of the tensor need to be transmitted. Semantic information M_i is mapped into a complex symbol stream T_i in the semantic encoder to overcome channel distortion and noise. On the receiver side, the received symbol T_i' is first passed through a semantic decoder $\Psi_d(\cdot)$, which demaps the complex symbols into the recovered semantic information R_i for further fusion.

Feature fusion. Through the feature fusion module, such as attention-based feature fusion, the recovered semantic information R_i is fused with the feature of the ego vehicle F_i' into a fused feature tensor that contains the semantic information from both the CAV and the ego vehicle. The output of the feature fusion module D_i can be represented as

$$D_i = \chi(R_i, F_i'), \tag{4.44}$$

where the spatial relationship between R_i and F_i' can be captured through the self-attention fusion network $\chi(\cdot)$.

Detection result generation. Upon receiving the final fused feature tensor, the prediction header would be employed for the tasks of classification and box regression. The regression output consists of the position, size, and yaw angle of the predefined anchor boxes; the classification output is the confidence score assigned to each anchor box, indicating the probability of an object or background.

To minimize the average communication distortion, the semantic encoder and decoder can optimize the MSE between the feature tensor and its reconstruction produced by the decoder. Meanwhile, to measure the performance of the cooperative

---

[4] The primary sensing vehicle in the CAV group.

perception, the smooth ℓ_1 loss is adopted for regression [177], and the focal loss is employed for classification [178]. The total loss can be represented by

$$\mathcal{L}_{total} = \mu \mathcal{L}_{rec} + \mathcal{L}_{per}, \tag{4.45}$$

where μ represents the weighting ratio between the reconstruction loss and the perception loss in the overall loss function. $\mathcal{L}_{rec}$ and $\mathcal{L}_{per}$ represent the reconstruction loss and perception loss, respectively. $\mathcal{L}_{rec}$ is defined as

$$\mathcal{L}_{rec} = \frac{1}{N} \sum_{i=1}^{N} ||M_i - R_i||_2^2, \tag{4.46}$$

where N is the number of samples. L_{per} is defined as

$$\mathcal{L}_{per} = \frac{1}{N} \sum_{i=1}^{N} \ell(Y_i, \hat{Y}_i), \tag{4.47}$$

where $\ell(\cdot)$ denotes the perception loss function [176]. Y_i and $\hat{Y}_i$ represent the label and the model detection output, respectively.

The performance of the cooperative perception is evaluated in terms of average precision (AP) at an intersection-over-union (IoU) threshold of 0.5. AP is a commonly used evaluation metric to assess the performance of object detection algorithms. It quantifies the perception accuracy and trade-off between precision and recall by calculating precision at different recall levels and taking their average. The AP is defined as

$$\text{AP@0.5} = \int_0^1 \max \left\{ p(r'|r' > r) \right\} dr, \tag{4.48}$$

where $p(r)$ is the precision-recall curve at an IoU threshold of 0.5.

As Figure 4.21 illustrates, the semantic communication method is compared with traditional baseline schemes under the assumption of a fast Rayleigh fading channel. The upper bound can be obtained assuming perfect communication with CR $= 1$. For traditional baseline schemes, $1/2$-rate LDPC coding with code length of 1000 is considered for channel coding, and 16-QAM or 256-QAM is applied for modulation after channel coding. In terms of AP@0.5, the semantic communication method outperforms the separate coding schemes and approaches the performance upper bound constrained by the cooperative perception module at SNR $= 6$ dB. This can be attributed to the use of JSCC in semantic communication, which is designed and trained to jointly optimize the entire system, thereby preserving more semantic information given the same communication resources. Meanwhile, semantic communication exhibits smoother performance in low-SNR regimes without the cliff effect that is commonly observed in traditional communication systems. Hence, the proposed method can prevent catastrophic perception performance degradation in low-SNR regimes, which is critical to autonomous driving.

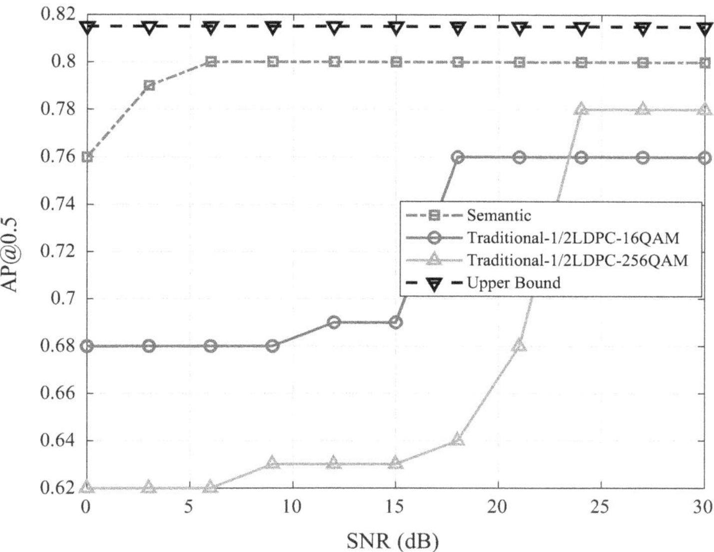

Figure 4.21 Performance of semantic communication compared with baseline schemes at different SNRs over the Rayleigh fading channel. The baseline schemes and semantic communication method exploit the same channel uses for fairness [175].

4.6 Exercises

Exercise 4.1 What are the advantages of learned end-to-end communication systems compared with conventional communication systems?

Exercise 4.2 What are the potential applications of learned end-to-end communication systems in practical scenarios?

Exercise 4.3 What challenges might learned end-to-end communication systems encounter when applied to practical scenarios?

Exercise 4.4 Section 1.2.3 introduces the basic concepts of RL, whereas Section 4.2.2 outlines an RL-based framework for learned end-to-end communication systems without channel models. Compare the two sections and consider the following questions:

(a) In the RL-based framework described in Section 4.2.2, which specific components correspond to the agent, state, action, and reward in RL?
(b) In the RL-based framework described in Section 4.2.2, is the next state determined based on the previous state? Additionally, can the decision process described in Section 4.2.2 be considered a Markov decision process?

Exercise 4.5 What are the three layers of communication systems defined by Shannon and Weaver? Explain these three layers with practical examples from real-world communication scenarios.

Exercise 4.6   Section 4.5 introduces applications of deep learning-based semantic communication systems across various domains. What are the differences in the architecture of neural networks between these different domains? Why do different domains require distinct neural network models?

Exercise 4.7   Section 4.2.2 introduces an RL-based framework that can derive the gradient without $p(y|x)$. Derive the gradient of the loss function in (4.15), i.e.,

$$\widehat{\mathcal{L}}(\boldsymbol{\theta}_T, \boldsymbol{\theta}_R) \triangleq \mathbb{E}_s \left\{ \int \hat{\pi}_{f_{\boldsymbol{\theta}_T}^{(T)}(s), \sigma}(\mathbf{x}) \cdot \int l\left(f_{\boldsymbol{\theta}_R}^{(R)}(\mathbf{y}), s\right) p(\mathbf{y}|\mathbf{x})\, d\mathbf{y} d\mathbf{x} \right\}.$$

Hint: Use the chain rule and the logarithmic trick, given by

$$\nabla_{\mathbf{x}} \log(g(\mathbf{x})) = \frac{\nabla_{\mathbf{x}} g(\mathbf{x})}{g(\mathbf{x})}.$$

Exercise 4.8   Let $\mathcal{X}$ and $\mathcal{Y}$ denote the domains of the functions $\hat{\pi}_{\bar{\mathbf{x}}, \sigma}$ and $p(\mathbf{y}|\mathbf{x})$, respectively. Prove that the gradient of the actual loss function with respect to the transmitter parameters in (4.14), denoted by $\nabla_{\boldsymbol{\theta}_T} \mathcal{L}$, can be accurately approximated by the gradient of a surrogate loss function in (4.16), $\nabla_{\boldsymbol{\theta}_T} \widehat{\mathcal{L}}$, with an error that becomes negligible. Specifically, we aim to prove the following limit,

$$\lim_{\sigma \to 0} \nabla_{\boldsymbol{\theta}_T} \widehat{\mathcal{L}}(\boldsymbol{\theta}_T, \boldsymbol{\theta}_R) = \nabla_{\boldsymbol{\theta}_T} \mathcal{L}(\boldsymbol{\theta}_T, \boldsymbol{\theta}_R),$$

under each of the following two distinct assumptions.

(a) Prove that the limit above holds under the following five assumptions:

- $\mathcal{X}$ and $\mathcal{Y}$ are compact.[5]
- if $\mathbf{x} \sim \hat{\pi}_{\bar{\mathbf{x}}, \sigma}$, $\mathbf{x} + \varepsilon \sim \hat{\pi}_{\bar{\mathbf{x}}+\varepsilon, \sigma}$.
- $\hat{\pi}_{\bar{\mathbf{x}}, \sigma}$ converges to the Dirac distribution as σ decreases to 0.
- $\nabla_{\mathbf{x}} p(\mathbf{y}|\mathbf{x})$ is continuous jointly in $\mathbf{x}$ and $\mathbf{y}$.
- $\hat{\pi}_{\bar{\mathbf{x}}, \sigma}$ uniformly converges to the Dirac distribution as σ decreases to 0, i.e.,

  for all $\eta > 0$, there exists $\sigma_0 > 0$, for all $\sigma \le \sigma_0$, for all $\bar{\mathbf{x}}$, for all $\mathbf{y}$:

  $$\left| \int_{\S} \hat{\pi}_{\bar{\mathbf{x}}, \sigma}(\mathbf{x}) g(\mathbf{x}, \mathbf{y}) d\mathbf{x} - g(\bar{\mathbf{x}}, \mathbf{y}) \right| \le \eta.$$

(b) Prove that the limit above holds under the following four assumptions:

- $\mathcal{X}$ and $\mathcal{Y}$ are not compact.
- if $\mathbf{x} \sim \hat{\pi}_{\bar{\mathbf{x}}, \sigma}$, $\mathbf{x} + \varepsilon \sim \hat{\pi}_{\bar{\mathbf{x}}+\varepsilon, \sigma}$.
- $\hat{\pi}_{\bar{\mathbf{x}}, \sigma}$ converges to the Dirac distribution as σ decreases towards 0, i.e., given a continuous and bounded function g: $\lim_{\sigma \to 0} \mathbb{E}_{\mathbf{x} \sim \hat{\pi}_{\bar{\mathbf{x}}, \sigma}} \{g(\mathbf{x}, \mathbf{y})\} = g(\bar{\mathbf{x}}, \mathbf{y})$.
- A function $h : (\mathbf{x}, \mathbf{y}) \mapsto \mathbb{R}^{2N}$ exists that is integrable with respect to $\mathbf{y}$, and it satisfies the following condition:

  for all $\bar{\mathbf{x}}$, for all $\mathbf{y}$, $\left| \mathbb{E}_{\mathbf{x} \sim \hat{\pi}_{\bar{\mathbf{x}}, \sigma}} \{\nabla_{\mathbf{z}} p(\mathbf{y}|\mathbf{z})|_{\mathbf{z}=\mathbf{x}}\} \right| \le h(\bar{\mathbf{x}}, \mathbf{y}),$

---

[5] A compact set, within the context of a metric or topological space, is defined as a set that is both closed, meaning it includes all of its limit points, and bounded, indicating it can be confined within a finite region.

where the inequality is true for each element, and the variable z is introduced to improve readability.

Exercise 4.9   Consider a learned end-to-end communication system over an AWGN channel based on Figure 4.1 and Table 4.1. Assume the code rate is 4/7. Please use Python and the Pytorch deep learning framework to complete the following programming tasks. The code snippet has been provided in https://github.com/le-liang/wcmlbook/blob/main/ch4/Exercise_4.9/.

(a) Calculate the M and n in Table 4.1.
(b) Generate the entire neural network based on Figure 4.1 and Table 4.1.
(c) Calculate the loss based on (4.2).
(d) Train the entire network at $E_b/N_0 = 7$ dB using the Adam optimizer with a learning rate of 0.001. After training, evaluate the network's performance over a range of E_b/N_0 values from -4 to 8 dB, with increments of 0.5 dB. Compare the simulation results with Figure 4.2, and analyze how the training E_b/N_0 affects the performance.
(e) Train and test the whole network without the power normalization. Please compare the simulation results with Figure 4.2 and analyze the impact of the power normalization on the performance.

Exercise 4.10   Consider a learned end-to-end communication system over a multipath fading channel. Use Python and the Pytorch deep learning framework to complete the following programming tasks. The code snippet has been provided in https://github.com/le-liang/wcmlbook/blob/main/ch4/Exercise_4.10/.

(a) Consider a time-varying multipath fading channel, which is modeled based on the tapped delay line (TDL) model in 3GPP [78]. Generate the channel model without trainable parameters.
(b) Extend the channel model above to the OFDM scheme according to Table 4.6.

Table 4.6 Simulation parameters for OFDM scheme.

Parameters	Values
Number of subcarriers	2048
Number of OFDM symbols	14
Subcarrier spacing	15 kHz
Carrier frequency	3.5 GHz

Exercise 4.11   Consider a learned end-to-end communication system over a Rayleigh fading channel based on Figure 4.3 and Table 4.2. Assume the code rate is 1. Use Python and the Pytorch deep learning framework to complete the following programming tasks. The code snippet has been provided in https://github.com/le-liang/wcmlbook/blob/main/ch4/Exercise_4.11/.

(a) Train the entire network with SNR $= 7$ dB and different channel uses of 32, 64, and 128, respectively, using the Adam optimizer. After training, evaluate their performance over the range of $[0, 20]$ dB with steps of 3 dB. Compare the simulation results under different channel uses, and analyze the impact of the channel uses on the performance.

(b) Train the entire network with channel uses $n = 128$ and different SNR of 1 dB, 7 dB, and 20 dB, using the Adam optimizer. After training, evaluate their performance over the range of $[0, 20]$ dB with steps of 3 dB. Compare the simulation results under different training SNRs, and analyze the impact of the training SNR on the performance.

Exercise 4.12   Section 4.2.2 establishes that the distributions of channel outputs can be learned through a conditional GAN [159]. Consider a learned end-to-end communication system over a Rayleigh fading channel with a conditional GAN. Use Python and the Pytorch deep learning framework to construct a neural network based on Table 4.5, and train the whole network based on Algorithm 4.1. Plot the curve of the BER performance over the SNR range of $[0, 8]$ dB. The code snippet has been provided in https://github.com/le-liang/wcmlbook/blob/main/ch4/Exercise_4.12/.

Exercise 4.13   Section 4.5.1 describes how the transformer can transform a sentence into a vector. Consider the following two sentences:

(a) She enjoys reading books.
(b) She loves to read.

Use Python and the *transformers* package to compute their semantic similarity based on (4.35). The code snippet has been provided in https://github.com/le-liang/wcmlbook/blob/main/ch4/Exercise_4.13/.

Exercise 4.14   Consider a semantic communication system for images over an AWGN channel as illustrated in Exercise 4.9. Use CIFAR10 or CIFAR100 to train and test the semantic communication network. The code snippet has been provided in https://github.com/le-liang/wcmlbook/blob/main/ch4/Exercise_4.14_4.15/.

(a) Derive the new loss function, and design a novel network structure based on the CNN.
(b) Calculate the transmission ratio k/n.
(c) Train the entire network with different SNRs of 1 dB, 7 dB, and 20 dB, using the Adam optimizer. After training, evaluate their performance over the range of $[0, 20]$ dB with steps of 2 dB. Calculate the PSNR metric under different SNRs according to (4.40).

Exercise 4.15   In Section 4.5.3, we present a semantic communication system for images that focuses on the task of semantic recovery of images based on the loss function in (4.38) and (4.39). However, most of the time we are interested in the task performance. For example, for the image classification task, the classification performance is more important than the reconstruction performance of the images. Therefore, the loss function needs to be modified according to the specific task.

(a) Design the loss function for a classification task and train the network in Exercise 4.14 using the CIFAR10 or CIFAR100 dataset.
(b) Compare its performance with the network in Exercise 4.14 in terms of PSNR in (4.40) and classification accuracy. The classification accuracy δ is defined as $\delta = N_{correct}/N$, where $N_{correct}$ is the number of correctly predicted samples, and N is the total number of all samples.

5   Learning Resource Allocation in Wireless Networks

In this chapter, we explore the application of deep learning techniques for resource allocation in wireless networks. Resource allocation refers to the dynamic assignment of communication resources, including frequency spectrum, transmitted power, time slots, and others, among many users in a wireless network to optimize network performance and improve resource utilization efficiency. Traditionally, resource allocation is addressed by explicitly formulating an optimization problem and then exploiting mathematical programming techniques to solve the problem to a certain level of optimality. Despite its widespread adoption, it turns out that many of the optimization problems formulated for resource allocation are difficult to solve. Moreover, with myriad new applications to support, conventional methods face increasing difficulty in accurately balancing and modeling the diverse service requirements in a mathematically precise way.

To overcome these limitations, we first investigate supervised and unsupervised learning approaches that help solve resource optimization problems in a computationally efficient way. Supervised learning methods rely on traditional optimization algorithms to generate training labels and shift the heavy computation load to the offline training stage, thus enabling a low-complexity online deployment with performance close to traditional methods. Unsupervised learning methods approach the optimization objective directly by performing gradient descent over a loss function that is closely related to the objective, thus holding the potential to outperform any existing optimization algorithms. To enhance the scalability and generalization capability of these resource optimization methods, we further present graph neural network (GNNs), which incorporate the structures of the target task into the network architecture. Finally, we introduce reinforcement learning (RL)-based resource allocation methods, which provide native support for addressing sequential decision-making under uncertainty and optimize the network performance in response to environment changes.

5.1   Resource Allocation Problems in Wireless Networks

Wireless resource allocation encompasses a broad range of topics and challenges. In this chapter, we concentrate on several quintessential examples, including link scheduling, power control, and spectrum allocation, which are key to the operation of wireless networks but are notoriously difficult to solve due to their non-convex (and sometimes combinatorial) nature. Such problems are described and formulated to help readers understand the basic form of resource allocation in wireless networks, and they pave the way for the algorithm development in the rest of this chapter.

Link Scheduling in Wireless Networks

link scheduling is crucial to maximizing network throughput while minimizing interference. Consider a wireless network with N communication links sharing a single spectrum sub-band, where link k is characterized by transmit power p_k (fixed and known) and channel state h_{kj} from the transmitter of link j to the receiver of link k. For the sake of simplicity, it is assumed that the sub-band is frequency flat. The noise power at each receiver is σ^2. An indicator variable x_k is introduced for each link k, which is equal to 1 if the link is activated and 0 otherwise. The scheduling problem can be formulated as a combinatorial optimization problem that seeks to maximize the weighted sum rate

$$\max_{\mathbf{x}} \quad \sum_{k=1}^{N} \omega_k \log \left(1 + \frac{|h_{kk}|^2 p_k x_k}{\sum_{j \neq k} |h_{kj}|^2 p_j x_j + \sigma^2} \right), \tag{5.1}$$

$$\text{s.t.} \quad x_k \in \{0, 1\}, \quad \text{for all } k,$$

where ω_k denotes the weight for link k, and $\mathbf{x} = [x_1, \ldots, x_N]^T$ denotes the vector of the indicator variables for link scheduling. This formulation represents scheduling in a single time slot and frequency sub-band, and can be seen as a special case of the power allocation problem (where the power selection is binary) to be discussed next. We note that link scheduling can also happen across different time slots or frequency sub-bands, where only one link, or a subset of links, is scheduled during each orthogonal resource unit in the time and frequency domain.

Power Allocation in Wireless Networks

power allocation in wireless networks involves the selection of transmit powers to optimize the system metrics of interest. Consider the same wireless network with N communication links as above. Link k is characterized by transmit power P_k and channel state h_{kj} from the transmitter of link j to the receiver of link k. Then the received signal-to-interference-plus-noise ratio (SINR) of link k is

$$\gamma_k(\mathbf{p}) = \frac{p_k |h_{kk}|^2}{\sum_{j \neq k} p_j |h_{kj}|^2 + \sigma^2}, \tag{5.2}$$

where $\mathbf{p} = [p_1, \ldots, p_N]^T$ is the transmit power vector for the N links. Consequently, a power allocation problem to optimize a generic weighted sum rate is formulated as

$$\max_{\mathbf{p}} \quad \sum_{k=1}^{N} \omega_k \log\left(1 + \gamma_k(\mathbf{p})\right),$$

$$\text{s.t.} \quad 0 \le p_k \le P_{\max}, \quad k \in \mathcal{N}, \tag{5.3}$$

where $P_{\max}$ is the maximum transmit power for all links, and ω_k is the nonnegative weight of link k, which can be adjusted to prioritize either sum rate maximization or fair scheduling.

Joint Spectrum and Power Allocation

The design and optimization of multi-carrier communication systems often involve maximizing total throughput under system resource constraints, such as available spectrum and transmit power. Specifically, we consider K users sharing N sub-bands in an orthogonal frequency division multiple access (OFDMA) system, which can be formulated as

$$\max_{\mathbf{p}} \quad \sum_{k=1}^{K} \omega_k \sum_{n=1}^{N} \log\left(1 + \frac{p_k^n |h_{kk}^n|^2}{\sum_{j \ne k} p_j^n |h_{kj}^n|^2 + \sigma^2}\right),$$

$$\text{s.t.} \quad \sum_{n=1}^{N} p_k^n \le P_k^{\max}, \quad \text{for all } k, \tag{5.4}$$

$$p_k^n \ge 0, \quad \text{for all } k, n,$$

where K is the number of users, p_k^n is the power of the kth user, and h_{kj}^n is the channel state from the transmitter of user j to the receiver of user k over the nth sub-band. ω_k is the weight for user k, and $P_k^{\max}$ is the total power constraint for user k. Conventional optimization algorithms transform problem (5.4) into the dual domain by forming its Lagrangian dual as

$$\max_{\mathbf{p}} \quad \sum_{k=1}^{K} \omega_k \sum_{n=1}^{N} \log\left(1 + \frac{p_k^n |h_{kk}^n|^2}{\sum_{j \ne k} p_j^n |h_{kj}^n|^2 + \sigma^2}\right) +$$

$$\sum_{k=1}^{K} \lambda_k (P_k^{\max} - \sum_{n=1}^{N} p_k^n), \tag{5.5}$$

$$\text{s.t.} \quad p_k^n \ge 0, \quad \text{for all } k, n,$$

where $\lambda_1, \ldots, \lambda_K$ are non-negative dual variables. The transformed problem is (almost) an unconstrained optimization problem, except for the non-negative power constraint, and is much easier to solve. On that basis, a low-complexity dual algorithm is shown to achieve near-optimality when N is large in light of the duality theory as proved in [179].

5.2 Supervised Learning Approach for Optimization

In this section, we first introduce a conventional optimization-based power control method, the weighted minimum mean square error (WMMSE) algorithm, to give the ground truth for supervised training and then demonstrate how to use deep neural networks (DNNs) and convolutional neural networks (CNNs) to deal with the wireless resource optimization problem.

5.2.1 WMMSE Algorithm

The WMMSE algorithm [180] is a commonly used conventional algorithm for the power allocation problem in (5.3), which takes an iterative procedure to adjust each optimization variable in the problem in turn and is guaranteed to converge to a stationary point. It is proved in [178] that the weighted-sum rate maximization problem in (5.3) can be equivalently reformulated as the following weighted-sum mean squared error minimization problem, hence the name WMMSE, as the following WMMSE problem:

$$
\min_{\{w_k, u_k, v_k\}_{k=1}^{N}} \quad \sum_{k=1}^{N} \omega_k (w_k e_k - \log(w_k)),
\tag{5.6}
$$

$$
\text{s.t.} \quad 0 \le v_k \le \sqrt{P_k^{\max}}, k = 1, 2, \ldots, N,
$$

where w_k is a positive variable and

$$
e_k = (u_k |h_{kk}| v_k - 1)^2 + \sum_{j \ne k} (u_k |h_{kj}| v_j)^2 + \sigma^2 u_k^2.
\tag{5.7}
$$

The problem (5.6) is defined over the variables $\{w_k, u_k, v_k\}_{k=1}^{N}$ and is convex while holding other variables fixed. This makes it easier to solve compared with problem (5.3), whose objective function is non-convex and difficult to deal with. In particular, WMMSE optimizes one set of variables while keeping the rest fixed, using the block coordinate descent method [181]. At the beginning of the algorithm, we initialize v_k^0 as

$$
0 \le v_k^0 \le \sqrt{P_k^{\max}}, \quad \text{for all } k,
\tag{5.8}
$$

and

$$
u_k^0 = \frac{|h_{kk}| v_k^0}{\sum_{j=1}^{N} |h_{kj}|^2 (v_j^0)^2 + \sigma^2}, \quad \text{for all } k,
$$

$$
w_k^0 = \frac{1}{1 - u_k^0 |h_{kk}| v_k^0}, \quad \text{for all } k,
\tag{5.9}
$$

given channel parameters $\{|h_{kj}|, \sigma^2\}$. The iterations are repeated as

$$
v_k^t = \left[\frac{\omega_k w_k^{t-1} u_k^{t-1} |h_{kk}|}{\sum_{j=1}^{N} \omega_j w_j^{t-1} (u_j^{t-1})^2 |h_{kj}|^2} \right]_0^{\sqrt{P_k^{\max}}}, \quad \text{for all } k,
$$

$$
u_k^t = \frac{|h_{kk}| v_k^t}{\sum_{j=1}^{N} |h_{kj}|^2 (v_j^t)^2 + \sigma^2}, \quad \text{for all } k,
\tag{5.10}
$$

$$
w_k^t = \frac{1}{1 - u_k^t |h_{kk}| v_k^t}, \quad \text{for all } k,
$$

until some stopping criteria is met, and the final output of WMMSE is $p_k = (v_k)^2$, for all k. Note that the first equation means clipping v_k^t to the range of $(0, \sqrt{P_k^{max}})$.

Example 5.1 As shown in Table 5.1, the channel power gains of three users at a certain time are given (consider small-scale fading only for simplicity). Assume that the maximum transmit power and σ per user is 1, and the weight $\omega_k = 1$, for all k. Show how to derive the WMMSE optimization solution.

Table 5.1 Channel power gains between transmitter j and receiver k.

Transmitter/Receiver	1	2	3
1	1.623	0.254	0.704
2	0.699	0.723	0.676
3	0.150	1.546	1.206

Solution
Set $v_k^0 = 1$, for all k. The initial u_1^0 and w_1^0 of user 1 are

$$u_1^0 = \frac{|h_{11}|v_1^0}{\sum_{j=1}^{3} |h_{1j}|^2 (v_j^0)^2 + \sigma^2}$$

$$= \frac{1.623}{1.623^2 + 0.699^2 + 0.150^2 + 1}$$

$$= 0.39,$$

$$w_1^0 = \frac{1}{1 - u_1^0 |h_{11}| v_1^0} = 8.11.$$

Similarly, we can obtain the initial u_2^0, w_2^0, u_3^0, and w_3^0. Repeat the iterations given by (5.10) until convergence. It usually converges after about a hundred iterations.

5.2.2 Supervised Learning Approach for Optimization

For supervised learning, DNNs are applied to learn the mapping from the network conditions to the solution of a given optimization algorithm. The most straightforward way to leverage deep learning for resource allocation is treating the given optimization problem as a black box and using various deep learning techniques to learn its input–output relationship. In this case, a traditional optimization method usually acts as a supervisor whose output will serve as the ground truth when training DNNs. With the universal approximation property of DNNs, the mapping from the network conditions to the solution of the given optimization algorithm can be well approximated.

The training and testing stages are shown in Figure 5.1. During the training stage in Figure 5.1a, the labeled training dataset is constructed by first generating

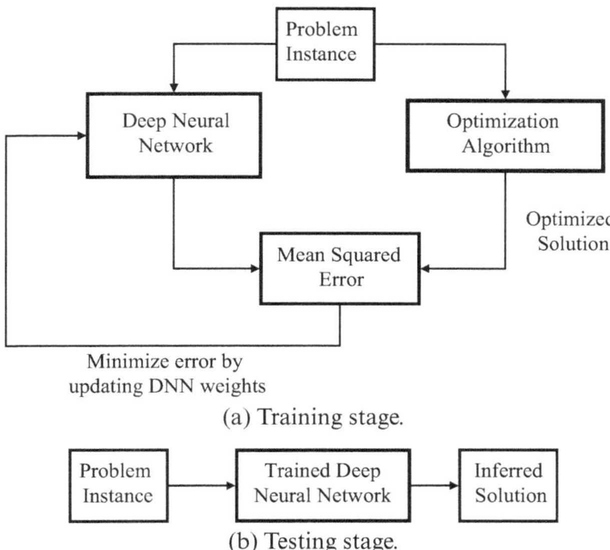

(a) Training stage.

(b) Testing stage.

Figure 5.1 Illustration of supervised learning for optimization.

random samples of the parameters of the optimization problems, e.g., the channel parameters, according to certain distributions, and then running existing optimization algorithms whose outputs will serve as the training labels. We then leverage the obtained training datasets to minimize the discrepancy between the DNN outputs and the optimized solutions by updating the DNN parameters with stochastic gradient descent methods. In the testing stage in Figure 5.1b, we pass the parameters of a new problem instance as the input to the trained network and obtain the inferred solution almost instantly. In doing so, the computationally heavy model training process can be performed offline, while the online deployment only requires a simple forward pass through the trained network, which is usually an order-of-magnitude faster than running traditional optimization algorithms directly [18]. Moreover, the quality of training labels can be ensured by executing optimization algorithms offline that are deemed to be computationally complex and thus not suitable for online deployment in the traditional paradigm.

Example 5.2 Here we give an example to illustrate how to allocate transmit power by supervised learning and compare this approach with the conventional WMMSE algorithm. We consider problem (5.3) with a Gaussian interference channel model that generates the channel coefficients according to the Rayleigh fading distribution with zero mean and unit variance. This has been widely used to evaluate the performance of various resource allocation algorithms.

In the training stage, we first randomly generate a few realizations of the channels $\{|h_{kj}^{(i)}|\}_{i \in \mathcal{T}}$ according to the Rayleigh distribution and store them in a dataset, where $\mathcal{T}$ includes the indices for the training set. Then we run the WMMSE algorithm to compute the corresponding optimized power vectors $\{p_k^{(i)}\}_{i \in \mathcal{T}}$. The resulting training dataset consists of pairs $(\{|h_{kj}^{(i)}|\}, \{p_k^{(i)}\})_{i \in \mathcal{T}}$. A fully connected neural network is used, and the loss function is set as the squared error between the ground truth $\{p_k^{(i)}\}$ and the output of the network.

In the testing stage, we process the testing datasets through the trained network to obtain the optimized power. We then compare the sum rates generated by the trained network and WMMSE to show the difference. Figure 5.2 compares the performance of the following schemes: 1) WMMSE; 2) the random power allocation strategy, which allocates the power according to the uniform distribution $p_k \sim$ U$(0, P_{\max})$, for all k; 3) maximum power allocation, which sets the power $p_k = P_{\max}$, for all k. In the figure, $P_{\max}$ and σ^2 are set to 1, and the number of links, N, is set to 10. From the figure, we see the learning-based method performs measurably close to the WMMSE method, outperforming the two heuristic benchmarks by a large margin. Experiments also show that the learning method runs an order-of-magnitude faster than the WMMSE algorithm, demonstrating its potential in practical deployment.

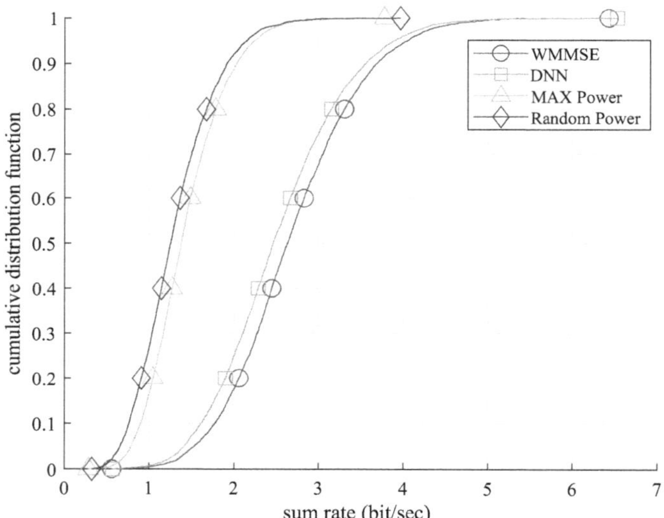

Figure 5.2 The cumulative distribution function (CDF) describes the rates achieved by different algorithms, over 5000 randomly generated testing data points with $N = 10$.

5.2.3 CNN for Optimization

The fully connected neural network has several limitations when applied to optimization tasks. First, the architecture of a fully connected neural network is not scalable. For the power allocation problem with N transmitter–receiver pairs, the network has N^2 different input channel coefficients and N nodes in the output layer, which require at least $O(N^3)$ interconnect weights (and most likely many more). Thus, the complexity of the training and testing grows rapidly with N. Second, the neural network necessitates retraining in the event of alterations to the number of transmitter–receiver pairs, indicating a lack of generalization.

To solve these problems, geographic location information (GLI) is proposed to serve as input in [21], which is linear with the number of links. This approach enables link scheduling without relying on explicit CSI. The link-scheduling problem focuses on device-to-device (D2D) communications, where the devices in proximity communicate with each other directly, without relying on the base station to relay their data as is done in traditional cellular networks. The goal of link scheduling is to maximize the network utility by activating a subset of links at any given time, which can be formulated as the combinatorial optimization problem in (5.1). Traditional methods rely on mathematical optimization techniques and require accurate CSI. For a network with N D2D links, N^2 channel coefficients need to be estimated, which is both time- and resource-consuming. To bypass the channel estimation stage, the feature-embedded paradigm is used for the link scheduling problem.

In fact, in many propagation environments, the CSI is largely a function of the distance-dependent path loss and shadowing. Thus, the CSI can be considered as a stochastic function of GLI,

$$\text{CSI} = f(\text{GLI}). \tag{5.11}$$

As a result, it is possible to bypass the channel estimation and directly find a mapping from the GLI to the link scheduling decisions,

$$\mathbf{x} = g(\text{CSI}) = g(f(\text{GLI})), \tag{5.12}$$

where $\mathbf{x} = [x_1, \ldots, x_N]^T$ denotes the vector of the indicator variables for link scheduling. To construct the network input based on GLI, the continuous $(\mathbf{d}_k^{\text{tx}}, \mathbf{d}_k^{\text{rx}})$ are quantized into a grid form, where $\mathbf{d}_k^{\text{tx}} \in \mathbb{R}^2$ and $\mathbf{d}_k^{\text{rx}} \in \mathbb{R}^2$ are the transmitter and the receiver locations of the kth link, respectively. As illustrated in Figure 5.3, a square deployment area is divided into M^2 cells. The density grid matrices are constructed by directly counting the total number of transmitters and receivers in each cell.

As illustrated in Figure 5.4, the network structure includes two stages: a convolution stage and a fully connected stage. The former captures the interference patterns that each link causes to its neighbors and each link receives from its neighbors. The transmitter and receiver density grids are operated in parallel by a spatial convolutional filter whose coefficients are optimized in the training process. Given that different neighboring links of different ranges have multiple components of the interference pattern, employing multiple convolutional layers can provide more information to describe the interference pattern. Thus, we construct three convolution filters with a small range, medium range, and full range, respectively, to improve the feature vector. The computed convolutions are referred to as TxINT_k and RxINT_k, for link k. The fully connected stage following the convolution stage produces an output, with the feature vector extracted for each link as input:

$$x_k \leftarrow F_{fc}(\text{TxINT}_k, \text{RxINT}_k, \|\mathbf{d}_k^{\text{tx}} - \mathbf{d}_k^{\text{rx}}\|_2), \tag{5.13}$$

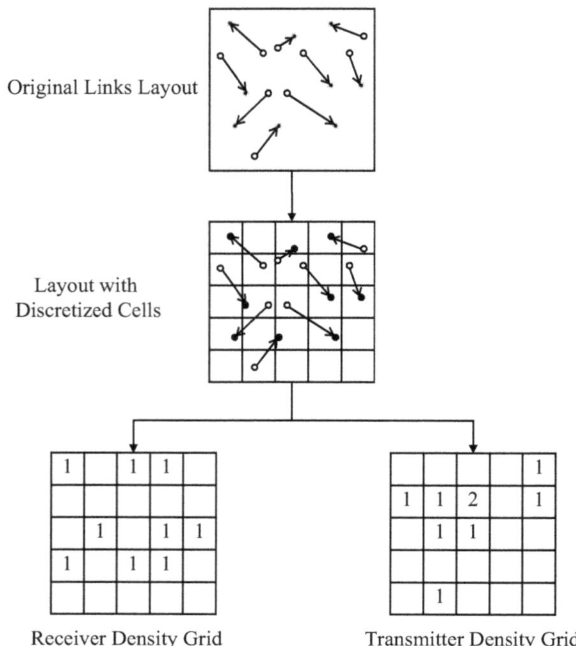

Figure 5.3 Transmitter and receiver density grids.

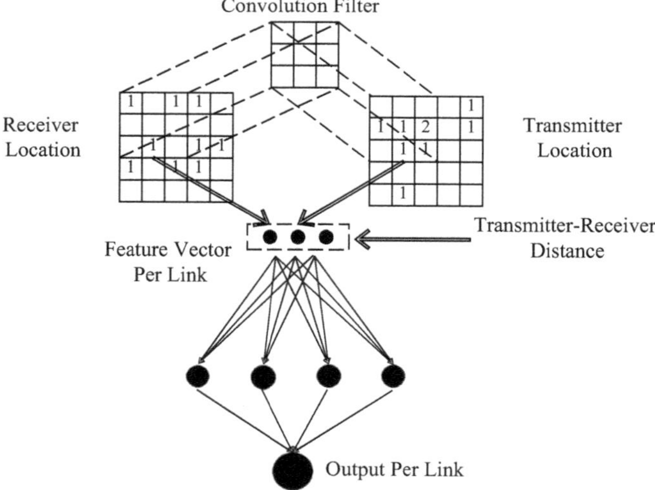

Figure 5.4 Example of a forward computation path for a single link with spatial convolutions and link distance as input to a neural network.

where $F_{fc}(\cdot)$ is the mapping dictated by a fully connected neural network, and $\|\mathbf{d}_k^{tx} - \mathbf{d}_k^{rx}\|_2$ is the distance between the transmitter and receiver. In the training stage, the network is trained in a supervised learning paradigm with the ground truth generated by either the WMMSE or the FPLinQ algorithm [182]. The trained neural network can produce a scheduling pattern close to the FPLinQ output and exhibits robust generalizability to layouts of larger dimensions [21].

5.3 Unsupervised Learning Approach for Optimization

The supervised learning paradigm has several limitations. First, since the ground truth must be provided by conventional optimization algorithms, the performance of the deep learning approach is inherently bounded by these conventional algorithms. Second, a large number of labeled samples are usually required to obtain a good model, as demonstrated in [18] and [20], which use 1,000,000 and 50,000 labeled samples for training, respectively. However, high-quality labeled datasets are difficult to generate in wireless resource allocation due to several factors, including inherent problem difficulty and computational resource constraints. These factors further limit the scalability of supervised learning methods. In this section, we introduce an objective-oriented unsupervised learning paradigm for resource allocation in wireless networks. The optimization objective is employed as the loss function, which is optimized directly during training.

5.3.1 Sum Rate Maximization

We consider the sum rate maximization problem in (5.3), where the weights ω_k are all set to one. Since there is no data label to guide learning, the unsupervised learning methods need to design an appropriate loss function and train the neural network with backpropagation to establish a mapping from the CSI to the optimal power-control vector. It is natural to use the objective function from the optimization problem as the loss function because the only constraint is that each transmit power must be in the range $[0, P_{\max}]$. This constraint can be guaranteed by using the sigmoid function defined in Section 1.2.2 at the network output layer, followed by a linear scaling of $P_{\max}$. Therefore, we can define the loss function for the sum rate maximization problem as

$$\ell_{\mathrm{SRM}} = -R(\mathbf{h}; \boldsymbol{\theta}) \triangleq -\sum_{k=1}^{N} \log\left(1 + \gamma_k(\mathbf{p}(\mathbf{h}; \boldsymbol{\theta}))\right), \tag{5.14}$$

where $\mathbf{p}(\mathbf{h}; \boldsymbol{\theta})$ is the power vector determined by the power allocation network with trainable parameters $\boldsymbol{\theta}$ and the channel $\mathbf{h}$, and $\gamma_k(\mathbf{p}(\mathbf{h}; \boldsymbol{\theta}))$ is the corresponding SINR as defined in (5.2). The loss function (5.14) is differentiable with respect to $\boldsymbol{\theta}$ and can be used to train the network via minimizing the empirical risk

$$\mathcal{L}_{\mathrm{SRM}}(\boldsymbol{\theta}) = -\frac{1}{|\mathcal{H}|} \sum_{\mathbf{h} \in \mathcal{H}} R(\mathbf{h}; \boldsymbol{\theta}), \tag{5.15}$$

where $|\mathcal{H}|$ denotes the number of samples.

Due to the non-convex nature of the loss function, the network training may converge to a local optimum. Thus, even with sufficient data we still cannot guarantee

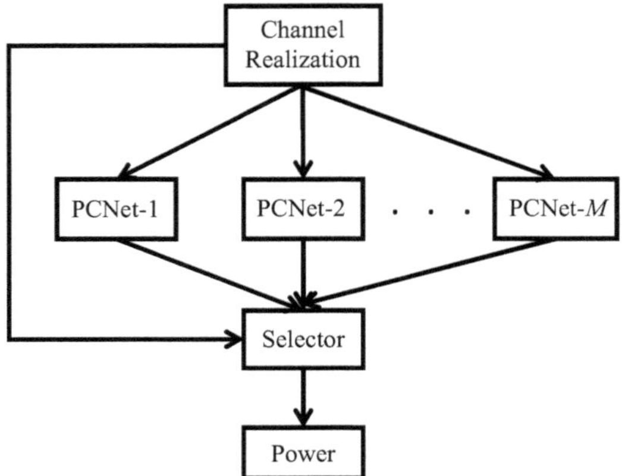

Figure 5.5 Illustration of ePCNet with an ensemble of M PCNets.

that the network outputs the globally optimal power profile $\mathbf{P}^*(\mathbf{h})$ for given channel information $\mathbf{h}$. To address this issue, a novel family of methods called the ensemble power control network (ePCNet) has been proposed in [183], which enhances the abovementioned unsupervised learning-based network training with ensemble learning. As shown in Figure 5.5, the proposed ePCNet consists of multiple parallel copies of power control networks (PCNets), which are trained with the loss in (5.15). These PCNets share the same network structure but are trained with different initial parameters and independently generated training data for diversity. During deployment, the channel information $\mathbf{h}$ is fed into all PCNets, which generate power control vectors $\{\mathbf{p}(\mathbf{h}; \boldsymbol{\theta}_\mathbf{m}), \text{for all } m\}$. A selector then computes the sum rate for each PCNet, $\{\mathbf{p}_m(\mathbf{h}), \text{for all } m\}$, and selects the largest sum rate according to

$$\max_{m} \; R(\mathbf{h}; \boldsymbol{\theta}_m), \tag{5.16}$$

where $R(\mathbf{h}; \boldsymbol{\theta}_m)$ denotes the sum rate given the channel $\mathbf{h}$ and the power control vector determined by the mth PCNet with parameters $\boldsymbol{\theta}_m$.

Ensemble learning enhances the neural network performance by combining multiple weak learners. For instance, although each individual PCNet is not powerful enough, it can output a universally better power profile by carefully forming an ensemble of these weak local learners. Moreover, the overall running time can be reduced with ensemble learning since all networks are trained independently and in parallel.

We note that unsupervised learning can automatically uncover underlying patterns and structures from data without requiring manually labeled examples. This is especially useful for resource allocation problems, where the relationships between resources and the optimal allocation scheme are unclear. Through unsupervised learning, it is possible to discover the internal relationships and make better allocation decisions. Moreover, the unsupervised learning models have superior adaptability and can be updated and adjusted when faced with new data. This flexibility is crucial for addressing wireless resource allocation problems where the propagation environment experiences strong dynamics. However, it is unclear in general whether unsupervised learning will outperform supervised learning approaches or vice versa. Unsupervised learning might get stuck at some local-quality local optima, and supervised learning approaches may converge faster when high-quality labels are provided. Therefore, selecting the appropriate learning method should be based on the specific characteristics of the problem and the availability of data.

5.3.2 Sum Rate Maximization with QoS Constraints

Let us consider the sum rate-maximization problem with quality-of-service (QoS) constraints, which maximizes the sum rate while satisfying the minimum rate requirement for each link, i.e.,

$$\max_{\mathbf{p}} \quad \sum_{k=1}^{N} \log(1 + \gamma_k(\mathbf{p})),$$
$$\text{s.t.} \quad \log(1 + \gamma_k(\mathbf{p})) \geq r_{k,\,\min}, \quad \text{for all } k,$$
$$0 \leq p_k \leq P_{\max}, \quad \text{for all } k, \tag{5.17}$$

where $r_{k,\,\min}$ is the minimum rate requirement of the kth link, and $P_{\max}$ is the maximum transmit power. We define the minimum rate and maximum transmit power vector as

$$\mathbf{r}_{\min} = [r_{1,\,\min}, \cdots, r_{N,\,\min}],$$
$$\mathbf{p}_{\max} = [P_{\max}, \cdots, P_{\max}]. \tag{5.18}$$

It is evident that problem (5.17) degenerates into the weighted sum rate-maximization problem defined in (5.3) when we set $\mathbf{r}_{\min} = \mathbf{0}$. In the following, we describe two methods to address the problem in (5.17), and they both center around designing appropriate loss functions.

Loss Function Design with Penalty Terms

For the sum rate-maximization problem, the maximum power constraint can be easily met by using the sigmoid function properly. However, the minimum rate constraints in (5.17) are more difficult to handle since no neural network structure can directly capture and enforce them. To address this issue, a new loss function [183] is introduced that encourages sum rate maximization and in the meantime penalizes constraint violation, given by

$$\ell_{\text{SRM-QoS}} = -R(\mathbf{h}; \theta) + \lambda \sum_{k=1}^{N} \text{ReLU}(r_{k,\,\min} - r_k(\mathbf{p}(\mathbf{h}; \theta))), \qquad (5.19)$$

where $\mathbf{p}(\mathbf{h}; \theta)$ is the power vector determined by the power allocation network with trainable parameters θ and the channel $\mathbf{h}$. $R(\mathbf{h}; \theta)$ is the sum rate defined in (5.14), and $r_k(\mathbf{p}(\mathbf{h}; \theta))$ is the rate of the kth link, given by

$$r_k(\mathbf{p}(\mathbf{h}; \theta)) = \log\left(1 + \gamma_k(\mathbf{p}(\mathbf{h}; \theta))\right), \qquad (5.20)$$

where $\gamma_k(\mathbf{p}(\mathbf{h}; \theta))$ is the SINR of the kth link as defined in (5.14). The loss function in (5.19) introduces penalty terms to incentivize the network output to meet the minimum rate constraints. For example, if the rate constraint is satisfied, $\text{ReLU}(r_{k,\,\min} - r_k(\mathbf{p}(\mathbf{h}; \theta))) = 0$, and the loss function (5.19) degenerates to the loss function (5.14), which does not influence the network training. However, if the rate constraint is not satisfied, i.e., $r_{k,\,\min} > r_k(\mathbf{p}(\mathbf{h}; \theta))$, and $\text{ReLU}(r_{k,\,\min} - r_k(\mathbf{p}(\mathbf{h}; \theta))) > 0$, the network parameters are updated to decrease this term, which in turn gradually alleviates the constraint violation. The scaling factor λ is a hyperparameter to balance the sum rate and constraints, which needs to be tuned carefully. If it is too large, the network will focus on the minimum rate constraints and sacrifice the sum rate performance; if it is too small, the penalty terms do not work for training, and the network may not output a feasible power profile. A more fine-grained tuning of this parameter will be introduced shortly based on (Lagrangian) primal-dual learning.

It should be noted that PCNet may still output an infeasible solution even with the new loss function. This might be because the weighted sum maximization problem with QoS constraints is infeasible under this specific parameter setting or because the PCNet fails to produce a feasible solution, though one may in fact exist.

Primal-Dual Learning Formulation

A more principled approach to the constrained optimization problem in (5.17) can be developed using Lagrangian duality theory and is presented in this subsection. To begin with, we define the non-negative multiplier dual variables $\lambda \in \mathbb{R}_+^N$, $\mu \in \mathbb{R}_+^N$, and $\tau \in \mathbb{R}_+^N$, respectively associated with the constraints $\mathbf{r}(\mathbf{h}; \theta) \geq \mathbf{r}_{\min}$, $\mathbf{p}(\mathbf{h}; \theta) \geq \mathbf{0}$, and $\mathbf{p}(\mathbf{h}; \theta) \leq \mathbf{p}_{\max}$, where $\mathbf{r}(\mathbf{h}; \theta) = [r_1(\mathbf{h}; \theta), \cdots, r_N(\mathbf{h}; \theta)]$ is the rate vector. The Lagrangian of the optimization problem (5.17) is then given by

$$\mathcal{L}(\mathbf{p}(\mathbf{h}; \theta), \lambda, \mu, \tau) := \sum_{k=1}^{N} r_k(\mathbf{p}(\mathbf{h}; \theta)) + \lambda^T \left(\mathbf{r}(\mathbf{p}(\mathbf{h}; \theta)) - \mathbf{r}_{\min}\right)$$
$$+ \mu^T \mathbf{p}(\mathbf{h}; \theta) + \tau^T (\mathbf{p}_{\max} - \mathbf{p}(\mathbf{h}; \theta)). \qquad (5.21)$$

The dual function $D(\lambda, \mu, \tau)$ is defined as the maximum Lagrangian value achieved by the overall power control schemes

$$D(\lambda, \mu, \tau) := \max_{\theta} \mathcal{L}(\mathbf{p}(\mathbf{h}; \theta), \lambda, \mu, \tau). \qquad (5.22)$$

It is important to note that the problem in (5.22) is a penalized version of (5.17), in which the constraints are not enforced but penalized by the terms $\lambda^T (\mathbf{r}(\mathbf{p}(\mathbf{h};\theta)) - \mathbf{r}_{\min}))$, $\mu^T \mathbf{p}(\mathbf{h};\theta)$, and $\tau^T(\mathbf{p}_{\max} - \mathbf{p}(\mathbf{h};\theta))$. Similar to (5.19), this unconstrained optimization problem can be well approximated by training a neural network via stochastic gradient descent. Since for any choice of $\lambda \geq \mathbf{0}$, $\mu \geq \mathbf{0}$, and $\tau \geq \mathbf{0}$, we have $D(\lambda,\mu,\tau) \geq \sum_{k=1}^{N} r_k(\mathbf{p}(\mathbf{h};\theta))$ when the constraints are satisfied. Thus, a Lagrangian dual problem is formulated to search for the multipliers that make $D(\lambda,\mu,\tau)$ as small as possible to narrow the duality gap between $D(\lambda,\mu,\tau)$ and $\sum_{k=1}^{N} r_k(\mathbf{p}(\mathbf{h};\theta))$, i.e.,

$$D^* := \min_{\lambda,\mu,\tau \geq \mathbf{0}} D(\lambda,\mu,\tau), \tag{5.23}$$

which is also an unconstrained optimization problem and can be optimized via gradient descent.

With the unconstrained optimization formulations in (5.22) and (5.23), we proceed to develop a primal-dual learning framework [22] where both the primal variable θ and dual variables λ, μ, τ are updated iteratively using stochastic gradient descent. At the tth iteration, we update current primal variables θ_t by adding the corresponding partial gradients $\nabla_\theta \mathcal{L}$ of the Lagrangian in (5.21), i.e.,

$$\theta_{t+1} = \theta_t + \gamma_{\mathbf{p},t}\nabla_\theta \mathcal{L}(\mathbf{p}(\mathbf{h};\theta),\lambda,\mu,\tau), \tag{5.24}$$

where $\gamma_{\mathbf{p},t} > 0$ is the scalar step size. Likewise, λ, μ, τ are operated in a similar manner,

$$\begin{aligned}
\lambda_{t+1} &= \lambda_t + \gamma_{\lambda,t}\nabla_\lambda \mathcal{L}(\mathbf{p}(\mathbf{h};\theta),\lambda,\mu,\tau), \\
\mu_{t+1} &= \mu_t + \gamma_{\mu,t}\nabla_\mu \mathcal{L}(\mathbf{p}(\mathbf{h};\theta),\lambda,\mu,\tau), \\
\tau_{t+1} &= \tau_t + \gamma_{\tau,t}\nabla_\tau \mathcal{L}(\mathbf{p}(\mathbf{h};\theta),\lambda,\mu,\tau),
\end{aligned} \tag{5.25}$$

with associated step sizes $\gamma_{\lambda,t}, \gamma_{\mu,t}, \gamma_{\tau,t} > 0$ and constraints $\lambda,\mu,\tau \geq \mathbf{0}$. The iterative updates in (5.24) and (5.25) are performed on each batch of data sampled from certain channel distributions, e.g., Rayleigh fading, and repeated until convergence. We note that since the original problem in (5.18) is non-convex, the above stochastic primal-dual descent does not guarantee convergence to a global optimum, although its excellent performance is empirically verified in [22] for wireless resource allocation problems.

5.4   Resource Management with GNNs

Although deep learning-based algorithms have demonstrated satisfactory results for resource allocation problems, these methods often struggle as the wireless network scales. To address this challenge, we introduce a novel deep learning architecture – GNNs. In this section, we first introduce some foundational concepts of graphs, followed by an overview of several classic GNN models, including the graph convolutional network (GCN), graph attention network (GAT), and graph isomorphism network (GIN). Finally, we demonstrate how to leverage the spatial structures of wireless networks for resource allocation through the graph embedding process.

5.4.1 Basics of Graphs

The graph is an important class of data structures to describe complex systems with many objects interacting with each other. It consists of a set of nodes representing these objects and a collection of edges that connect these nodes and capture their relationships. Compared to other structures like matrices, tensors, and sequences, graphs are particularly effective for modeling and solving problems that require special consideration of the relationships between different objects of interest, such as social networks, traffic systems, grammatical structures, and citation networks.

As shown in Figure 5.6, a graph can be represented as $G = (V, E)$, where $V = \{v_i \mid i = 1, 2, ..., N\}$ represents the set of nodes and N denotes the number of nodes. $E = \{e_{ij} \mid v_i, v_j \in V\}$ is the set of edges between nodes where the edge e_{ij} is drawn from node j to node i. $N(i)$ denotes a group of neighboring nodes connected to node i, i.e., $N(i) = \{v_j \mid e_{ij} \in E\}$. For a graph, each node and edge have their own properties (weight, rank, type, sign, etc.), denoted by $\mathbf{x}_i$ and $\mathbf{e}_{ij}$, respectively. Depending on whether the edges have directions or not, graphs can be classified as undirected or directed graphs. Additionally, graphs can be classified as unweighted or weighted graphs depending on whether the edges have weights or not.

The connectivity of edges in the graph can be represented by an adjacency matrix, a square $N \times N$ matrix populated with binary elements. In particular, an undirected graph is represented by a symmetric matrix. The adjacency matrix can be defined as follows:

$$\mathbf{A} \in \{0, 1\}^{N \times N}, \quad \text{with} \quad A_{ij} = \begin{cases} 1, & \text{if } e_{ij} \in E, \\ 0, & \text{otherwise.} \end{cases} \tag{5.26}$$

However, constructing and storing such an adjacency matrix can be expensive and inefficient, especially when the graph is sparsely connected. To efficiently describe the edge connectivity, deep learning frameworks like PyTorch Geometric (PyG) and Deep Graph Library (DGL) adopt edge list $\mathbf{E}$, a two-dimensional matrix where the first line indicates the serial number of the source nodes, and the second line indicates

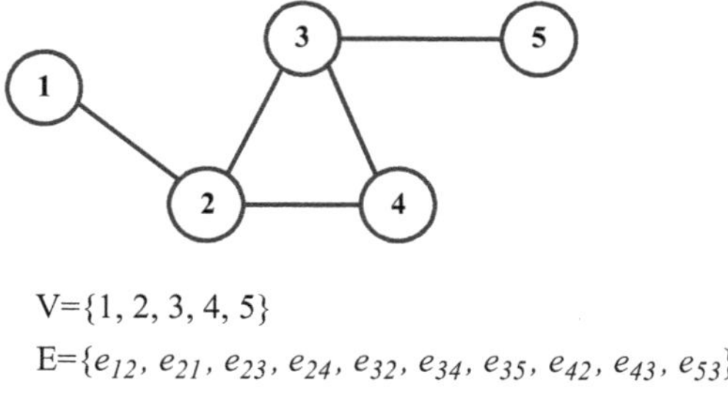

$V=\{1, 2, 3, 4, 5\}$

$E=\{e_{12}, e_{21}, e_{23}, e_{24}, e_{32}, e_{34}, e_{35}, e_{42}, e_{43}, e_{53}\}$

Figure 5.6 A simple example of an undirected graph.

the corresponding destination nodes. Figure 5.6 illustrates an undirected graph, and its adjacency matrix and edge list are expressed respectively by

$$\mathbf{A} = \begin{bmatrix} 0 & 1 & 0 & 0 & 0 \\ 1 & 0 & 1 & 1 & 0 \\ 0 & 1 & 0 & 1 & 1 \\ 0 & 1 & 1 & 0 & 0 \\ 0 & 0 & 1 & 0 & 0 \end{bmatrix}, \quad \mathbf{E} = \begin{bmatrix} 1 & 2 & 2 & 2 & 3 & 3 & 3 & 4 & 4 & 5 \\ 2 & 1 & 3 & 4 & 2 & 4 & 5 & 2 & 3 & 3 \end{bmatrix}.$$

Another important matrix for describing the properties of a graph is the degree matrix, a diagonal matrix where the diagonal elements represent the degrees of each node. The degree of a node $d(v_i)$ indicates the number of edges connected to that node. In an undirected graph, the degree of a node equals the number of its neighbors. In contrast, in a directed graph, the degree of a node is divided into out-degree and in-degree, representing the number of edges directed away from the node and the number of edges directed toward the node, respectively.

Generally, there are three types of deep learning tasks on graphs: node-level, edge-level, and graph-level. In node-level tasks, the goal is to predict properties or perform classification on individual nodes. Edge-level tasks focus on predicting relationships between nodes, e.g., in the form of link prediction or graph completion, where the primary goal is to predict whether there exists a link or edge between two nodes given the set of nodes and an incomplete set of edges in a graph. In graph-level tasks, we aggregate the features of all nodes in the graph to represent the entire graph and make predictions at the graph level.

5.4.2 Basics of GNNs

GNNs are deep learning architectures designed specifically for processing graph-structured data. Unlike traditional neural networks, such as CNNs for grid-structured inputs or recurrent neural networks for sequences, GNNs are specialized for handling complex relational networks that are usually described by a graph as discussed above.

Spectral-Based and Spatial-Based GNNs

In GNNs, graph convolution is a crucial operation for extracting features from graph-structured data, enabling more efficient data representation. This operation can be implemented in various ways, with the most common approaches being spectral-based and spatial-based GNNs. spectral-based GNNs use the eigenvalues and eigenvectors of the Laplacian matrix of the graph to perform the convolution operation. In contrast, spatial-based GNNs directly leverage the graph structure and node features to carry out the convolution operation.

As the earliest GNNs, spectral-based GNNs have achieved impressive results in many graph-related tasks. However, spectral-based GNNs become less efficient as the graph size increases due to the high computational cost of calculating eigenvectors

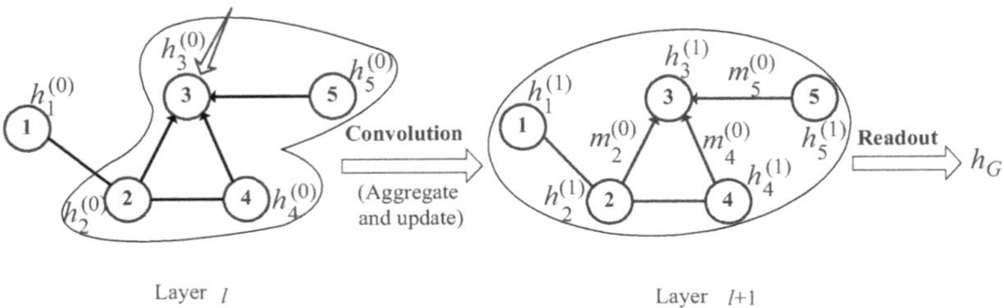

Figure 5.7 Overview of the message-passing neural network structure. The message-passing phase aims to learn the node-embedding vector.

or processing the entire graph. Consequently, spectral-based GNNs struggle with scalability and are less suited for large graphs. Besides, spectral-based GNNs assume a fixed graph, thus making it difficult to incorporate new nodes or generalize to new graphs. Moreover, such fixed graphs are limited to being undirected since the Laplacian matrix on directed graphs is not well-defined.

On the contrary, spatial-based GNNs perform convolution directly in the graph domain by aggregating information from all neighbors at each node. Essentially, a form of neural message-passing is used in which vector messages are exchanged between nodes and are updated by neural networks [184].

As shown in Figure 5.7, messages m_j are generated for each node $j \in V$ based on the node features and are passed according to the graph's topology. After aggregating the information from its neighbors and itself, each node obtains its embedding vector $\mathbf{h}_j$. For graph-level tasks, an additional read-out stage generates a global representation $\mathbf{h}_G$ that describes the features of the entire graph based on the node features.

Figure 5.8 illustrates a two-layer message-passing model. During each message-passing iteration l, a hidden embedding $\mathbf{h}_i^{(l)}$ is updated for each node $i \in V$ according to the information aggregated from i's neighborhood $N(i)$. The different iterations of message passing are also known as the different layers of the GNN model, and we use superscripts to index the iteration and subscripts to differentiate different nodes.

The message-passing stage consists of two core functions, AGGREGATE($\cdot$) and UPDATE($\cdot$) [185], and can be expressed as

$$
\begin{aligned}
\mathbf{m}_{N(i)}^{(l)} &= \text{AGGREGATE}^{(l)}\left(\left\{\mathbf{h}_j^{(l)}, \text{for all } j \in N(i)\right\}\right), \\
\mathbf{h}_i^{(l+1)} &= \text{UPDATE}^{(l)}\left(\mathbf{h}_i^{(l)}, \mathbf{m}_{N(i)}^{(l)}\right).
\end{aligned}
\tag{5.27}
$$

For each node $j \in V$, AGGREGATE($\cdot$) takes as input the embeddings of its neighborhood $N(i)$ and generates a message $\mathbf{m}_{N(i)}^{(l)}$ at each iteration l. Subsequently, the UPDATE($\cdot$) function combines the previous embedding of the node itself $\mathbf{h}_i^{(l)}$ with the message $\mathbf{m}_{N(i)}^{(l)}$ that is aggregated from neighborhood information and generates the updated hidden embedding $\mathbf{h}_i^{(l+1)}$. In particular, the initial embeddings at $l = 0$ are the node features, i.e., $\mathbf{h}_i^{(0)} = \mathbf{x}_i$, for all $i \in V$. Since each iteration aggregates 1-hop neighborhood information, the message-passing process can be iterated several times to obtain more comprehensive information about the graph structure, and

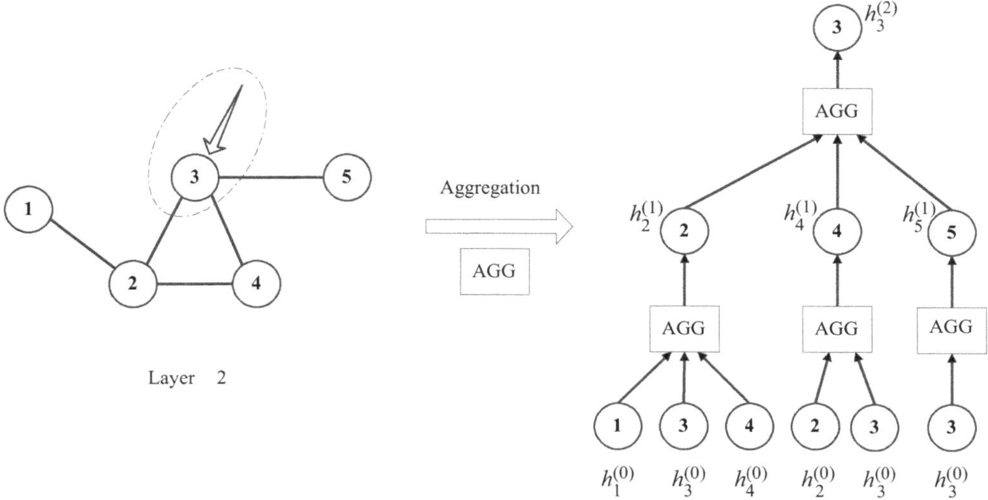

Figure 5.8 The process of message passing for node 3. The message-passing model aggregates the messages from node 3's local graph neighbors (i.e., 2, 4, and 5). In turn, the messages coming from these neighbors are based on information aggregated from their respective neighborhoods.

the final output of GNNs is the embedding vector of each node at the last layer. Generally, different types of GNNs can be realized by designing various UPDATE($\cdot$) and AGGREGATE($\cdot$) functions. In the following, we will describe three classic GNNs.

GCNs

Spatial-based GCNs are inspired by the convolution operation used in conventional CNNs for images. The key difference is that the spatial-based GCNs define graph convolution according to spatial relationships between nodes [186] rather than neighboring pixels in images.

As shown in Figure 5.9a, an image can be considered as a particular form of a graph, where each pixel represents a node and is directly connected to pixels in its neighborhood. Using a 3×3 window, also called a receptive field, the neighborhood of each pixel consists of the eight surrounding pixels. A filter is then applied to this 3×3 window, mapping each pixel to its unique feature using the weighted average of the pixel values of both the center node and its neighboring nodes. Notably, it is possible to share the trainable weights at different positions, a distinctive feature of CNNs.

Similarly, for general graphs, a spatial-based GNN aggregates the center node embedding and the adjacent node embeddings to obtain a new vector for the center node, as shown in Figure 5.9b. The update of the representation for node i is given by

$$\mathbf{m}_{N(i)}^{(l)} = \mathbf{W}^{(l)} \sum_{j \in N(i)} \frac{\mathbf{h}_j^{(l-1)}}{|N(i)|}$$

$$\mathbf{h}_i^{(l)} = \mu\left(\mathbf{m}_{N(i)}^{(l)}\right),$$

(5.28)

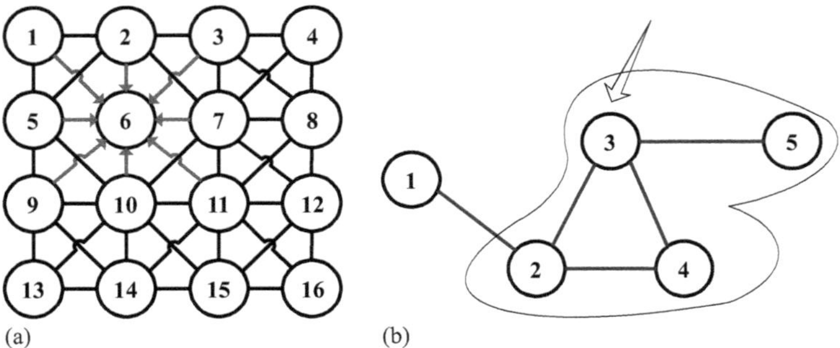

(a)   (b)

Figure 5.9 Similarities and differences between CNNs and GCNs.
(a) Two-dimensional convolution. Similar to a graph, each pixel in the image
can be considered as a node, and its neighbors depend on the filter size. The
convolution takes a weighted average (depending on the filter and its
neighbors, and each node's neighbors are ordered. (b) Graph convolution.
To generate the hidden embedding of node 3, the GCN model takes the
average value of the features of node 3 and its neighbors. Unlike images,
each node's neighbors are unordered and differ in size.

where node embeddings of node i's neighborhood in the previous $(l-1)$th layer are
normalized by the node degree and multiplied by the weight matrix of the lth layer.
These embeddings are then summed together to generate the current message $\mathbf{m}_{N(i)}^{(l)}$
for each node, and the node representation can be obtained by applying the non-
linear activation function $\mu(\cdot)$ to the message to enhance the expressiveness.

GATs

It is important to note that GCNs treat all neighboring nodes equally during the
convolution process without assigning different weights based on the relative importance of the nodes. To solve this problem, GATs [187] have been developed. GATs
introduce a self-attention mechanism that assigns different weights to neighboring
nodes based on their features when calculating node embeddings. These two types of
GNNs are compared in Figure 5.10.

The graph convolution process in GATs is defined as

$$\mathbf{h}_i^{(l)} = \mu \left(\sum_{j \in N(i)} \alpha_{ij} \mathbf{W}^{(l)} \mathbf{h}_j^{(l-1)} \right), \text{for all } i \in V, \tag{5.29}$$

where $\alpha_{ij}, j \in V$ are the attention weights given by

$$\alpha_{ij} = \frac{\exp\left(\varpi_{ij}\right)}{\sum_{k \in N(i)} \exp\left(\varpi_{ik}\right)},$$

$$\varpi_{ij} = a\left(\mathbf{W}^{(l)} \mathbf{h}_j^{(l-1)}, \mathbf{W}^{(l)} \mathbf{h}_i^{(l-1)}\right), \tag{5.30}$$

where $a(\cdot, \cdot)$ is an attention score function defined over pairs of nodes j, i based
on their embeddings, and parameters of $a(\cdot, \cdot)$ are trained jointly together with
weight matrices $\mathbf{W}^{(l)}$ in an end-to-end fashion. For instance, we can use a simple

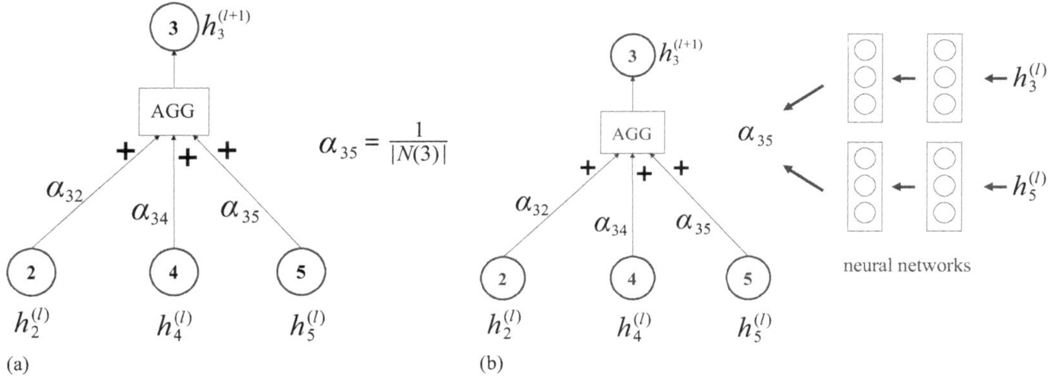

Figure 5.10 Similarities and differences between GCNs and GATs. (a) During the aggregation, GCNs assign equal weights to all neighboring nodes. (b) GATs assign different weights (based on the self-attention (GCN) mechanism commonly realized by neural networks) to the nodes' neighborhoods so that important nodes obtain higher weights.

single-layer neural network to parameterize $a(\cdot, \cdot)$. Here, ϖ_{ij} indicates the importance of node j to node i, and the final attention weights α_{ij} are obtained by normalizing ϖ_{ij} with the softmax function, so that $\sum_{j \in N(i)}(\alpha_{ij}) = 1$. Furthermore, GATs introduce a multi-head attention mechanism to stabilize the learning process of the attention mechanism. Let k denote the kth attention head. The GAT computes the attention coefficients by various $\mathbf{W}_k^{(l)}$ and aggregates different attention coefficients by concatenation for an update or average for the final output:

$$
\text{Update}: \quad \mathbf{h}_i^{(l)} = \text{CONCAT}\left\{ \mu \left(\sum_{j \in N(i)} \alpha_{ij}^k \mathbf{W}_k^{(l)} \mathbf{h}_j^{(l-1)} \right) \right\},
$$

$$
\text{Output}: \quad \mathbf{h}_i^{(l)} = \mu \left(\frac{1}{K} \sum_{k=1}^{K} \sum_{j \in N(i)} \alpha_{ij}^k \mathbf{W}_k^{(l)} \mathbf{h}_j^{(l-1)} \right).
$$

(5.31)

GINs

GINs have proved to be as expressive as the Weisfeiler–Lehman (WL) test in distinguishing isomorphic graphs [188]. The expressiveness of GNNs can be characterized by their message and aggregation functions. Injective message and aggregation functions, as employed by GINs, enable them to achieve the most expressive GNNs. It follows from the universal approximation theorem that multilayer perceptrons (MLPs) with sufficiently large hidden dimensionality and appropriate non-linearity can approximate any continuous injective function to arbitrary accuracy. The injective function for GIN update can be modeled as

$$
\mathbf{h}_i^{(l+1)} = \text{MLP}_a \left((1 + \epsilon) \cdot \text{MLP}_f \left(\mathbf{h}_i^{(l)} \right) + \sum_{j \in N(i)} \text{MLP}_f \left(\mathbf{h}_j^{(l)} \right) \right),
$$

(5.32)

where ϵ is a learnable scalar, and message computation will be performed through an MLP named MLP_f, after which another MLP named MLP_a will aggregate the messages for each node i from its neighbors $N(i)$ and itself. The embedding result of node i that contains the information of its l-hop neighborhoods is represented by $\mathbf{h}_i^{(l+1)}$.

Example 5.3 Consider the message-passing neural network structure shown in Figure 5.8. Assume that the AGGREGATE$(\cdot)$ and the UPDATE$(\cdot)$ are Mean$(\cdot)$ and Sum$(\cdot)$, respectively. Compute the updated hidden embedding $\mathbf{h}_3^{(2)}$ of node 3 with the inital hidden embedding $\left[\mathbf{h}_1^{(0)}, \mathbf{h}_2^{(0)}, \mathbf{h}_3^{(0)}, \mathbf{h}_4^{(0)}, \mathbf{h}_5^{(0)}\right] = [1.0, 2.0, 3.0, 4.0, 5.0]$.

Solution

Layer 0:

This is the input layer, where each node has its initial features.

Layer 1:

Node 2 has three child nodes: 1, 3, and 4. The updated embedding of node 2 becomes

$$\mathbf{h}_2^{(1)} = \text{Sum}\left(\mathbf{h}_2^{(0)}, \text{Mean}\left(\left\{\mathbf{h}_1^{(0)}, \mathbf{h}_3^{(0)}, \mathbf{h}_4^{(0)}\right\}\right)\right)$$
$$= \text{Sum}\left(\mathbf{h}_2^{(0)}, 2.67\right) = 4.67.$$

Similarly, the embeddings of nodes 4, 5, and 3 become

$$\mathbf{h}_4^{(1)} = \text{Sum}\left(\mathbf{h}_4^{(0)}, \text{Mean}\left(\left\{\mathbf{h}_2^{(0)}, \mathbf{h}_3^{(0)}\right\}\right)\right) = 6.5,$$
$$\mathbf{h}_5^{(1)} = \text{Sum}\left(\mathbf{h}_5^{(0)}, \text{Mean}\left(\left\{\mathbf{h}_3^{(0)}\right\}\right)\right) = 9.0,$$
$$\mathbf{h}_3^{(1)} = \text{Sum}\left(\mathbf{h}_3^{(0)}, \text{Mean}\left(\left\{\mathbf{h}_2^{(0)}, \mathbf{h}_4^{(0)}, \mathbf{h}_5^{(0)}\right\}\right)\right) = 6.67.$$

Layer 2:

Node 3 has three child nodes: 2, 4, and 5. The embedding of node 3 becomes

$$\mathbf{h}_3^{(2)} = \text{Sum}\left(\mathbf{h}_3^{(1)}, \text{Mean}\left(\left\{\mathbf{h}_2^{(1)}, \mathbf{h}_4^{(1)}, \mathbf{h}_5^{(1)}\right\}\right)\right) = 13.39.$$

5.4.3 GNN-Based Resource Management

Consider a wireless network with N transmitter–receiver pairs in Figure 5.11a, where each transmitter communicates with its intended receiver and causes interference to other receivers. This is an example of the interference channels discussed in Section 5.1. We use h_{ii} and h_{ij} to denote the channel gain of the ith link and the interference link from the jth link's transmitter to the ith link's receiver, respectively. We consider the power allocation problem in (5.3), where all links share the same spectrum and adjust their transmit power to maximize the sum rate. Specifically, let p_i denote the transmit power of the ith link, collectively represented by $\mathbf{p}^{\text{T}} = [p_1, p_2, ..., p_N]$; the SINR of the ith link follows

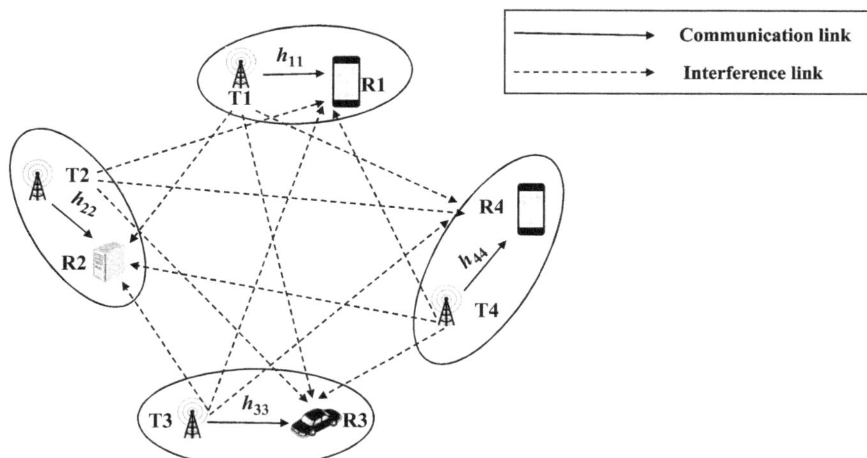

(a) An interference channel with four transmitter–receiver pairs.

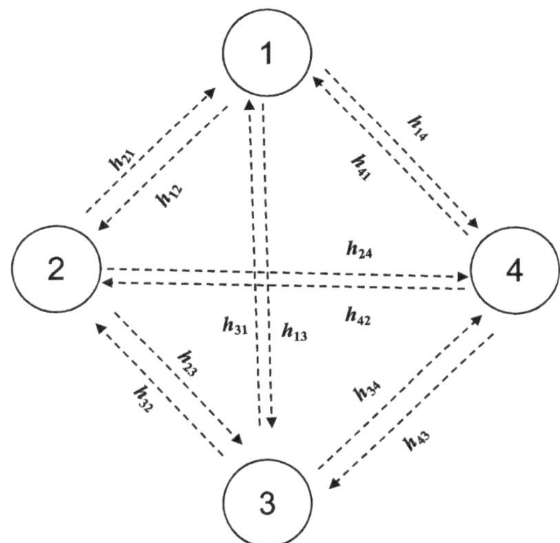

(b) Graph representation of the interference channel.

Figure 5.11 Illustration of an interference channel.

$$\mathrm{SINR}_i(\mathbf{p}) = \frac{p_i \mid h_{ii} \mid^2}{\sigma^2 + \sum_{j \neq i} p_j \mid h_{ij} \mid^2}, \tag{5.33}$$

where σ^2 denotes the power of the additive noise that is assumed to follow the complex Gaussian distribution $\mathcal{CN}(0, \sigma^2)$. Then the power allocation problem can be rewritten as

$$\max_{\mathbf{p}} \ \sum_{i=1}^{N} \log\left(1 + \mathrm{SINR}_i(\mathbf{p})\right), \tag{5.34}$$

$$\text{s.t.} \quad 0 \leq p_i \leq P_{\max}, \quad \text{for all } i \in \{1, \ldots, N\},$$

where $P_{\max}$ denotes the maximum transmit power. The formulated optimization problem is difficult to solve due to its non-convex nature, particularly when the network size is large. Existing optimization algorithms, such as the WMMSE method discussed in Section 5.2, usually involve a lengthy iterative procedure. Meanwhile, emerging deep learning-based resource allocation methods have demonstrated satisfactory performance in small-scale networks as demonstrated so far in this chapter. Unfortunately, these methods exhibit limited scalability and generalization. To address these issues, we turn to GNNs to exploit the spatial structures of wireless networks and obtain scalability for large-scale wireless networks.

The wireless network in Figure 5.11a can be modeled as a directed graph with node and edge features, as illustrated in Figure 5.11b. Specifically, we treat each transmitter–receiver pair as a node, and the edge e_{ji} can be drawn from node j to node i, indicating there exists an interference link from the jth link's transmitter to the ith link's receiver. The node features $\mathbf{X} = \left[\mathbf{x}_1, \cdots, \mathbf{x}_N \right] \in \mathbb{R}^{n \times N}$ indicate the communication condition of the links, where $\mathbf{x}_i$ denotes an n-dimensional feature vector for node i. The edge features $\mathbf{W} = \left[w_{ji} \right]_{i,j \in V, i \neq j} \in \mathbb{R}^{N \times N}$ incorporate the properties of the corresponding interference channels, where $w_{ji} = h_{ij}$.

For power control, we introduce the maximum transmit power $P_{\max}$ as a supplementary part of the initial node feature and leverage the GINs to perform the graph embedding process. The update rule of GINs for the ith node in the lth layer can be expressed as

$$\mathbf{x}_i^{(l)} = \mathrm{MLP}_2 \left(\mathbf{x}_i^{(l-1)}, \mathrm{MAX}_{j \in N(i)} \left\{ \mathrm{MLP}_1 \left(\mathbf{x}_j^{(l-1)}, w_{ji} \right) \right\} \right), \tag{5.35}$$

where $\mathbf{x}_i^{(l-1)}$ represents the node embedding at the $(l-1)$th layer. After the message computation in MLP_1, MLP_2 aggregates the messages for each node i from its neighbors $N(i)$, determined by graph topology. The embedding vector of node i that contains the information of its l-hop neighborhoods is represented by $\mathbf{x}_i^{(l)}$.

With each link represented by a low-dimensional vector, the network can then be embedded into the feature space. Power control can subsequently be conducted based on the graph embedding results from the final GIN layer, which are expected to capture the network interference topology and are essential for power control designs. Letting f_{out} denote the prediction head for our node-level task and L denote the number of GIN layers, the final embedding result, $\mathbf{x}_i^{(L)}$, can be mapped into the power control decision for the ith communication link given by

$$p_i = f_{\mathrm{out}}(\mathbf{x}_i^{(L)}). \tag{5.36}$$

We simulate a simple interference channel where each channel gain is generated according to a standard normal distribution with zero mean and unit variance. We set the maximum transmit power to $P_{\max} = 40$ dBm and the noise power to -102 dBm (assuming 5 MHz bandwidth and noise power spectral density of -169 dBm/Hz). We generate wireless networks with N transmitter–receiver pairs located randomly within a square area of side length $R = 1,000$ m. Specifically, we drop the transmitters randomly within the network area. For each transmitter, we ensure a

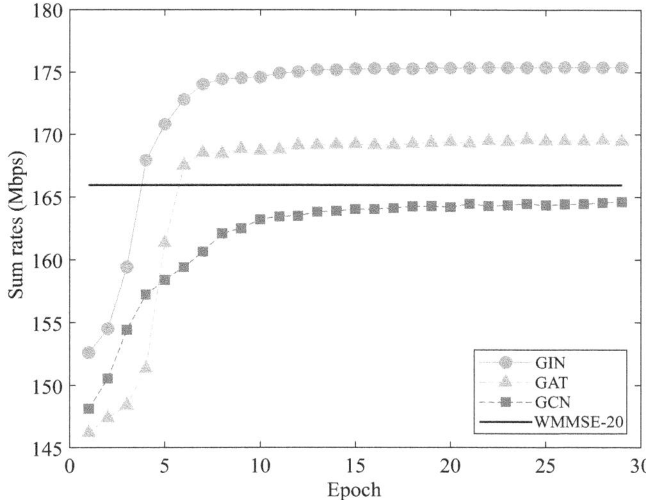

Figure 5.12 Convergence behavior of GCN, GAT, and GIN methods when $N = 50$.

distance within [2 m, 65 m] between the transmitter and its associated receiver and drop the receivers randomly.

Figure 5.12 shows the performance comparison between several GNN methods, including GCN, GAT, and GIN, compared to the WMMSE-20 method, with $N = 50$ pairs. The GNN-based methods are trained for 30 epochs on 1280 samples with a batch size of 320, using the Adam optimizer with a decaying learning rate. WMMSE-20 indicates that the iterative procedure of the WMMSE method is repeated 20 times, and we take the best one for performance comparison. We see that all the GNN methods converge in less than 10 epochs, and the GIN-based method outperforms the other two GNNs and the WMMSE benchmark. Moreover, there is a noticeable performance gap of around 6% between the GCN-based and GAT-based methods. Such a performance drop can be attributed to the fact that the GCNs do not consider the importance of different neighboring nodes when aggregating messages.

We further investigate the generalization of GNN-based methods to scenarios with varying numbers of links, N. We train all GNNs with 50 pairs in a 1000 m $\times$ 1000 m area and change the number of transmitter–receiver pairs in testing but with a fixed link density. Figure 5.13 shows the average rate of each link with an increasing number of links. It can be observed that as the number of communication links grows, the average rate of all methods drops since the increased communication density will give rise to more severe interference. Moreover, GAT-based and GIN-based methods maintain their satisfactory performance, and the performance of the three GNN-based methods is stable and competitive to the WMMSE baseline, which confirms the excellent generalization of GNN-based methods to larger wireless network scales.

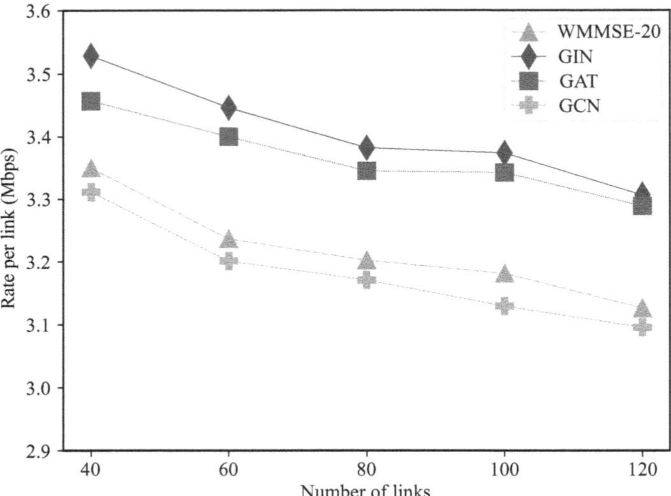

Figure 5.13 Generalization to different network scales but with a fixed density.

5.5 Learning Resource Allocation Based on RL

We have discussed various effective approaches to addressing wireless resource allocation based on deep learning, but several challenges still remain. On the one hand, fast-changing channel conditions in wireless networks cause substantial uncertainty for resource allocation, e.g., in terms of performance loss induced by the inaccuracy of acquired channel information. On the other hand, increasingly diverse service requirements dictated by new applications in wireless networks are emerging, such as simultaneously maximizing throughput and reliability for a mix of different links in vehicular networks. Such requirements are often difficult to model in a mathematically precise way, let alone finding a systematic approach to obtain optimal solutions.

Fortunately, RL has been shown effective in addressing decision-making under uncertainty [29]. It provides a robust and principled way to treat environmental dynamics and perform sequential decision-making under uncertainty, thus representing a promising method to handle challenging environmental dynamics in wireless networks. In addition, the hard-to-optimize objective issues can also be nicely addressed in an RL framework by designing training rewards that align with the final objective. Subsequently, the learning algorithm can autonomously devise an intelligent approach to pursuing the ultimate objective effectively. Another potential advantage of using RL for resource allocation is that it enables distributed algorithms, as demonstrated in [189], where each decision-making entity is treated as an agent that learns to refine its resource-sharing strategy through interactions with the unknown environment.

We will start with formulating the resource allocation in an RL framework and then show an extended example of applying this RL-based framework to address the spectrum-sharing problem in vehicular networks. Meta-RL will be further investigated to enable the trained RL agent to quickly adapt to a new environment and thus make RL-based resource allocation more applicable in practice.

5.5.1 RL-based Resource Allocation Framework

In the RL formulation of the resource sharing problem, each decision-making entity acts as an agent, concurrently exploring the unknown environment. For instance, the base station usually serves as an agent for centralized resource allocation, while each device can be treated as an agent for distributed resource allocation in, e.g., D2D communications. Mathematically, the resource allocation problem can be modeled as a Markov decision process (MDP). At each time step t, given the current environment state s_t, the agent receives an observation z_t of the environment, determined by the observation function $O(\cdot)$, i.e., $z_t = O(s_t)$, after which the agent takes an action a_t. Thereafter, the agent receives a reward r_{t+1}, and the environment evolves to the next state s_{t+1} with probability $p(s_{t+1}, r_{t+1} \mid s_t, a_t)$. The new observations z_{t+1} are then received by the agent.

RL Elements Design

The environment state, in theory, should include whatever information is available and useful for decision-making. In the case of wireless resource allocation, this state could include global channel conditions, the queue length of all links, real-time QoS requirements, behaviors of all agents, etc. However, this is usually not available to each individual link (agent) in distributed resource allocation, where each agent can only acquire knowledge of the underlying environment through the lens of an observation function. In the resource-sharing scenario, the observation z_t commonly consists of partial CSI and other extra information related to the formulated optimization goal. To better address the service requirements, extra information can be added for decision-making. For example, we can include the time budget and remaining payload as part of the observation when seeking high-reliability or low-latency transmission.

Typically, the resource-sharing design of wireless networks comes down to spectrum selection, transmit power control, or both. For spectrum selection, the action options can be a set of available frequency spectra. For power control, the transmit power, a_t, takes a continuous value in most existing power control literature. A more intricate scenario arises when both spectrum and power need to be jointly allocated.

What makes RL particularly appealing for solving problems with hard-to-optimize objectives is the flexibility in its reward design. Recall that the goal of RL is to find an optimal policy π_*, a mapping from a state to probabilities of selecting each action. It maximizes the expected return of any initial state s, where the return, denoted by G_t, is defined as the cumulative discounted rewards with a discount factor γ, $0 \le \gamma \le 1$, i.e.,

$$G_t = \sum_{k=0}^{\infty} \gamma^k r_{t+k+1}.$$

The system performance can be improved when the designed reward signal r_t at each step correlates with the desired objective. In this way, we can handle resource allocation problems involving sequential decision-making across multiple coherence time slots, which are difficult to solve using conventional optimization methods. Taking the optimization of the packet delivery rate of wireless communication as an example: For pure goal-directed consideration, we can design a reward function that is equal to zero at each step until the payload is delivered, beyond which point the reward is set to one [190]. Mathematically, this design will encourage all agents to finish delivering all payloads within each episode as early as possible. However, this sparse-reward design will hinder the learning process, since the agent can hardly learn anything useful at the beginning of each episode as it always receives a reward of zero for this period. One way to deal with this problem is to impart some prior knowledge into the reward, i.e., higher transmission rates should be helpful in improving the payload delivery rate. To be specific, the reward at each step can be set to the effective transmission rate, and when the payload is delivered successfully, the reward is set to a constant number slightly greater than the largest possible transmission rate.

Furthermore, as diverse service requirements are brought up to support new wireless devices, conventional optimization methods have difficulties in mathematically modeling the resource allocation problem in an exact form. In contrast, in RL, the diverse service needs can be balanced by setting the reward r_t as a weighted sum of various requirements (e.g., sum capacities, packet delivery rate).

Training Procedure

Similar to most deep learning algorithms, there are two stages in the RL-based resource allocation method, i.e., the training and the testing stages. The training and testing data are generated from the interactions of the environment simulator and the agents.

Figure 5.14 illustrates a general training stage of the RL-based resource allocation methods. First, the agents in the considered communication systems should be properly chosen, such as the vehicles in vehicle-to-everything (V2X) systems or the devices in the cellular D2D communications. Each agent learns a policy, which is usually a neural network parameterized by θ. Generally, RL-based resource allocation problems have an episodic setting, with each episode consisting of multiple time steps. Each episode starts with a randomly initialized environment or updated state following the last episode. The change in time steps indicates a transition of the environment state and causes the agents to adjust their actions. At time step t, the agents receive their local observations of the environment and select actions according to their policies.

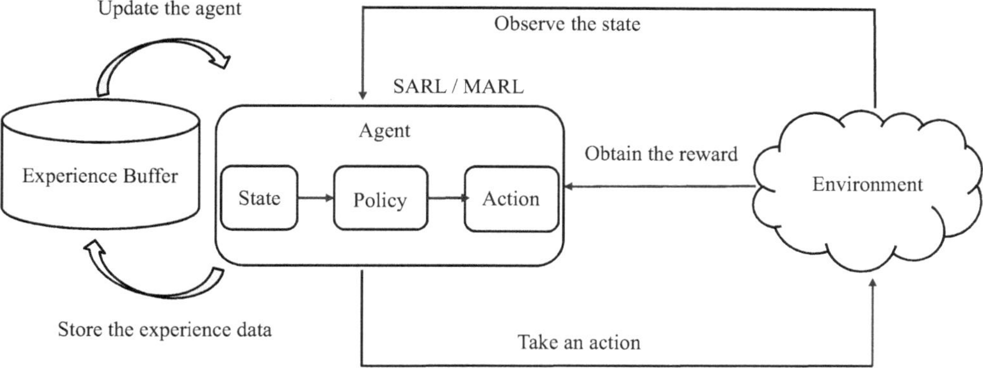

Figure 5.14 The training procedure of RL-based resource allocation.

There are two main mechanisms for training RL models, i.e., single-agent (SARL) and multi-agent (MARL). SARL means that there is only one agent in the system, and the experience data collected by the agent is used to train a neural network model. Indeed, as the action is selected independently based on the agent's local observation, the agent has no knowledge of actions selected by other agents. If the actions are updated simultaneously, the states observed by each agent cannot fully characterize the environment. To mitigate this issue, the agents are set to update their actions asynchronously in SARL, where only one or a small subset of agents will update their actions at each time step. In this way, the environmental changes caused by actions from other agents will be observable. Specifically, agents are selected one by one to change their actions. In MARL, each agent has its own policy, which is trained with the experience data collected by itself. At each time step, all agents update their actions simultaneously. To overcome the instability of the environment (i.e., the non-stationarity issue in MARL literature), additional information about other agents is needed to expand the agent's observation, with which the agent can predict other agents' actions and stabilize its training process.

In this section, we focus on the SARL mechanism. At time step t, the selected agent k obtains the current observation z_t^k. For deterministic policies (e.g., deep Q-network), the agent selects the action a_t^k with the largest Q-values $Q_{\pi_\theta}(z_t^k, a_t^k)$ calculated by the policy π_θ. For stochastic policies (e.g., policy gradient), the agent will output the probabilities of all actions according to its policy and then sample an action a_t^k from this distribution with probability $\pi_\theta(a_t^k | z_t^k)$. Afterward, the reward r_{t+1} will be generated by the environment simulator and sent to the agent, and the next local observation of environment z_{t+1}^k is obtained. The transition tuple, $(z_t^k, a_t^k, r_{t+1}, z_{t+1}^k)$, will be stored in the experience buffer. The experience data collected from one complete episode makes up one trajectory. After collecting several trajectories, this data is sampled to update the parameters. For on-policy algorithms, this data should be discarded after each update of the policy network or value network. In contrast, off-policy algorithms can reuse this collected data during the whole training stage, which improves sample efficiency. This training procedure is repeated until convergence. Algorithm 5.1 illustrates the whole training stage.

Algorithm 5.1 The Training Stage of Resource Allocation based on RL

1: Initialize the parameters of the policy θ, the environment simulator. Clear the experience buffer.

2: **while** repeat until convergence **do**

3:     **for** episode $= 1, 2, \cdots, N$ **do**

4:         **for** time step $t = 1, 2, \cdots, T$ **do**

5:             Choose an agent k in order and get its local observation z_t^k.

6:             Sample an action a_t^k according to the agent's policy π_θ.

7:             Get the reward r_{t+1}. The environment changes to the next state.

8:             Get the local observation of the next environment state z_{t+1}^k.

9:             Store $(z_t^k, a_t^k, r_{t+1}, z_{t+1}^k)$ into the experience buffer.

10:        Sample a batch of data from the experience buffer.

11:        Update the policy parameters via SGD with the gradient of the corresponding objective functions.

12:        If the algorithm is on-policy, clear the experience buffer after updating.

13: Output the trained parameters, θ.

In the testing stage, the policy of the agent is initialized with the trained parameters $\boldsymbol{\theta}$. Then, at each time step t, the agent is chosen in order and obtains its local observation of the environment z_t as input to the policy. Afterward, the agent selects an action according to the policy $\pi_\theta(a_t|z_t)$, and subsequently, the environment will change to the next state. This procedure is repeated until the testing is finished. Finally, the performance of the trained policy can be evaluated with these state–action pairs. In the next section, we will provide an extended example to show how to apply this RL framework to develop a distributed resource-allocation strategy in vehicular networks.

5.5.2   RL for V2X Spectrum Sharing: An Example

We consider spectrum-access design in vehicular networks, which in general comprise both vehicle-to-infrastructure (V2I) and vehicle-to-vehicle (V2V) connectivity [190]. Figure 5.15 shows a cellular-based vehicular communication network with M V2I and K V2V links that provides simultaneous support for mobile high-data-rate entertainment and reliable periodic safety message sharing for an advanced driving service, as discussed in 3GPP Release 15 for cellular V2X enhancement [191]. The V2I[1] links leverage cellular (Uu) interfaces to connect M vehicles to the base station for high-data-rate services, while the K V2V links disseminate periodically generated safety messages via sidelink (PC5) interfaces with localized D2D communications. We assume that all transceivers use a single antenna, and the set of V2I links and

---

[1] This V2I definition is in fact equivalent to the vehicle-to-network (V2N) definition in 3GPP documents.

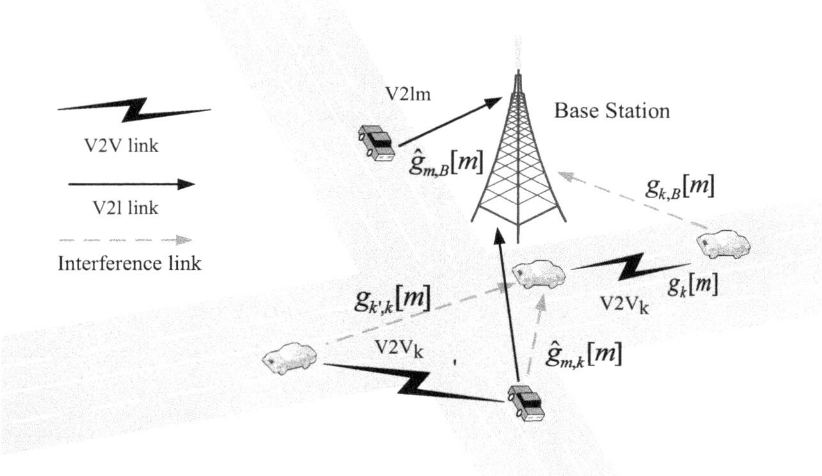

Figure 5.15 An illustrative structure of vehicular networks, where V2I and V2V links are indexed by m and k (or k'), respectively. Each of the V2I links is preassigned an orthogonal spectrum sub-band, and hence the sub-band is also indexed by m [190].

V2V links in the studied vehicular networks are denoted by $\mathcal{M} = \{1, \ldots, M\}$ and $\mathcal{K} = \{1, \ldots, K\}$, respectively.

We focus on Mode 4 for LTE-V2X or Mode 2 for NR-V2X specified in the 3GPP cellular V2X architecture [192], where V2I and V2V connections are supported through Uu and PC5 radio interfaces, respectively. In this mode, vehicles have a pool of radio resources that they can autonomously select from for V2V communications. Such resource pools can overlap with those of cellular V2I interfaces for better spectrum utilization, provided the necessary interference management design is in place. We further assume that the M V2I links (uplink considered) have been pre-assigned orthogonal spectrum sub-bands with fixed transmit power, i.e., the mth V2I link occupies the mth sub-band. As a result, the major challenge is to design an efficient spectrum-sharing scheme for V2V links such that both V2I and V2V links achieve their respective goals with minimal signaling overhead, given the strong dynamics underlying high-mobility vehicular environments.

Orthogonal frequency division multiplexing (OFDM) is exploited to convert the frequency-selective wireless channels into multiple parallel flat channels over different subcarriers. Several consecutive subcarriers are grouped to form a spectrum sub-band, and we assume channel fading is approximately the same within one sub-band and independent across different sub-bands. During one coherence time period, the channel power gain, $g_k[m]$, of the kth V2V link over the mth sub-band (occupied by the mth V2I link) follows $g_k[m] = \alpha_k h_k[m]$, where $h_k[m]$ is the frequency-dependent small-scale fading power component and assumed to be exponentially distributed with unit mean, and α_k captures the large-scale fading effect, including path loss and shadowing, assumed to be frequency independent. The interfering channel from the k'th V2V transmitter to the kth V2V receiver over the mth sub-band, $g_{k',k}[m]$, the interfering channel from the kth V2V transmitter to the BS over the mth sub-band, $g_{k,B}[m]$, the channel from the mth V2I transmitter to the BS over the mth sub-band, $\hat{g}_{m,B}[m]$, and the interfering channel from the mth V2I transmitter to the kth V2V receiver over the mth sub-band, $\hat{g}_{m,k}[m]$, are similarly defined.

The received SINRs of the mth V2I link and the kth V2V link over the mth sub-band are expressed as

$$\gamma_m^c[m] = \frac{P_m^c \hat{g}_{m,B}[m]}{\sigma^2 + \sum_k \rho_k[m] P_k^d[m] g_{k,B}[m]} \tag{5.37}$$

and

$$\gamma_k^d[m] = \frac{P_k^d[m] g_k[m]}{\sigma^2 + I_k[m]}, \tag{5.38}$$

respectively, where P_m^c and $P_k^d[m]$ denote the transmit powers of the mth V2I transmitter and the kth V2V transmitter over the mth sub-band, respectively. σ^2 is the noise power, and

$$I_k[m] = P_m^c \hat{g}_{m,k}[m] + \sum \rho_{k'}[m] P_{k'}^d[m] g_{k',k}[m] \tag{5.39}$$

denotes the interference power. $\rho_k[m]$ is the binary spectrum allocation indicator, with $\rho_k[m] = 1$ implying the kth V2V link uses the mth sub-band and $\rho_k[m] = 0$ otherwise. We assume that each V2V link only accesses one sub-band, i.e., $\sum_m \rho_k[m] = 1$.

Capacities of the mth V2I link and the kth V2V link over the mth sub-band are then obtained as

$$C_m^c[m] = W \log\left(1 + \gamma_m^c[m]\right) \tag{5.40}$$

and

$$C_k^d[m] = W \log\left(1 + \gamma_k^d[m]\right), \tag{5.41}$$

where W is the bandwidth of each spectrum sub-band. As described earlier, the V2I links are designed to support mobile high-data-rate entertainment services. Hence, an appropriate design objective is to maximize their sum capacity, defined as $\sum_m C_m^c[m]$, for smooth mobile broadband access. Meanwhile, V2V links are mainly responsible for the reliable dissemination of safety-critical messages that are generated periodically with varying frequencies depending on vehicle mobility for advanced driving services. We mathematically model such a requirement as the delivery rate of packets of size B within a time budget T as

$$\Pr\left\{\sum_{t=1}^{T} \sum_{m=1}^{M} \rho_k[m] C_k^d[m,t] \geq B/\Delta_T\right\}, k \in \mathcal{K}, \tag{5.42}$$

where B denotes the size of the periodically generated V2V payload in bits, Δ_T is channel coherence time, and the index t is added in $C_k^d[m,t]$ to indicate the capacity of the kth V2V link at different coherence time slots.

Formally, the resource allocation problem is stated as follows: Design the V2V spectrum allocation, expressed through binary variables $\rho_k[m]$ for all $k \in \mathcal{K}, m \in \mathcal{M}$, and the V2V transmit power, $P_k^d[m]$ for all $k \in \mathcal{K}, m \in \mathcal{M}$, to simultaneously maximize the sum capacity of all V2I links $\sum_m C_m^c[m]$ and the packet delivery rate of V2V links defined in (5.42).

Observation Space

In the SARL formulation of the resource-sharing problem, each V2V link k acts as an agent, and the observation space of an individual V2V agent k contains local channel information, including strength of its own signal channel, $g_k[m]$, for all $m \in \mathcal{M}$, interference channels from other V2V transmitters, $g_{k',k}[m]$, for all $k' \neq k, m \in \mathcal{M}$, the interference channel from its own transmitter to the BS, $g_{k,B}[m]$, for all $m \in \mathcal{M}$ and the interference channel from all V2I transmitters, $\hat{g}_{m,k}[m]$, for all $m \in \mathcal{M}$. The received interference power over all bands, $I_k[m]$, for all $m \in \mathcal{M}$, expressed in (5.39), can be measured at the V2V receiver and is also introduced in the local observation. In addition, the local observation space includes the remaining V2V payload, B_k, and the remaining time budget, T_k, to better capture the queuing states of each V2V link. As a result, the observation function for an agent k is summarized as

$$Z_t^k = \left\{ B_k, T_k, \{I_k[m]\}_{m \in \mathcal{M}}, \{G_k[m]\}_{m \in \mathcal{M}} \right\}, \tag{5.43}$$

with $G_k[m] = \left\{ g_k[m], g_{k',k}[m], g_{k,B}[m], \hat{g}_{m,k}[m] \right\}$.

Action Space

The resource-sharing design of vehicular links can be attributed to the spectrum sub-band selection and transmit power control for V2V links. While the spectrum naturally breaks into M disjoint sub-bands, each preoccupied with one V2I link, the V2V transmit power typically takes a continuous value in most existing power-control literature. Here, the power-control options are discretized to four levels, i.e., [23, 10, 5, −100] dBm, for the sake of both ease of learning and practical circuit restriction. It is noted that the choice of −100 dBm effectively means zero V2V transmit power. As a result, the dimension of the action space is $4 \times M$, with each action corresponding to one particular combination of a spectrum sub-band and power selection.

Reward Function

In this investigated V2X spectrum-sharing problem, the objectives are twofold: maximizing the sum V2I capacity while increasing the success probability of V2V payload delivery within a specific time constraint T.

In response to the first objective, the instantaneous sum capacity of all V2I links $\sum_{m \in \mathcal{M}} C_m^c[m, t]$ is simply included, as defined in (5.40), in the reward at each time step t. To achieve the second objective, for each agent k, the reward L_k is set as the effective V2V transmission rate until the payload is delivered, after which the reward is set to a constant number, β, that is greater than the largest possible V2V transmission rate. As such, the V2V-related reward at each time step t is set as

$$
L_k(t) = \begin{cases} \sum_{m=1}^{M} \rho_k[m] C_k^d[m, t], & \text{if } B_k > 0, \\ \beta, & \text{otherwise,} \end{cases}
\tag{5.44}
$$

where β can be approximated by rolling out several episodes and selecting the largest rate among these time slots.

As a result, the reward at each time step t is set as

$$
r_{t+1} = \lambda_c \sum_{m} C_m^c[m, t] + \lambda_d \sum_{k} L_k(t),
\tag{5.45}
$$

where λ_c and λ_d are positive weights to balance V2I and V2V objectives.

Learning Algorithm

Afterward, the proximal policy optimization (PPO) algorithm introduced in Section 1.2.3 is used to train the RL model for the spectrum-sharing problem. Each V2V link in this system is considered as an agent, and SARL is used to stabilize the training process. In both stages, we focus on an episodic setting, with each episode spanning the V2V payload delivery time constraint T. Each episode starts with a randomly initialized environment state (determined by the initial transmit powers of all vehicular links, channel state, etc.) and a full V2V payload of size B for transmission that lasts until the end of T. The change in small-scale channel fading triggers a transition of the environment state and causes each V2V agent to adjust its action. By simply modifying Algorithm 5.1, we can obtain the training procedure of the PPO-based resource-allocation method.

Performance Analysis

The simulator is custom built and adheres to the evaluation methodology for the urban case defined in [192] and [193], which describes in detail vehicle drop models, densities, speeds, the direction of movement, vehicular channels, V2V data traffic, etc. The number of vehicles in the system is set as $M = 4$, and the number of V2V links is set as $K = 4$. The time constraint for V2V payload transmission is $T = 100$ ms.

Each V2V agent's actor and critic networks comprise three fully connected hidden layers, with 500, 250, and 120 neurons, respectively. ReLU is utilized as the activation function. A softmax layer is added after the last layer of the actor network to obtain the probabilities of all actions. The discount rate is $\gamma = 0.99$. The cumulative rewards per training episode with increasing training iterations are shown in Figure 5.16 to demonstrate the convergence behavior of the PPO-based resource allocation method. From the figure, the average cumulative rewards improve as training continues, and when the training episode reaches approximately 3000, the performance gradually

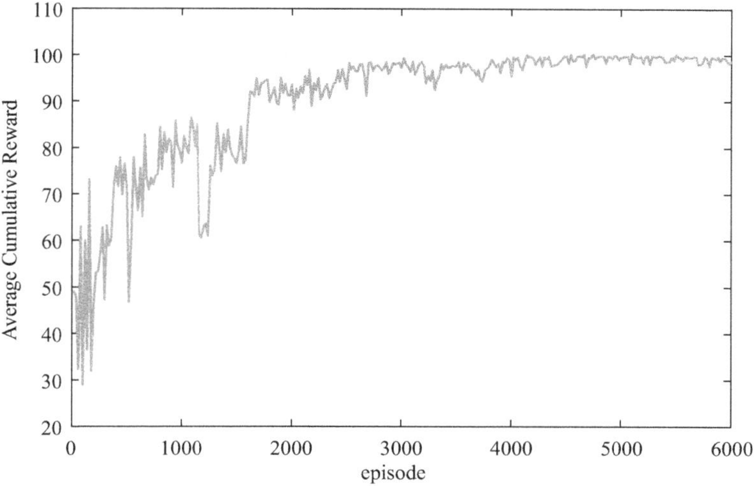

Figure 5.16 The convergence of the PPO algorithm.

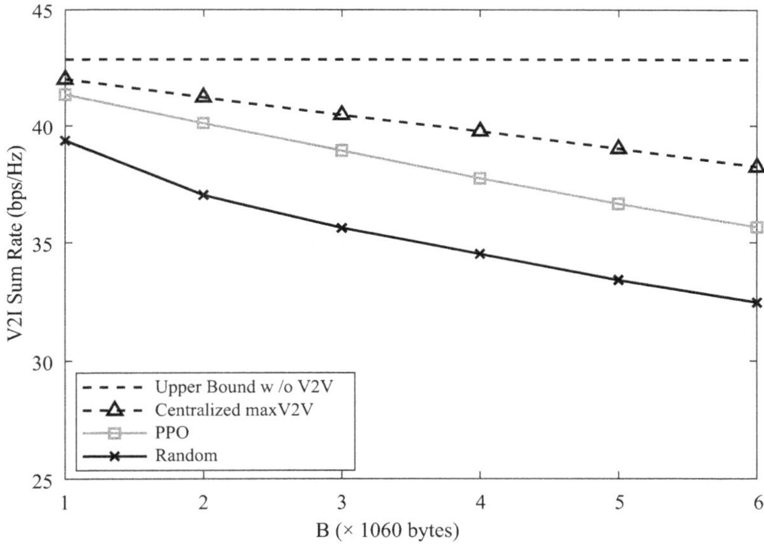

Figure 5.17 Sum capacity performance of V2I links with varying V2V payload sizes B.

converges despite some fluctuations due to mobility-induced channel fading in vehicular networks. In Figure 5.17 and 5.18, the PPO-based resource allocation scheme, termed PPO, is compared against the following three baseline methods.

1) Upper bound without V2V: The transmission of all V2V links is disabled to obtain the upper bound of V2I performance. In this case, the packet delivery rates for all V2V links are zero, thus not shown in Figure 5.18.

2) Centralized maxV2V: The action space of all K V2V agents in each step is exhaustively searched to maximize sum V2V rates. This method breaks the sequential decision-making of delivering B bytes over multiple steps within the time constraint T into separate optimizations of sum V2V rates over each step. Apart from the complexity, this scheme needs to be performed in a centralized way with accurate global CSI available.

3) Random baseline: The spectrum sub-band and transmit power for each V2V link are chosen in a random fashion at each time step.

Figure 5.17 shows the V2I performance with respect to increasing V2V payload sizes B for different resource allocation designs. From the figure, the performance drops for all schemes (except the upper bound) with growing V2V payload sizes. An increase of V2V payload leads to longer V2V transmission duration and possibly higher V2V transmit power to improve V2V payload transmission success probability. This will inevitably cause stronger interference to V2I links for a more extended period and thus jeopardize their capacity performance. The PPO-based method performs better than the random baseline and performs measurably close to the V2I performance upper bound, within 16.7% degradation in the worst case of 6×1060 bytes of payload. The centralized maxV2V scheme attains remarkable performance in terms of V2I performance. This could be attributed to the packet delivery rates of V2V links have been substantially enhanced with centralized maxV2V, and the V2V links incur no interference to V2I links once their payload delivery has finished. The PPO-based scheme performs close to the centralized maxV2V scheme within 6.72% degradation in the worst case for the V2I sum rate, demonstrating its efficiency. Besides, the PPO-based scheme only requires local CSI, and the computational complexity is relatively low in deployment as it merely needs a forward pass through the trained policy network.

Figure 5.18 shows the probability of V2V payload delivery against growing payload sizes B. As the V2V payload size grows, the transmission success probabilities drop for the PPO-based method and the random baseline, while the idealistic centralized maxV2V can achieve 100% packet delivery throughout the tested cases. The PPO-based method performs significantly better than the random baseline and stays close to the centralized maxV2V scheme. In conjunction with the observations from Figure 5.17 and Figure 5.18, we conclude that the PPO-based method achieves better performance across different V2V payload sizes though it is trained with a fixed size of 2×1060 bytes, demonstrating its robustness to a range of V2V payload sizes.

5.5.3 Meta-RL for Fast Resource Management

Although RL presents a promising solution for addressing resource allocation problems, several challenging issues remain unresolved. First, a simulation environment is typically needed as training a well-performing policy necessitates extensive interactions with the environment. When the real and simulation environments are identical, the policy trained in simulators can attain desirable performance. However,

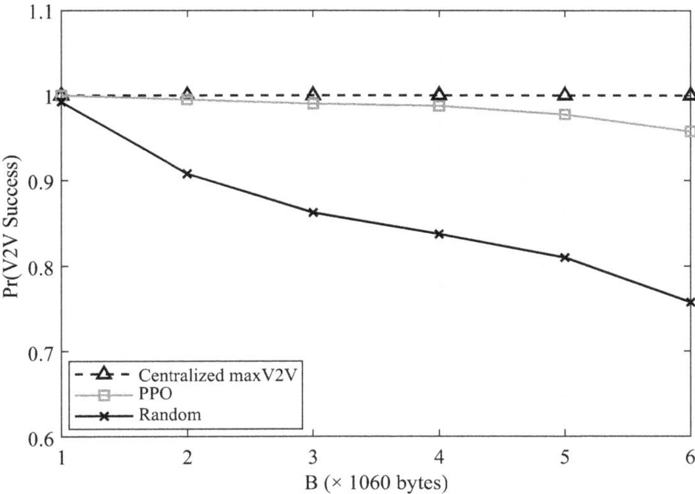

Figure 5.18 V2V payload transmission success probability with varying payload sizes B.

differences between these two environments, known as the reality gap, may reduce the efficacy of a trained agent that is high performing in simulation environments. Second, training a policy directly in the real environment is often infeasible due to the requirement of an extensive number of interactions. Furthermore, an immature policy is inappropriate for collecting the experience data as it may occasionally lead to catastrophic consequences in practice.

Inspired by the human ability to rapidly learn new tasks through experience with similar ones, meta-learning offers a powerful approach to improving policy generalization. Gradient-based meta-learning is one of the most popular meta-learning algorithms, aiming to obtain an initialization that enables the policy to adapt to a new task with minimal gradient descents by training across many similar tasks. In this section, we will introduce two widely used meta-learning algorithms and then combine meta-learning with RL to solve the resource allocation problem.

Model-Agnostic Meta-Learning

One of the most popular gradient-based meta-learning algorithms is model-agnostic meta-learning (MAML), proposed in [194]. This method can learn the parameters of neural networks via meta-learning for fast adaptation. The intuition of this method is that some internal representations are transferable.

Assume that in the new task, the neural networks will be fine-tuned with limited samples using gradient-based learning rules. The target of meta-learning is to learn the parameters that can serve as a good initialization for new tasks, allowing the model to make rapid improvements without overfitting. Consider a neural network in which the parameters are represented by θ. When adapting to a new task (such as regression problems or RL problems) $\mathcal{T}_i$, the model's parameters become θ_i', i.e.,

$$\theta_i' = \theta - \alpha \nabla_\theta \mathcal{L}_{\mathcal{T}_i}(\theta), \tag{5.46}$$

where α is the step size. $\mathcal{L}_{\mathcal{T}_i}(\theta)$ denotes the objective function of the task $\mathcal{T}_i$ that is calculated using the result generated by the neural networks parameterized by θ. Particularly, it denotes the reward function in RL problems. Moreover, using multiple gradient steps is also feasible. This procedure is called the inner loop in meta-learning. The parameters are trained via optimizing for the performance of θ_i' across tasks sampled from the task distribution, $p(\mathcal{T})$. Therefore, the optimization problem of MAML can be formulated as

$$\min_\theta \sum_{\mathcal{T}_i \sim p(\mathcal{T})} \mathcal{L}_{\mathcal{T}_i}(\theta_i'). \tag{5.47}$$

Through this objective, MAML aims to achieve maximally effective behavior on a new task using just one or a small number of gradient steps. The model parameters are updated via stochastic gradient descent,

$$\theta = \theta - \beta \nabla_\theta \sum_{\mathcal{T}_i \sim p(\mathcal{T})} \mathcal{L}_{\mathcal{T}_i}(\theta_i'), \tag{5.48}$$

where β is the learning rate of meta-learning. This procedure is called the outer loop in meta-learning. MAML requires an additional backward process as the meta-optimization is performed over the parameters θ, while the objective is computed using the updated parameters θ'. Therefore, the computational complexity is relatively high. By ignoring the second-order derivative terms, first-order MAML (FOMAML) uses the first-order approximation at the expense of losing some gradient information. In FOMAML, the model parameters are updated as

$$\theta = \theta - \beta \sum_{\mathcal{T}_i \sim p(\mathcal{T})} \nabla_{\theta_i'} \mathcal{L}_{\mathcal{T}_i}(\theta_i'). \tag{5.49}$$

It is shown in [194] that FOMAML works nearly as well as MAML in regression and classification tasks.

MAML can also be applied to RL to enable an agent to obtain a policy for a new task with only a small amount of experience data. Generally, policy gradient methods are used to evaluate the gradient for the inner and outer loops. In the outer loop, the additional gradient step of MAML requires new samples from the updated policy θ_i'. The detailed algorithm is shown in Algorithm 5.2.

Reptile

Another first-order gradient-based meta-learning method proposed in [195], called Reptile, also learns an initialization for the parameters of a neural network model, such that the model can adapt fast with a small number of examples in new tasks. As a variant of MAML, Reptile changes the update rule of the parameters to

$$\theta = \theta + \beta \frac{1}{N} \sum_{i=1}^{N} (\tilde{\theta}_i - \theta), \tag{5.50}$$

Algorithm 5.2 MAML for RL

    Input: a meta-training task set: $\{\mathcal{T}\} \sim p(\mathcal{T})$, the learning rates for inner and outer loops: (α, β).
    Output: meta-parameters: θ.
1: Randomly initialize the meta-parameters θ.
2: **while** not done **do**
3:     Sample a batch of tasks $\mathcal{T}_i$ from $\{\mathcal{T}\}$.
4:     **for** each $\mathcal{T}_i$ **do**
5:         Sample K trajectories $\mathcal{D}$ using π_θ in $\mathcal{T}_i$.
6:         Compute $\nabla_\theta \mathcal{L}_{\mathcal{T}_i}(\theta)$ using $\mathcal{D}$.
7:         Update the model parameters: $\theta_i' = \theta - \alpha \nabla_\theta \mathcal{L}_{\mathcal{T}_i}(\theta)$.
8:         Sample new trajectories $\mathcal{D}_i'$ using $\pi_{\theta_i'}$ in $\mathcal{T}_i$.
9:         Compute $\mathcal{L}_{\mathcal{T}_i}(\theta_i')$ using $\mathcal{D}_i'$.
10:    Update meta-parameters: $\theta = \theta - \beta \nabla_\theta \sum_{\mathcal{T}_i} \mathcal{L}_{\mathcal{T}_i}(\theta_i')$.

---

where $\tilde{\theta}_i$ represents the updated parameters on the ith task, which are updated k times using the data sample from $\mathcal{T}_i$. N denotes the number of sampled tasks.

According to [195], when the number of gradient steps in the inner loop is large enough, Reptile can achieve similar performance to MAML while significantly decreasing the computational complexity. Compared to FOMAML, Reptile can benefit from taking many inner loop steps because FOMAML only uses information of the final gradient, while Reptile utilizes the information of all gradient steps. Applying Reptile to RL is similar to Algorithm 5.2, and the main difference is that extra sampling for the additional gradient computation is unnecessary.

Resource Allocation based on Meta-RL

Meta-RL aims to obtain the parameters of an RL policy network through training the agent across a large number of tasks to help the agent adapt faster when deployed in the real world. Many similar tasks can be generated by adjusting the values of some key factors of the environment, and the task mentioned here refers to learning an efficient policy for resource allocation in a specific environment.

Meta-RL-based resource allocation consists of two stages: the meta-training stage, which learns the meta parameters across a variety of similar tasks, and the adaptation stage, which commences by initializing the policy network with the meta parameters and then adapts for a few episodes in the new environment. The meta-training stage includes two loops: the inner and outer loops. The inner loop updates the policy parameters with the experience data from a specific environment. The inner loop typically addresses a standard RL problem, focusing on a specific resource allocation problem. Policy-gradient algorithms, for example, the PPO algorithm discussed in Section 1.2.3, can be used directly in the inner loop. The outer loop amalgamates the updated parameters from different inner loops and updates the meta-parameters.

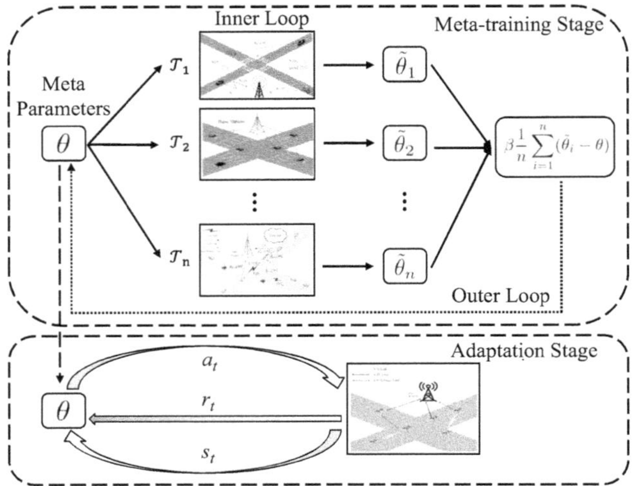

Figure 5.19 Illustration of meta-RL-based resource allocation.

Through this amalgamation, the generalization of the meta-parameters for this class of resource-allocation problems is improved. MAML, Reptile, or other meta-learning algorithms can be used in the outer loop to obtain the gradient of the meta-optimization. Figure 5.19 shows the process of a meta-RL-based resource allocation that uses Reptile to amalgamate the results of multiple inner loops.

Next an example will be provided to show how the spectrum-sharing design for vehicular networks can be enhanced with meta-RL.

Example 5.4 Consider the spectrum-sharing problem in vehicular networks described in Section 5.5.2, but the training and testing environment are different here, leading to the so-called "reality gap." Five key factors of the vehicular environment are considered: the number of V2V links that each vehicle constructs with its neighboring vehicles, the size of the V2V payload, the average velocity of the vehicles, and the fast-fading types of these two links. For the fast-fading type, only Ricean fading is considered for simplicity, of which the Ricean factor is adjustable. The simulators and neural network models are identical to those in Section 5.5.2. Before meta-training, a meta-training task set is generated by adjusting the values of these five factors. During training, a batch of tasks will be sampled from this task set for each outer loop. After training for enough outer loops, the meta-parameters will be learned.

Figure 5.20 investigates the performance of a meta-RL-based resource allocation scheme with varying numbers of gradient descents. Here, only the number of gradient steps in the adaptation changes, while the number of gradient steps for meta-training remains fixed. This illustrates that a model trained with meta-RL can continue to improve with additional gradient steps, as is the case in [194]. Two baselines are considered:

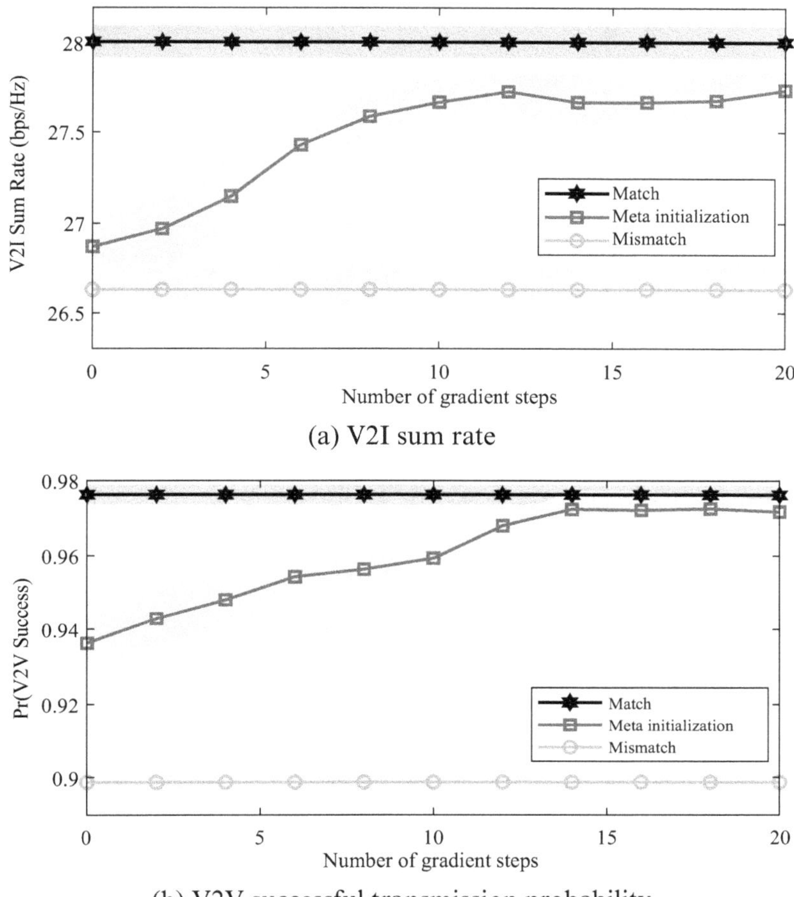

(a) V2I sum rate

(b) V2V successful transmission probability

Figure 5.20 Performance with varying gradient steps in adaptation.

- Matched policy: a policy trained for 3000 episodes in this new task.
- Mismatched policy: a policy trained for 3000 episodes in a separate task, of which 5 main factors are different from this new task.

Figure 5.20 indicates that an increase in the number of gradient descents during adaptation leads to improvement in both the sum rate of the V2I links and the successful transmission probability of the V2V links. For a policy initialized with meta parameters, about 14 gradient descents are needed in this new task to attain performance close to the matched policy.

To assess the performance within a new environment, a comparison of the meta-RL-based method with several baselines is shown in Figure 5.21. The meta-RL-based resource allocation method is designated as meta initialization (20 episodes), indicating that the policy is initialized with the meta parameters, followed by 20 episodes for adaptation to the new task. In addition to matched and mismatched policies, randomly initialized policies are also considered, which are trained using PPO methods. The number of episodes for training is shown inside the bracket.

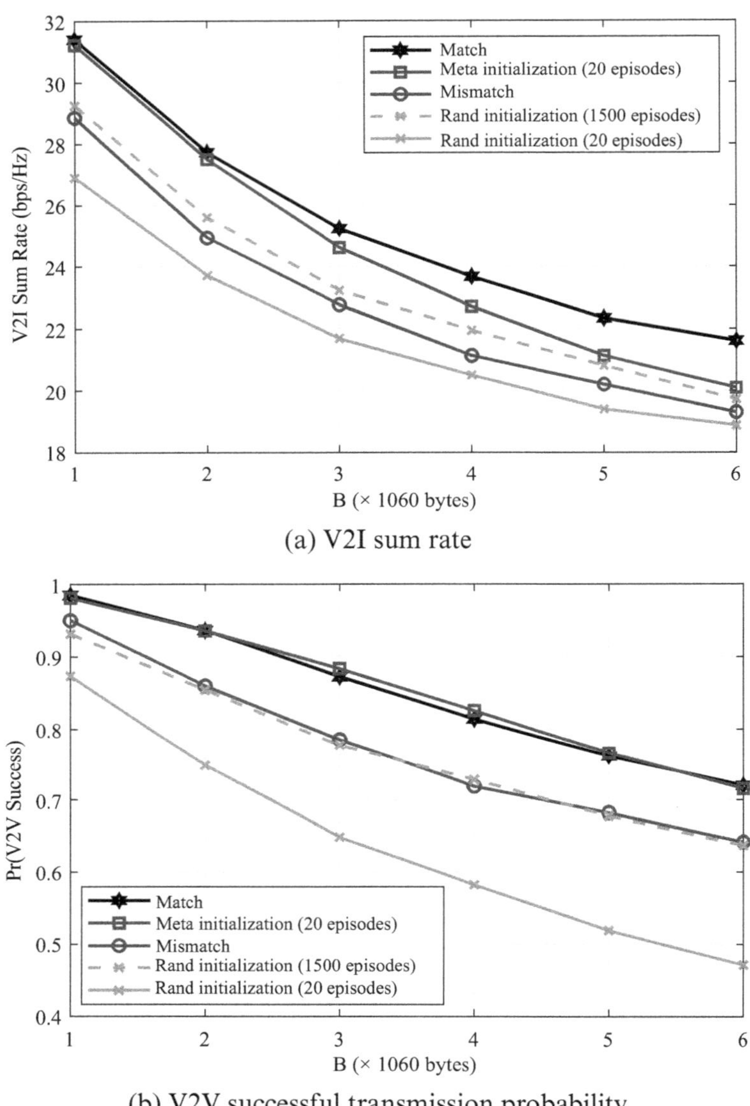

(a) V2I sum rate

(b) V2V successful transmission probability

Figure 5.21 Performance with varying size of V2V payload B (The number of V2V links is $K = 8$).

According to Figure 5.21, the meta-parameter-initialized policy, which subsequently rolled out 20 episodes for adaptation, closely follows the performance of the matched policy that has been trained in the same environment for 3000 episodes. It suffers about 6.8% degradation in the sum rate of V2I links compared with the matched policy even in the worst case (6 $\times$ 1060 bytes of payload) and still outperforms the mismatched policy by about 4.95%. In terms of successful transmission probability of V2V links, the meta-RL-based method attains a performance close to the matched policy. The mismatched policy suffers about 9% degradation in both performance metrics, demonstrating the necessity of adopting meta-RL to tackle the reality-gap problem. In the case of a random-initialized policy, 20 episodes of adaptation are grossly insufficient for adapting to the new environment effectively. For the random-initialized policy, 1500 training episodes are necessary to achieve a reasonable performance in the current setting. To conclude, an efficient resource-allocation scheme can be obtained with fewer interactions with this meta-RL method, which is much faster than letting the policy learn from scratch in the new environment.

5.6   Exercises

Exercise 5.1   Consider a wireless network with $N = 3$ links indexed as $k = 1$, 2, 3. The link k is characterized by its transmit power P_k. The channel power gains are given in Table 5.2. The maximum transmit power for each link is $P_{\mathrm{max}} = 10$ dBm, and the noise power at each receiver is $\sigma^2 = 0$ dBm. The weights for all links are $w_1 = w_2 = w_3 = 1$.

(a) If two links can be activated simultaneously, calculate the sum rate for each possible pair of active links. Identify which pair of links, when activated together, achieve the maximum sum rate.
(b) Calculate the sum rate when all three links are activated simultaneously. Compare this with the previous results to determine the most efficient link-scheduling strategy to maximize the sum rate.

Table 5.2 Channel power gains between transmitter j and receiver k.

Transmitter j/Receiver k	1	2	3
1	2.0	1.0	0.5
2	1.5	3.0	0.8
3	0.7	1.2	4.0

Exercise 5.2   The link-scheduling problem can be viewed as a specific instance of the power-allocation problem. Can a trained power-allocation network be used for link scheduling? If so, how is this implemented?

Exercise 5.3   Discuss how supervised learning, unsupervised learning, and RL can optimize the resource-allocation problem in wireless communication. What are the advantages and disadvantages of each approach?

Exercise 5.4   In the Lagrange multiplier method, why is the primal problem transformed into a dual problem for solution? What is the relationship between the solution of the dual problem and the solution of the primal problem? Under what conditions are the two solutions equal?

Exercise 5.5   Assume that the channel fading obeys Rayleigh distribution with zero mean and unit variance. As shown in Table 5.3, the channel power gains of four users

Table 5.3 Channel power gains between transmitter j and receiver k.

Transmitter j/Receiver k	1	2	3	4
1	1.616	0.284	0.704	1.919
2	0.699	0.723	0.636	1.403
3	0.150	1.346	1.006	0.287
4	0.778	1.457	1.215	1.397

are given (consider small-scale fading only for simplicity). Assume that the maximum transmit power per user is normalized to 1. Program the WMMSE algorithm to solve the user power allocation problem. The code snippet has been provided in https://github.com/le-liang/wcmlbook/blob/main/ch5/Exercise_5.5/.

Exercise 5.6 Sections 5.2 and 5.3 introduce supervised learning and unsupervised learning approaches for optimization in wireless networks. Complete the following programming tasks using the Python and Pytorch deep learning framework. The code snippet has been provided in https://github.com/le-liang/wcmlbook/blob/main/ch5/Exercise_5.6/.

(a) Use the supervised learning approach to solve the power-allocation problem in (5.3). Randomly generate Rayleigh-distributed channel coefficients for 10 pairs of D2D users. Use these as training data for a fully connected network. Compare the results with the WMMSE algorithm.
(b) Use the unsupervised learning method to solve the above task with and without QoS constraints. Retrain the model and compare the results with the supervised learning approach.

Exercise 5.7 In GNN, each iteration of the message-passing process aggregates 1-hop neighborhood information. Therefore, more comprehensive information on the graph structure can be obtained through several iterations. Discuss whether or not the GNN model performs better with more iterations.

Exercise 5.8 Section 5.3 considers a fully connected graph and utilizes GINs for power control in wireless networks. However, such a method invariably requires accurate CSI, which is impractical for a densely deployed network due to the significant CSI acquisition and feedback overhead. A simple solution is neglecting the interference caused by transmitter j to receiver i if the distance between them is above a certain threshold. Consider the wireless network in Figure 5.11b where an edge is drawn from node j to node i if there is a direct communication or interference link with node j as the transmitter and node i as the receiver. Given the threshold $D = 50$ m, try to regenerate the graph model according to Table 5.4 and determine the following:

(a) Adjacency matrix and edge list of the graph.
(b) Node degree of each node in the graph.

Table 5.4 Distances (m) between transmitter j and receiver i.

Transmitter j/Receiver k	1	2	3	4
1	/	26.96	24.38	52.96
2	26.96	/	34.96	37.96
3	24.38	34.96	/	38.43
4	52.96	37.96	38.43	/

Exercise 5.9   According to the generated graph in Exercise 5.8, try to use the message-passing framework in PyG to implement the GAT architecture and conduct resource management. The channel model and the training code snippet has been provided in https://github.com/le-liang/wcmlbook/blob/main/ch5/Exercise_5.9/.

(a) Compare your results with the performance of the same GNN architecture but equipped with complete CSI.
(b) Change the threshold D, regenerate the graph model, and retrain the GAT network for power control. Draw its performance with varying threshold D.

Exercise 5.10   The spectrum-sharing problem considered in this chapter is modeled as a discrete RL problem, as the power control options are limited to four discrete levels. PPO can tackle continuous learning because it is a kind of policy-gradient method. Therefore, the spectrum-sharing problem can be extended to a hybrid optimization problem, which chooses a discrete sub-band and a continuous power value from $(-100, 23)$ dBm to transmit. The code snippet has been provided in https://github.com/le-liang/wcmlbook/blob/main/ch5/Exercise_5.10/.

(a) Design the policy architecture properly to handle this situation.
(b) Compare your results with Figure 5.17 and Figure 5.18. Draw your conclusion.

Exercise 5.11   What problems in the field of deep RL can be addressed by meta-RL? In meta-RL, what are the inner loop and outer loop used for, respectively?

Exercise 5.12   Refer to Algorithm 5.2 to understand how MAML is applied to RL, and write the pseudo-code of Reptile for RL.

Exercise 5.13   As shown in the formulation (5.48), MAML requires an extra backward process as the meta-optimization is performed over the meta-parameters θ, while the objective is obtained using the parameters θ' updated in the inner loop. Show that by ignoring the second-derivative terms, the formulation (5.48) can be simplified to (5.49).

Exercise 5.14   Section 5.4.3 introduces the basic gradient-based meta-learning methods and utilizes the Reptile algorithm to solve the resource-allocation problems in vehicular networks. However, many other efficient meta-learning algorithms are used in resource allocation problems. Two recommended papers are listed below, and the reader can utilize these meta-learning algorithms to solve the spectrum sharing problem described in Section 5.5.2. Refer to Example 5.4 to explore how spectrum sharing methods based on these meta-learning methods can reduce the number of interactions or the number of training samples. You can draw figures such as Figure 5.21 to demonstrate the effect of these methods. The code snippet has been provided in https://github.com/le-liang/wcmlbook/blob/main/ch5/Exercise_5.14/.

(a) The context-based meta-learning method CAVIA is provided in https://proceedings.mlr.press/v97/zintgraf19a.html [196].
(b) The modular meta-learning method is provided in https://proceedings.mlr.press/v87/alet18a.html [197].

6   Wireless for AI: Distributed and Federated Learning

Traditional machine learning methods typically involve gathering training data, performing model training on a central server, and then providing users with access to the trained model to support various applications. However, with the widespread use of mobile devices, vast amounts of data are distributed across various devices, making centralization for model training a significant challenge. This centralized approach not only incurs high communication costs due to the need to transfer large volumes of data but also raises serious privacy concerns, as sensitive data must be transmitted to the server. Federated learning addresses these challenges by enabling multiple devices, referred to as clients in this context, to collaboratively train a shared global model while keeping the raw data stored locally on each device. Instead of sharing raw data, federated learning exchanges only the updated local models between clients and the server during training. This decentralized approach preserves data privacy while significantly reducing communication overhead, offering a more efficient and secure alternative to centralized machine learning.

This chapter provides an overview of federated learning over wireless networks, beginning with the foundational concepts and classic algorithms in federated learning. Subsequently, we explore how federated learning integrates with communication design, focusing on optimizing client selection and resource allocation to improve communication efficiency. We then progress to decentralized federated learning scenarios where no central server is available. Furthermore, we delve into over-the-air computation for efficient model aggregation in federated learning, leveraging the superposition property of multiple access channels. Finally, we discuss the application of the alternating direction method of multipliers (ADMM) framework in federated learning.

6.1   Wireless Federated Learning

This section explains the basic principles and commonly used algorithms in federated learning. We start by outlining the key concepts of federated learning, including its structure and training methods. Then, we provide an overview of three classic algorithms: federated stochastic gradient descent (FedSGD), federated averaging (FedAvg) [46], and federated proximal (FedProx) [198].

6.1.1 Basic Concept

First introduced by Google in 2016 [46], federated learning represents a paradigm shift in collaborative machine learning by enabling distributed model training across clients without requiring them to share their local data. In this approach, each client updates the model locally using its own data, and only the updated model parameters or gradients are sent to a central server. The server aggregates these updates to create a unified global model, which is then shared back with the clients.

The core principles of federated learning are distributed training and model aggregation. Distributed training allows clients to process data locally, ensuring data privacy by keeping sensitive information on the clients. The central server performs model aggregation, combining the locally computed updates into a single global model that benefits from the diverse data held by different clients. By combining these principles, federated learning enhances the training performance of the machine learning task while maintaining strict data privacy, making it a powerful solution for data-sensitive applications.

How Federated Learning Works

Federated learning operates through a structured process comprising several key steps:

1. **Initialization:** The central server initializes a global model. This model serves as the starting point for all participating clients in the federated learning network.
2. **Global model broadcasting:** The central server broadcasts the current global model to all participating clients, ensuring a uniform starting point for training.
3. **Local training and uploading:** Each participating device trains the global model locally using its private data. This localized training allows the model to adapt to the specific data distribution of each device. After local training is completed, only model updates (e.g., gradients or weights) are transmitted back to the central server.
4. **Global model aggregation:** Upon receiving model updates from all participating clients, the central server aggregates these updates to refine and improve the global model.

Steps 2–4 are repeated until the model converges or reaches the desired accuracy, as illustrated in Figure 6.1.

Challenges in Federated Learning

Federated learning, as a subset of distributed learning, enables model training across distributed clients while preserving data privacy by avoiding centralizing the data. Unlike traditional distributed learning, federated learning introduces distinct challenges for optimization due to its unique operational environment. These challenges

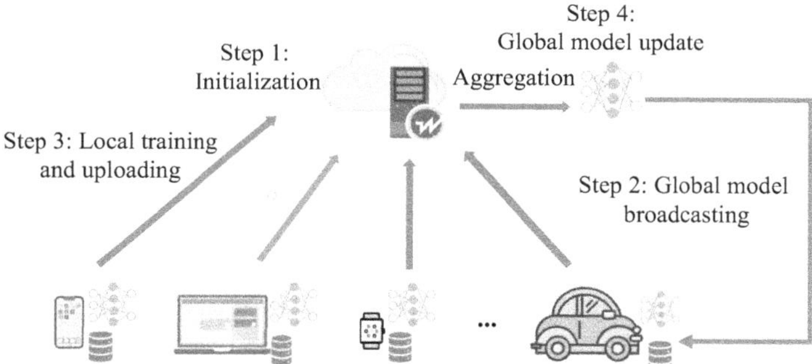

Figure 6.1 The basic workflow of federated learning.

arise from the heterogeneous data distributions, wide range of device connectivity, and limited communication resources. Below, we outline the primary challenges inherent in federated learning:

- **Non-i.i.d. data:** The data distribution among various clients is not independent and identically distributed (i.i.d.). Each local dataset can significantly deviate from the overall population distribution. This non-i.i.d. nature complicates model aggregation, making it challenging to derive unified insights from datasets that do not represent a common underlying distribution.
- **Unbalanced data volumes:** Owing to diverse usage patterns among users, the data volumes available across clients can vary significantly. Some clients may possess substantial amounts of data, while others may contribute only a negligible amount. This stark imbalance complicates model optimization as traditional methods often overlook the uneven contributions from different clients. Consequently, models can become biased toward clients with larger datasets, thereby limiting the training performance.
- **Massively distributed clients:** Federated learning can engage far more clients compared with traditional distributed machine learning. This massive scale of participants introduces significant scalability challenges, requiring robust optimization techniques to efficiently manage and coordinate such a large, diverse set of participants.
- **Limited communication resources:** In federated learning, clients are often spread across diverse geographical regions and connect to a central server via remote and sometimes unstable networks. This limitation increases the risk of communication errors or distortions, which may degrade the quality of model updates compared to traditional distributed learning systems that operate under more stable network conditions.

Federated optimization has been developed to address these unique challenges. Consequently, by employing tailored strategies for these problems, federated optimization ensures learning models are trained efficiently in a distributed and collaborative manner, thereby improving the learning performance.

Wireless Communication in Federated Learning

In the federated learning workflow, as shown in Figure 6.1, communication emerges as a pivotal component, which however often becomes the major bottleneck in practical implementations. In particular, reliance on wireless communication significantly affects the efficiency and reliability of model training in federated learning due to the unpredictable nature of wireless channels.

Wireless federated learning involves extensive data exchanges between clients and the central server, leading to substantial communication overhead. In environments with limited or shared bandwidth, effectively managing the amount of transmitted data is crucial to maintaining the speed and reliability of the learning process. To tackle these challenges, various optimization strategies are employed, such as compressing model updates to reduce data size, lowering the number of communication rounds, and selectively choosing which clients participate in each round. Additionally, the strong dynamics and unique challenges in wireless communications, such as noise, fading, interference, and mobility, can lead to unreliable model uploading. Consequently, such variability in communication quality can impede model convergence and degrade the training performance. To mitigate these issues, adaptive communication protocols can be designed to adjust transmission methods based on real-time network conditions. We will explore these topics in Section 6.2.

6.1.2 Classical Algorithms in Federated Learning

In this part, we review three classical algorithms: FedSGD, FedAvg [46], and FedProx [198]. FedSGD represents a basic approach to federated optimization but falls short in handling challenges such as communication efficiency, data heterogeneity, and system variability. In contrast, FedAvg introduces a model-averaging strategy to enhance communication efficiency, and FedProx incorporates regularization techniques to better handle data and system heterogeneity.

We consider a federated learning scenario with N clients, where each client i maintains a local dataset $\mathcal{D}_i$, consisting of $n_i = |\mathcal{D}_i|$ data samples. Each dataset is used to compute a local objective $f_i(\cdot)$, defined as

$$f_i(\mathbf{x}) \triangleq \mathbb{E}_{\xi_i \sim \mathcal{D}_i} \left[\ell_i(\mathbf{x}, \xi_i) \right], \tag{6.1}$$

where $\mathbf{x} \in \mathbb{R}^d$ represents the vector of parameters for a machine learning model. ξ_i denotes a mini-batch of size B, which is randomly sampled from the local dataset $\mathcal{D}_i$. The function $\ell_i(\cdot)$ specifies the loss function associated with the training task, used to evaluate the performance of the model parameters $\mathbf{x}$ based on the data in the mini-batch ξ_i.

The global objective, denoted by $f(\cdot)$, represents the weighted average of the local objectives $f_i(\cdot)$ from each of the N clients. It is mathematically formulated as

$$f(\mathbf{x}) = \sum_{i=1}^{N} w_i f_i(\mathbf{x}), \tag{6.2}$$

Algorithm 6.1 FedSGD: SGD-Based Federated Learning

1: Initialize $\mathbf{x}_i^0 = \mathbf{x}^0$, for all $i \in \{1, 2, \ldots, N\}$, a learning rate $\eta > 0$. Set $t \Leftarrow 0$.
2: **for** each round $t = 0, 1, 2, \ldots$ **do**
3:     Global model broadcasting: The central server broadcasts the model parameter $\mathbf{x}^t$ to every local client.
4:     **for** each client $i = 1, 2, \ldots, N$ **in parallel do**
5:         Local model updates: Update its model parameter with local mini-batch ξ_i^t by

$$\mathbf{x}_i^{t+1} = \mathbf{x}^t - \eta \nabla f_i(\mathbf{x}^t, \xi_i^t). \tag{6.3}$$

6:         Local model uploading: Send its model parameter $\mathbf{x}_i^{t+1}$ to the central server.
7:     Global model aggregation: The central server calculates the average model parameter $\mathbf{x}^{t+1}$ by

$$\mathbf{x}^{t+1} = \sum_{i=1}^{N} w_i \mathbf{x}_i^{t+1}. \tag{6.4}$$

---

where the weight w_i for each client i is given by $w_i = n_i/n$, and n represents the total data volume across all clients, i.e., $n = \sum_{i=1}^{N} n_i$.

FedSGD

FedSGD is an adaptation of the classic stochastic gradient descent (SGD) algorithm, specifically designed for federated learning environments. It represents the most basic approach to federated learning. As outlined in Algorithm 6.1, FedSGD involves clients performing local updates by computing stochastic gradients on their local data. These gradients are then periodically sent to a central server, where they are aggregated to update the global model. While being simple and easy to implement, FedSGD incurs high communication costs due to the frequent exchange of gradient updates, which makes it less efficient in large-scale federated learning scenarios.

FedAvg

FedAvg is a widely adopted algorithm in federated learning designed to efficiently train models on decentralized data. Rather than exchanging model updates after every computation, as in FedSGD, FedAvg allows each client to perform multiple local computation iterations (local epochs t_0) before sending the updated model parameters back to the server. This strategy reduces the frequency of communication with the central server, thereby enhancing the overall efficiency of the training process.

The FedAvg algorithm uses a synchronous update scheme organized into communication rounds. In each round, a random subset of clients is selected from the available pool, with the subset size denoted by M, where $M = C \cdot N$, and C is a parameter ranging from 0 to 1 that indicates the fraction of clients involved in each round.

At the beginning of each round, the central server distributes the current global model to the selected clients. Upon receiving this global model, each client performs local training using its own private dataset. To improve communication efficiency and fully use the computational capabilities of each client, two key strategies are applied. The first involves increasing parallelism by engaging more clients to work independently between communication rounds, thereby enhancing the scale and diversity of local updates. The second strategy focuses on deepening client-side computation by enabling each client to perform multiple SGD iterations within each communication round, rather than a single iteration.

After completing their local training, each selected client sends its model updates back to the server. The server then aggregates these updates, typically using simple averaging or weighted averaging, to produce an updated global model. This iterative process continues for a pre-defined number of communication rounds or until a convergence criterion is met. The pseudocode of the FedAvg algorithm is shown in Algorithm 6.2.

Algorithm 6.2 FedAvg: Communication-Efficient SGD-Based Federated Learning

1: Initialize $\mathbf{x}_i^0 = \mathbf{x}^0$, for all $i \in \{1, 2, \ldots, N\}$, an integer $t_0 > 0$, an integer $M = C \cdot N$, a learning rate $\eta > 0$. Set $t \Leftarrow 0$.

2: **for** each round $t = 0, 1, 2, \ldots$ **do**

3:     Client selection: Randomly select n clients to form a subset S_t.

4:     Global model broadcasting: The central server broadcasts the model parameter $\mathbf{x}^t$ to every selected client $i \in S_t$ and then overrides their parameters by $\mathbf{x}_i^{t,0} = \mathbf{x}^t$.

5:     **for** each selected client $i \in S_t$ **in parallel do**

6:         **for** local epoch $\tau = 0, 1, 2, \ldots, t_0 - 1$ **do**

7:             Local model updates: Update its model parameter with local mini-batch $\xi_i^{t,\tau}$ by

$$\mathbf{x}_i^{t,\tau+1} = \mathbf{x}_i^{t,\tau} - \eta \nabla f_i(\mathbf{x}_i^{t,\tau}, \xi_i^{t,\tau}). \tag{6.5}$$

8:         Local model uploading: Send its model parameter $\mathbf{x}_i^{t+1} = \mathbf{x}_i^{t+1,t_0}$ to the central server.

9:     Global model aggregation: The central server calculates the average model parameter $\mathbf{x}^{t+1}$ by

$$w_t = \sum_{i \in S_t} w_i,$$

$$\mathbf{x}^{t+1} = \sum_{i \in S_t} \frac{w_i}{w_t} \mathbf{x}_i^{t+1}. \tag{6.6}$$

Experiments are conducted on image classification tasks for performance evaluation. Specifically, a convolutional neural network (CNN) model is trained, which is designed for modified National Institute of Standards and Technology (MNIST) handwritten digit recognition [31]. The model includes two convolutional layers, each followed by rectified linear unit (ReLU) activation and max-pooling operations. Additionally, the network includes two fully connected layers with 50 hidden units each and output layers of size 10. The output from the last layer is processed through a softmax function to produce a probability distribution for classifying the input images.

To examine the impact of data distribution on model performance, two different strategies are employed for partitioning the MNIST dataset, which contains 60,000 samples, among 100 clients. In the i.i.d. scenario, the dataset is randomly shuffled and then evenly distributed, ensuring that each client receives an equal allocation of 600 samples. Conversely, in the non-i.i.d. scenario, the dataset is initially sorted by numerical labels and then divided into 1200 shards, with each shard containing 50 samples. Subsequently, these shards are randomly allocated across clients, leading to a variable number of shards per client, ranging from 1 to 30. This allocation introduces substantial variation in both label distribution and sample size, thereby enabling an assessment of model robustness under more realistic and heterogeneous data distributions.

Figure 6.2 compares the performance of the FedAvg algorithm, with parameters $B = 10$, $C = 0.1$, and $t_0 = 5$, against FedSGD, which is a special case of FedAvg with $B = \infty$, $C = 1$, and $t_0 = 1$. This analysis systematically examines two pivotal aspects of federated learning: the computational efficiency per client and the level of parallelism across clients.

1. **Enhancing computational efficiency per client:** The expected number of updates per client per round, denoted by $u = \mathbb{E}[n_i]t_0/B = nt_0/NB$, is calculated by averaging over the stochastic selection of clients. The experiments are designed to evaluate the impact of modifications in batch size B and the number of local epochs t_0 on the overall performance.

   - **Optimizing local epochs t_0:** Increasing t_0 from one to five significantly improves model performance, demonstrating the effectiveness of enhanced local training efforts.
   - **Optimizing batch size B:** Reducing B from ∞ to 10, while maintaining t_0 and C constant, improves performance and highlights the strategic advantage of optimizing batch sizes.
   - **Combined strategy:** Employing a combined strategy of both increasing t_0 and decreasing B, with C fixed, has been proved to be more effective than altering a single parameter in isolation.

2. **Amplifying client parallelism:** The experiment examines the impact of varying the participation ratio C, which determines the degree of multi-client parallelism, by setting it to $\{0.1, 0.2, 0.5, 1.0\}$.

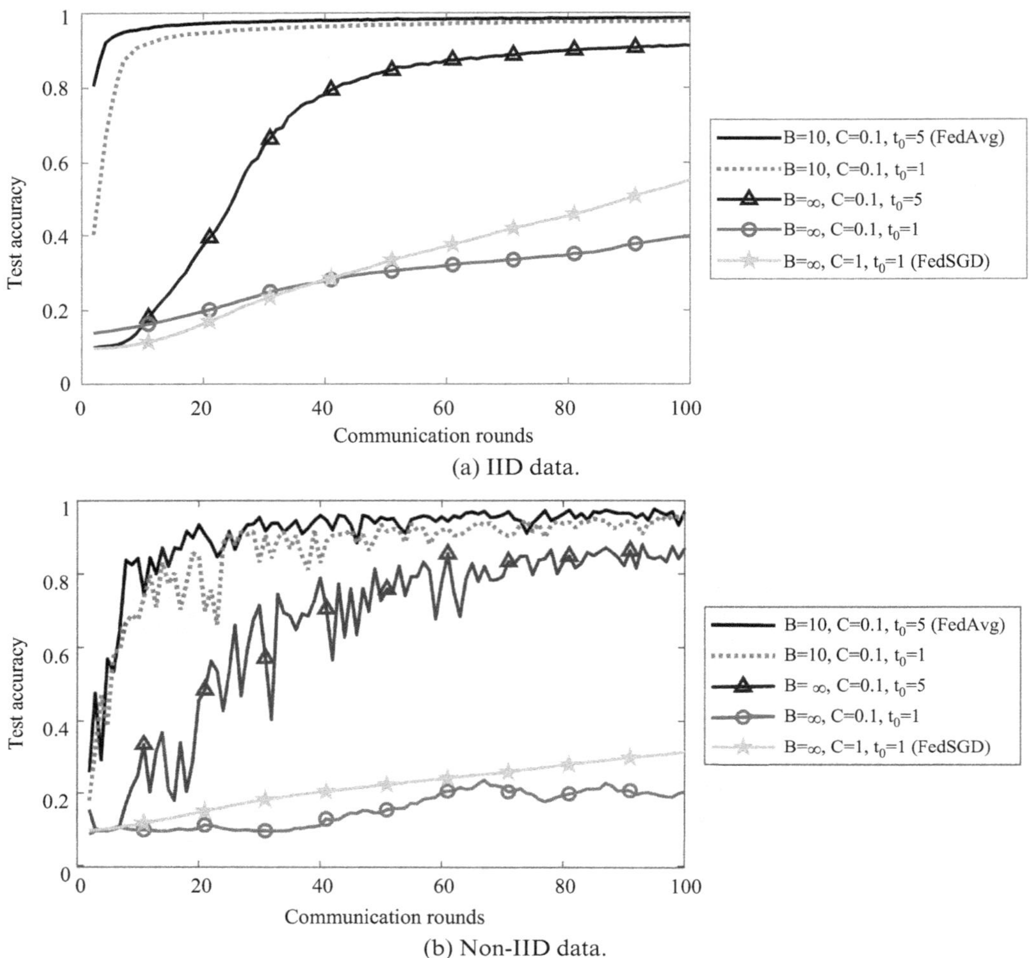

(a) IID data.

(b) Non-IID data.

Figure 6.2 Comparison of federated-learning performance with different parameters.

The results, as illustrated in Figure 6.3, reveal that higher participation ratios significantly boost convergence rates and improve model performance. This effect is particularly notable when dealing with non-i.i.d. data distributions, where involving more clients helps in better capturing the diverse data variations and stabilizes the global model updates more effectively.

FedProx

In federated learning, addressing heterogeneity among clients poses a significant challenge, manifesting in two primary aspects:

1. **Data heterogeneity:** The data across clients often deviates from the i.i.d. assumption, causing significant differences between the local models and the global model. This divergence leads to instability and poor generalization of the

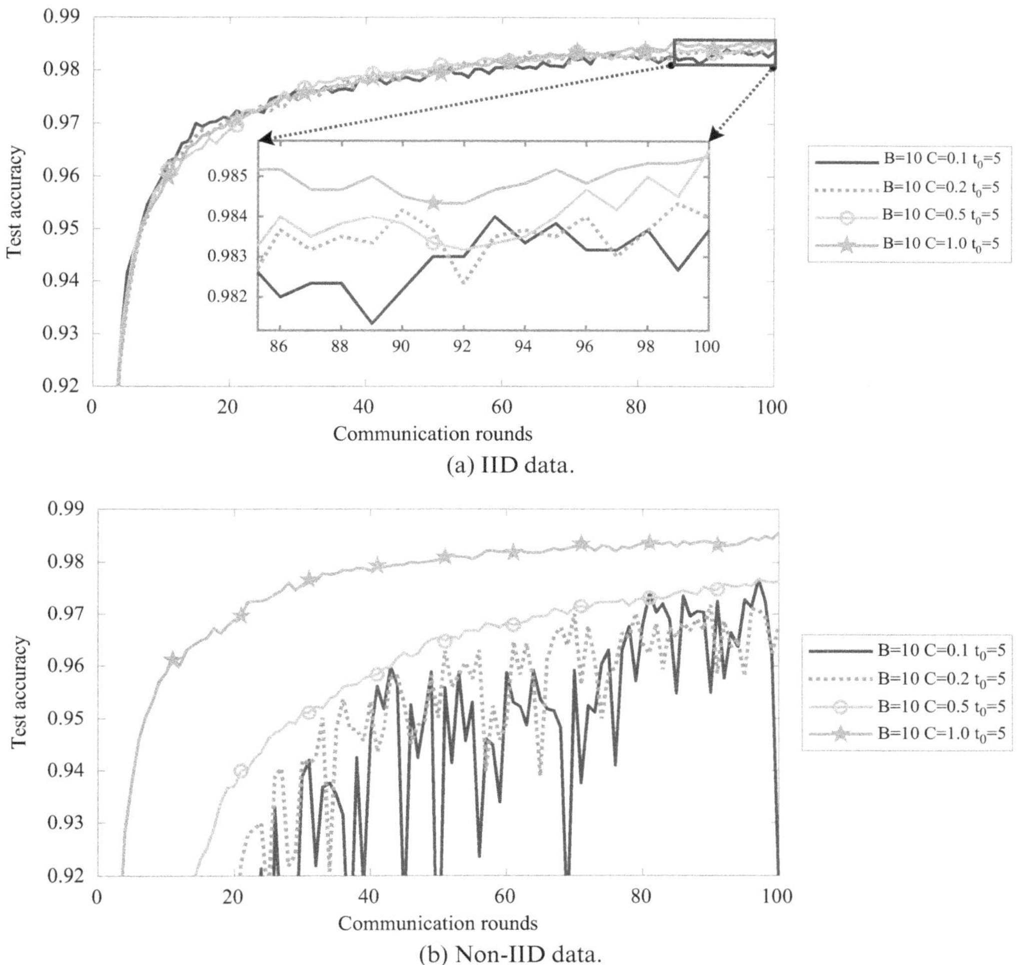

(a) IID data.

(b) Non-IID data.

Figure 6.3 Impact of varying fraction of clients C on FedAvg performance.

aggregated global model, as the model updates from clients may pull the global model in conflicting directions.

2. **System heterogeneity:** Clients have diverse communication and computational capabilities, adding complexity to the training process. This variability can lead to incomplete or delayed model updates from slower or less capable clients, raising questions about whether these updates should be integrated into the global model or discarded to maintain training efficiency and model quality.

Although FedAvg has shown empirical effectiveness in heterogeneous settings, it does not fully address these challenges. To mitigate these issues, FedProx is introduced as an enhanced variant of FedAvg, specifically designed to cope with heterogeneity. FedProx enhances FedAvg by incorporating two key modifications that more effectively address both data and system heterogeneity:

Proximal term: To mitigate the adverse effects of system heterogeneity and the potential divergence due to heterogeneous data, FedProx incorporates a proximal term into the local optimization problem. This term discourages large deviations from the global model during local updates, helping to stabilize the training process and ensure more consistent convergence across clients with varying data distributions and computational capabilities. Specifically, instead of merely minimizing the local objective $f_i(\cdot)$, each device i uses its preferred local solver to approximately minimize the enhanced objective $h_i(\cdot)$, defined as

$$h_i(\mathbf{x}; \mathbf{x}^t) \triangleq f_i(\mathbf{x}) + \frac{\mu}{2} \|\mathbf{x} - \mathbf{x}^t\|^2, \tag{6.7}$$

where $\mathbf{x}$ represents the vector of model parameters that is being optimized on each device, and $\mathbf{x}^t$ denotes the global model parameters at the tth round, serving as a reference point for local model updates. The hyperparameter μ controls the strength of the proximal term, determining how much local model updates $\mathbf{x}$ are penalized for deviating from the global model parameters $\mathbf{x}^t$. The inclusion of the proximal term mitigates the challenge of statistical heterogeneity by constraining local updates to remain close to the current global model.

ϵ**-inexact solution:** Unlike FedAvg, which requires all clients to perform the same number of local updates, FedProx allows for flexible computational workloads by permitting varying amounts of local computation. This flexibility recognizes the impracticality of enforcing the same number of local epochs for all clients. By aggregating both fully completed and partially completed updates, FedProx ensures the global model remains robust and benefits from all contributions, even from clients that provide partial or inexact solutions.

For (6.7), let $\mathbf{x}^*$ be an ϵ_i^t-inexact solution to the problem if

$$\|\nabla h_i(\mathbf{x}^*; \mathbf{x}^t)\| \leq \epsilon_i^t \|\nabla h_i(\mathbf{x}^t; \mathbf{x}^t)\|, \tag{6.8}$$

where $\nabla h_i(\mathbf{x}; \mathbf{x}^t) = \nabla f_i(\mathbf{x}) + \mu(\mathbf{x} - \mathbf{x}^t)$, and $\epsilon_i^t \in [0, 1]$. The parameter ϵ_i^t quantifies how much local computation is undertaken to solve the local subproblem on device i during the tth round, and a reduced ϵ_i^t corresponds to a more accurate solution.

In each communication round t, the server randomly selects a batch of clients, denoted by S_t. Subsequently, the server transmits the current parameters $\mathbf{x}^t$ to the selected clients. Each chosen client independently optimizes $h_i(\mathbf{x}; \mathbf{x}^t)$ to acquire an ϵ_i^t-inexact solution $\mathbf{x}_i^t$ and sends the obtained inexact solution back to the server. The server aggregates these parameters to obtain updated global model parameters $\mathbf{x}^{t+1}$, which will be used in the next round. The pseudocode of FedProx is shown in Algorithm 6.3.

An experiment focused on MNIST image recognition is conducted, operating under a global clock that governs the training process. Each client adjusts its local computation according to this global timing and its individual system constraints.

To better simulate real-world conditions in the MNIST image recognition task involving 1,000 clients, two primary forms of heterogeneity are introduced. First, data heterogeneity is implemented by allocating each device samples of only two digits, with the number of samples per device following a power-law distribution.

Algorithm 6.3 FedProx: Optimization-Based Federated Learning With Proximal Term

---

1: Initialize $\mathbf{x}_i^0 = \mathbf{x}^0$, for all $i \in \{1, 2, \ldots, N\}$, an integer $n = C \cdot N$. Set $t \Leftarrow 0$.

2: **for** each round $t = 0, 1, 2, \ldots$ **do**

3:     Client selection: Randomly select n clients to form a subset S_t.

4:     Global model broadcasting: The central server broadcasts the model parameter $\mathbf{x}^t$ to every selected client $i \in S_t$.

5:     **for** each selected client $i \in S_t$ **in parallel do**

6:         Local model updates: Calculate ϵ_i^t-inexact solution by

$$\mathbf{x}_i^{t+1} \approx \arg\min_{\mathbf{x}} h_i(\mathbf{x}; \mathbf{x}^t)$$
$$= \arg\min_{\mathbf{x}} f_i(\mathbf{x}) + \frac{\mu}{2} \|\mathbf{x} - \mathbf{x}^t\|^2. \tag{6.9}$$

7:         Local model uploading: Send its model parameter $\mathbf{x}_i^t$ to the central server.

8:     Global model update: The central server calculates the average model parameter $\mathbf{x}^{t+1}$ by

$$w_t = \sum_{i \in S_t} w_i,$$
$$\mathbf{x}^{t+1} = \sum_{i \in S_t} \frac{w_i}{w_t} \mathbf{x}_i^t. \tag{6.10}$$

Table 6.1 Simulation parameters for FedProx algorithm.

Parameters	Values	Parameters	Values
t_0	10	N	1000
μ	0.5	M	100
T	200	p	$\{0, 0.5, 0.9\}$
optimizer	SGD	η	0.01

This creates significant variability in the volume of data each device holds. Second, system heterogeneity affects the training process: Although all selected clients are expected to complete t_0 local epochs, system constraints imply that a certain percentage p of these clients, referred to as stragglers, can only complete a random number l of epochs, where l is uniformly chosen from the range $[1, t_0]$. The specific values of the parameters utilized in the simulation are outlined in Table 6.1.

The detrimental effects of system heterogeneity on convergence are evident in Figure 6.4, where different levels of system heterogeneity are simulated by assigning 0%, 50%, and 90% of clients as stragglers. In these simulations, FedProx consistently exhibits more stable convergence compared to FedAvg. This is due to the ability of FedProx to integrate partial solutions from stragglers, enhancing the robustness of the model training. In contrast, FedAvg discards the updates from these stragglers at the end of a global clock cycle. Figure 6.5 further highlights the importance of

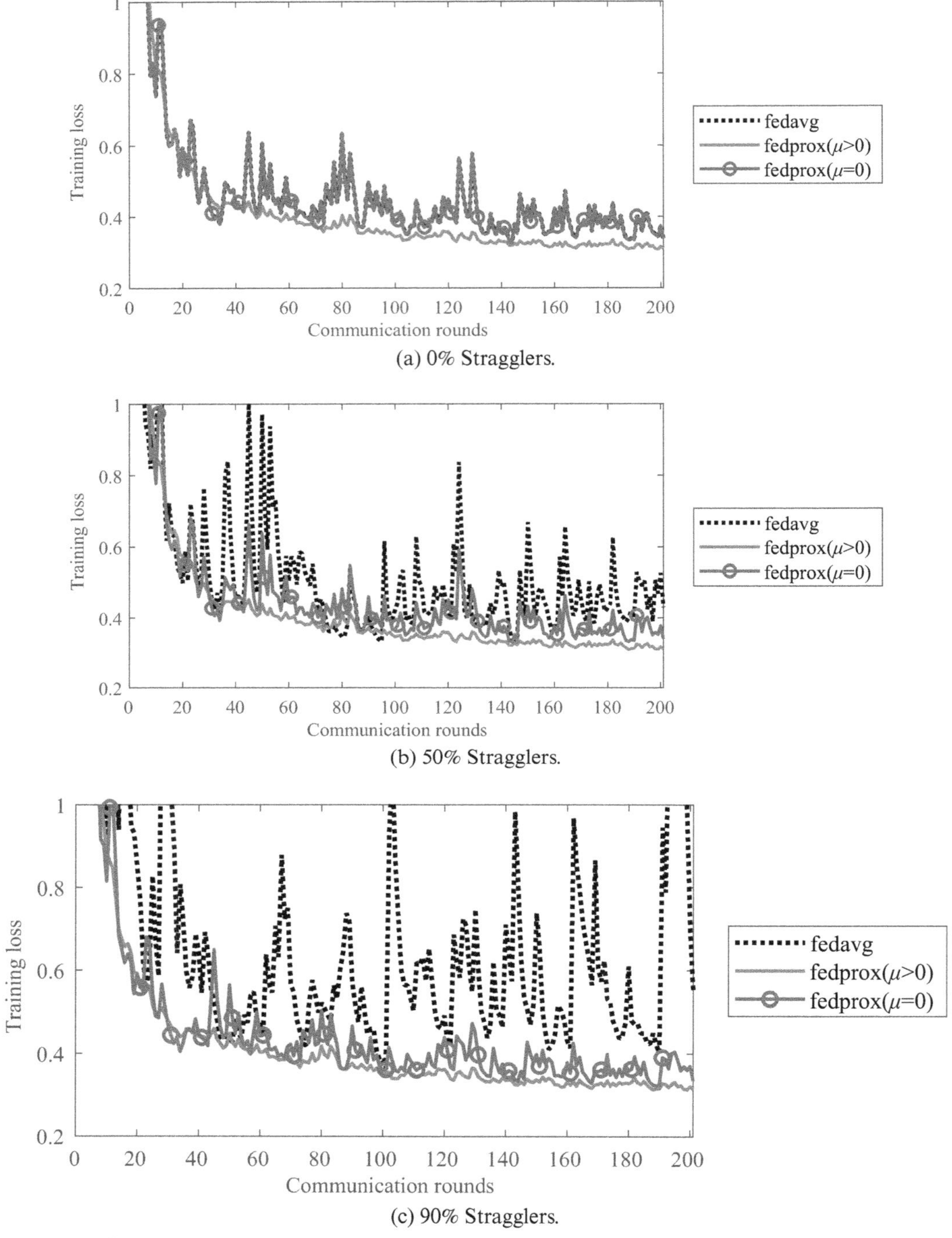

Figure 6.4 FedProx versus FedAvg in heterogeneous settings.

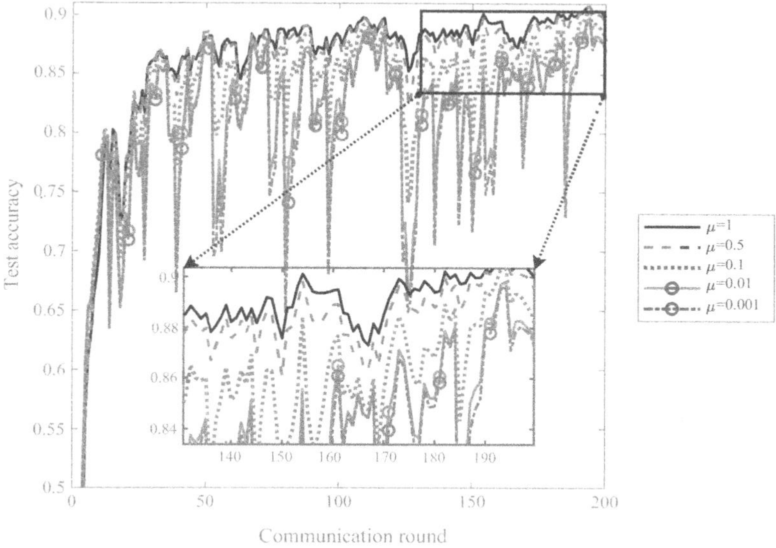

Figure 6.5 Impact of different regularization parameters μ on FedProx performance.

the hyperparameter μ, which plays a critical role in determining the performance of FedProx. Proper tuning of μ is essential to optimize the performance of the algorithm, and the appropriate value of μ varies depending on the specific task and the degree of heterogeneity present in the system.

6.2 Joint Federated Learning and Communication Design

In this section, we explore key advancements that aim to reduce communication overhead, improve client selection, and optimize resource allocation strategies, thereby enhancing the efficiency and effectiveness of federated learning over wireless networks.

6.2.1 Communication-Efficient Federated Learning Design

In both industry and academia, deep learning models, often composed of dozens to hundreds of layers and millions of parameters, are extensively used due to their ability to handle complex tasks. However, in federated learning, where numerous clients collaborate to train these models, the resulting communication overhead becomes a critical challenge. To improve the communication efficiency in federated learning, two primary strategies are employed:

- **Parameter reduction:** This strategy reduces the amount of data transmitted between clients and the central server using techniques like quantization and sparsification. Quantization lowers the precision of model parameters to shrink the data volume, while sparsification transmits only the most important elements, to minimize communication overhead without significantly deteriorating accuracy.

- **Communication management:** This strategy optimizes the timing and coordination of communication between clients and the central server to improve the efficiency of the training process. Techniques such as multiple local training epochs, asynchronous updates, hierarchical aggregation, and decentralized averaging help minimize communication rounds or better coordinate updates, thereby lowering communication costs and enhancing the overall efficiency of federated learning.

Parameter-Reduction Techniques

Parameter-reduction techniques are essential in federated learning to minimize the data transmitted between clients and the central server, particularly given the large volume of parameters in deep learning models. Techniques like quantization and sparsification play a key role in efficiently compressing model updates to address these communication challenges. Quantization reduces the size of transmitted data by converting continuous gradient values into low-precision discrete forms, such as binary representations. Meanwhile, sparsification minimizes communication overhead by eliminating gradient elements with minimal contributions and retaining only the most crucial ones. Together, these methods streamline the communication process and improve the efficiency of federated learning with marginal performance degradation.

For example, when client i needs to transmit model parameter $\mathbf{x}_i$, a compression operator $Q(\cdot)$ can transform it into a concise form $Q(\mathbf{x}_i)$, thereby reducing communication costs. However, compression inherently introduces errors in the transmitted model or gradient values, which can potentially degrade the learning performance. Therefore, it is crucial for compression methods to achieve a balance between preserving model performance and minimizing communication costs [199]. To this end, we formulate the following optimization problem:

$$\min_{Q} f_i^Q(\mathbf{x}_i) + \lambda \operatorname{Bit}(Q(\mathbf{x}_i)) + \mu \|\mathbf{x}_i - Q(\mathbf{x}_i)\|_2^2 , \tag{6.11}$$

where $f_i^Q(\mathbf{x}_i)$ represents the loss of the compressed neural network evaluated at the ith client, $\operatorname{Bit}(\cdot)$ denotes the total number of bits required for transmitting the compressed parameters, and the last term serves as a regularizer to constrain the compression-induced discrepancy. $\lambda, \mu > 0$ are hyperparameters to balance communication cost and compression accuracy. The objective of this optimization problem is to improve model performance while reducing the compression-induced errors through the regularization term.

To enhance quantization, clients transmit only the weight updates $\Delta_i^t = \mathbf{x}_i^t - \mathbf{x}_i^{t-1}$ instead of the full model weights, which is termed differential transmission [200]. Since these differential updates often have smaller magnitudes and lower variance, they are more suitable for quantization. Quantizing these updates reduces the number

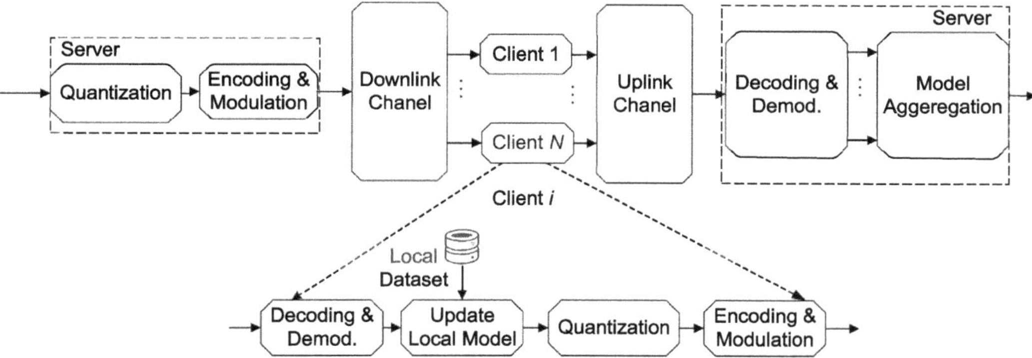

Figure 6.6 Wireless federated-learning system with quantization.

of bits required to represent them, significantly reducing the data volume. Subsequently, standard techniques such as source coding, channel coding, and modulation are applied to ensure reliable transmission, as shown in Figure 6.6. For downlink communication, where the server broadcasts the global model to clients, differential transmission is infeasible since newly participating clients lack the previous version of the global model. To overcome this limitation, layered quantization is employed, applying different quantization levels to various layers of the model based on their importance. This approach enhances communication efficiency and ensures that all clients, even those without prior model versions, can update their models effectively. By incorporating these tailored techniques into both uplink and downlink communications, wireless federated learning systems can achieve more robust and efficient performance, especially in environments with fluctuating network conditions and heterogeneous client capabilities.

Beyond quantization, sparsification is another effective method for compressing gradients by eliminating those with minimal impact on model updates. A common approach is top-k sparsification, which retains only the k largest gradient elements and sets the rest to zero, thereby significantly reducing communication overhead. Building on this idea, adaptive gradient sparsification (AGS), as proposed in [201], offers a more sophisticated solution. Unlike static sparsification techniques, AGS dynamically adjusts the sparsification level during training using an online learning algorithm, transmitting only the most significant gradients. This dynamic approach allows AGS to balance communication efficiency and model accuracy, particularly in environments with varying network conditions or client capabilities. By effectively managing the trade-off between communication cost and model performance, AGS demonstrates strong performance in large-scale federated learning applications.

Communication-Management Techniques

Communication-management techniques aim to optimize the communication process between clients and the central server by reducing the number of communication rounds or better-coordinating updates. Unlike parameter-reduction methods that focus on compressing data, these techniques improve when and how communication occurs, effectively reducing overhead and latency in federated learning systems.

Increasing the number of local epochs is one such technique that enhances interaction efficiency by allowing clients to perform several local training epochs before sending their updates to the central server for aggregation. In federated learning, clients upload their updated parameters to the server in a single communication round. This strategy reduces the number of communication rounds, significantly lowering the associated communication costs. The FedAvg algorithm serves as a prime example of this method, with clients performing local training over multiple epochs before transmitting their updates to the server, thereby minimizing the frequency of synchronization and reducing the overall communication overhead. Asynchronous update further optimizes the interaction efficiency by allowing clients to communicate with the server on their own schedule, rather than requiring all clients to synchronize their transmissions. In this setup, each client independently sends its updates to the server upon completing its local training, without waiting for others. This method is particularly advantageous in heterogeneous environments where clients differ in computational power and network conditions. Eliminating the need for strict synchronization reduces idle time and overall training latency, leading to more efficient resource utilization and faster convergence. This approach is especially well suited for wireless communication networks, where network conditions can be unpredictable and can vary greatly among clients, making federated learning more robust and scalable [202].

Hierarchical aggregation, as another effective communication-management technique, improves communication efficiency by organizing clients into a hierarchy where intermediate nodes such as edge servers aggregate local updates before uploading them to the central server. This multi-level approach reduces the frequency and volume of direct communication with the central server and lowers communication overhead and latency. In a typical setup, clients first send their updates to the nearest edge server, which aggregates the received updates locally. The edge servers then transmit the aggregated updates to the central server at a reduced frequency. This method balances the communication load, enhances scalability, and is particularly effective in large-scale networks or geographically distributed systems [203]. In contrast, decentralized averaging further improves communication efficiency by removing the need for a central server altogether and allowing clients to directly exchange model updates with their neighbors. This peer-to-peer model reduces the reliance on a central server, thereby minimizing communication bottlenecks and lowering overall latency. Clients can locally average their model parameters with nearby nodes, which not only reduces communication overhead but also enhances system flexibility and stability, especially in environments where centralized coordination is challenging or impractical. The details of this approach will be discussed further in Section 6.3.

6.2.2 Client Selection

In federated learning scenarios where the number of clients is limited, it is feasible to include all clients in every training round. However, as the number of clients increases, the communication overhead grows substantially, making it impractical to involve every client in each round. Moreover, in each training round, a subset of participants may drop out of the process voluntarily or involuntarily. This situation can lead to an insufficient number of participating clients and can result in poor generalization of the trained model. Therefore, it is crucial to design client-selection strategies that balance communication costs, adapt to varying client availability, and ensure fairness of clients. By implementing such strategies, we enhance the robustness and effectiveness of the federated learning process by maintaining model performance and generalization across different data distributions.

Designing effective client-selection strategies in federated learning involves addressing several key challenges [204], including:

- **Heterogeneity:** As discussed in Section 6.1.2, federated learning involves clients with non-i.i.d. local datasets and varying device capabilities. Differences in computation, communication, and storage capacities among client devices can significantly disrupt the training process. Slower clients may delay updates, and poor network conditions can cause asynchronous updates or dropped connections, resulting in uneven participation and asynchronous training. Additionally, the non-i.i.d. nature of data on edge devices often causes local data distributions to differ from the global population distribution. This mismatch can lead to biased model updates, ultimately deteriorating the accuracy and generalizability of the global model.

- **Dynamicity:** In federated learning, clients can be highly dynamic, frequently becoming unreliable or disconnecting due to high mobility, harsh network conditions, limited computational and communication resources, or limited energy supplies. This dynamic and volatile environment can significantly impact the convergence and stability of the federated learning process.

- **Fairness:** In federated learning, client devices with limited computational power or unreliable network connectivity are often underrepresented in the selection process, leading to unfair selection and reduced model generalization. Ensuring fairness in client selection involves providing all clients, regardless of their computational or network capabilities, with the opportunity to participate in the training process. This inclusivity helps minimize bias and improves the robustness of the global model.

- **Trustworthiness:** Ensuring the quality and reliability of data provided by clients is vital in federated learning, as unreliable or malicious data can compromise the integrity of the global model and lead to aggregation failures. To safeguard the learning process, it is crucial to identify and select trustworthy clients for participation. Effective client selection strategies can filter out unreliable or malicious participants, thereby enhancing the resilience of the system against adversarial attacks and maintaining the robustness and reliability of the global model.

Addressing these challenges is essential for optimizing client-selection mechanisms in federated learning. A well-designed client-selection scheme can enhance learning performance, ensure fair participation of clients, strengthen robustness against malicious clients, and reduce training overhead by eliminating unnecessary communication and computation.

The federated learning with client selection (FedCS) scheme proposed in [205] is one such example, which aims to improve the efficiency and speed of model training in federated learning by selecting clients based on their resource availability. Instead of focusing on client update frequency, FedCS prioritizes clients that can meet specific timing requirements based on their resource conditions. It evaluates the computational resource and network condition of each client to estimate training time, prioritizing those that can complete updates faster. This approach maximizes the number of clients that can participate within the deadline of each training round, thereby accelerating global model convergence. The selection process aims to maximize the number of clients capable of updating and transmitting their local models within a predefined deadline T_{round}, formulated as

$$
\begin{aligned}
&\max_{Q} \; |Q| \\
&\text{s.t.} \quad T_{\text{cs}} + T_{\text{gmb}} + T_{|Q|} + T_{\text{agg}} \leq T_{\text{round}},
\end{aligned}
\tag{6.12}
$$

where Q denotes the sequence of selected clients, and T_{cs}, T_{gmb}, $T_{|Q|}$, and T_{agg} correspond to the time costs associated with client selection, global model broadcasting, scheduled local updating and uploading, and global model aggregation, respectively, as illustrated in Figure 6.7. Since selected clients update their models in parallel but upload their updates sequentially, the total communication round time is given by $T_{|Q|} = T_{\text{ud}} + T_{\text{ul}}$, where T_{ud} is the longest client update time and T_{ul} is the cumulative upload time. To maximize client participation within the deadline T_{round}, a greedy

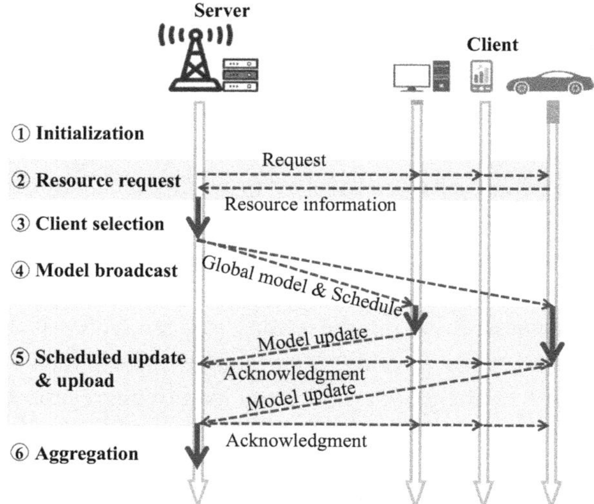

Figure 6.7 An overview of the FedCS protocol [205].

algorithm selects clients based on resource availability and estimated update and upload time. It prioritizes clients with shorter estimated update and upload time, adding them iteratively until the total time reaches T_{round}. This approach enhances federated learning efficiency by maximizing client participation and thereby reducing model convergence time.

While the FedCS strategy efficiently reduces model convergence time by prioritizing clients with faster updates and better network conditions, it may overlook data diversity and fairness. This focus on quick updates may exclude clients with valuable data but slower upload time, potentially compromising the diversity and representativeness of the global model. In contrast, reinforcement learning (RL), as discussed in Section 1.2.3 of Chapter 1, offers a more comprehensive solution by enabling dynamic adaptation to the evolving federated learning environment. RL-based approaches can optimize client selection policies by considering multiple objectives, such as maximizing test accuracy, minimizing communication costs, ensuring fairness among clients, and optimizing resource usage. By learning from real-time feedback and historical experiences, RL can prioritize clients with high-quality data, handle heterogeneity in client resources and data distributions, and adapt to dynamic network conditions, including unreliable connections and varying client availability. Moreover, RL can incorporate fairness as part of its reward function, promoting balanced participation and thus reducing bias and enhancing model robustness. This adaptability and ability to learn from experience make RL-based client-selection strategies particularly well suited for large-scale and complex federated learning systems. Next we provide an example to illustrate how RL-based client selection works in federated learning.

Example 6.1 The automated and quality-aware client selection (AUCTION) framework proposed in [206] demonstrates how RL can effectively refine client-selection strategies by balancing data quality, model performance, and budget constraints. Toward this goal, AUCTION considers a broader range of factors. For instance, the number of training samples each client provides, denoted by n_i, impacts convergence, with larger datasets offering more valuable updates. Data quality is also evaluated through metrics such as label distribution q_i^d and the number of mislabeled samples q_i^l. Moreover, the learning price b_i, which represents the participation expense of client i, is managed within a budget β to balance prices while maximizing valuable data contributions.

By leveraging RL, AUCTION makes more informed and efficient decisions about client participation in each training round, as illustrated in Figure 6.8. The RL model in AUCTION defines the following key components:

1. **State:** The state is denoted by $\mathbf{s} = \{\mathbf{s}_1, \mathbf{s}_2, \ldots, \mathbf{s}_N\}$, representing the features of all candidate clients for a given learning task. Each feature $\mathbf{s}_i$ is a four-dimensional vector represented as $\mathbf{s}_i = \{n_i, q_i^l, q_i^d, b_i\}$.
2. **Action:** Sequential actions are utilized to make client selection decisions one at a time. Each sequential action $a_j \in \{1, 2, \ldots, N\}$ selects one client from a maximum

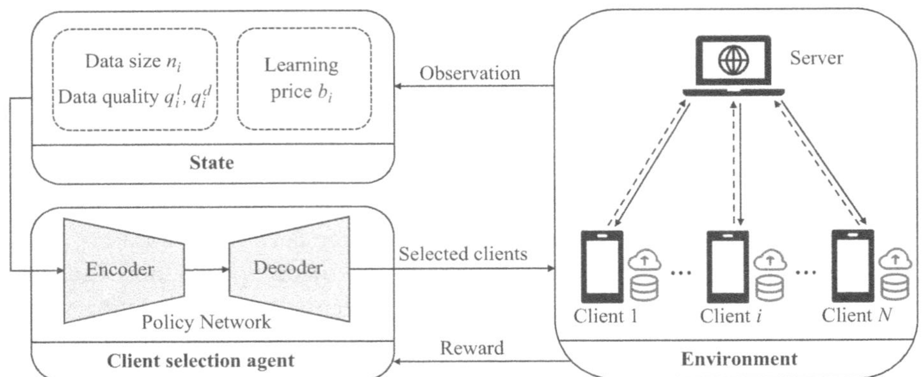

Figure 6.8 An overview of the AUCTION framework.

of N candidate clients, resulting in an action space of $\mathcal{O}(N)$. A series of these actions determine the final set of selected clients. The chosen clients are denoted by $\mathbf{a} = \{a_1, \ldots, a_j, \ldots\}$, where the total price $\sum_{a_j \in \mathbf{a}} b_{a_j} \leq \beta$, ensuring the selection adheres to the budget constraint β.

3. **Reward:** The reward function r is designed to accelerate the convergence of the global model and improve accuracy. It is formulated as $r = \sum_{t=0}^{T} \lambda^{\mathrm{acc}^t}$, where acc^t is the test accuracy of the global model on the test dataset after round t. λ is a constant that ensures reward r increases with the test accuracy.

4. **Policy:** The policy network generates a stochastic client selection policy $\pi(\mathbf{a}|\mathbf{s}, \beta)$ that determines a feasible action $\mathbf{a}$ based on the current state $\mathbf{s}$ and the learning budget β. The AUCTION policy network iteratively selects one client at each time step t, guided by the calculated probabilities $\{p_1^t, p_2^t, \ldots, p_N^t\}$. Each probability p_i^t represents the likelihood of adding client i to the chosen set. The sum of all probabilities at each time step t equals 1, i.e., $\sum_{i=1}^{N} p_i^t = 1$.

Training the policy network in the AUCTION framework focuses on optimizing its parameters to maximize the expected reward for client selection. This is achieved using the **REINFORCE** algorithm, as is introduced in Section 1.2.3 of Chapter 1, which performs gradient ascent on the expected reward:

$$\mathcal{J}(\theta|\mathbf{s}) = \mathbb{E}_{\mathbf{a} \sim \pi_\theta(\cdot|\mathbf{s}, \beta)}[r(\mathbf{a}|\mathbf{s})], \tag{6.13}$$

where θ denotes the parameters of the policy network, and $r(\mathbf{a}|\mathbf{s})$ is the reward obtained after taking action $\mathbf{a}$ in state $\mathbf{s}$. To optimize this objective, the policy gradient is calculated as

$$\nabla_\theta \mathcal{J}(\theta|\mathbf{s}) = \mathbb{E}_{\mathbf{a} \sim \pi_\theta(\cdot|\mathbf{s}, \beta)}[(r(\mathbf{a}|\mathbf{s}) - b(\mathbf{s}))\nabla_\theta \log \pi_\theta(\mathbf{a}|\mathbf{s}, \beta)], \tag{6.14}$$

where $b(\mathbf{s})$ is a baseline function that reduces variance in the gradient estimate. In the AUCTION framework, $b(\mathbf{s})$ is defined as the reward from a greedy rollout of the policy network, which always selects the action with the highest immediate reward. The training process involves multiple epochs, with each epoch comprising sampling states from the federated learning environment and performing policy and greedy rollouts to compute rewards and update the network parameters.

Figure 6.9 compares the learning performance of different client-selection mechanisms, trained on the Canadian Institute for Advanced Research-10 (CIFAR-10) dataset with 50 candidate clients. The AUCTION mechanism demonstrates superior performance by effectively balancing data volume, data quality, and price, making it particularly well suited for scenarios with massive clients. The greedy mechanism, which selects clients based on optimizing a combination of high data quality and low price, achieves reasonably good performance but is overshadowed by AUCTION. In contrast, the random mechanism, which selects clients at random, and the "price first" mechanism, which prioritizes clients with lower prices to maximize selection within a budget, both yield comparatively lower performance. These results corroborate the effectiveness of AUCTION in client selection.

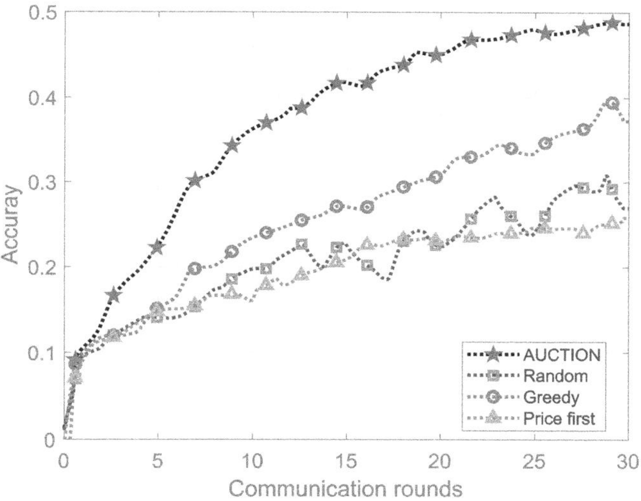

Figure 6.9 Performance comparison of client-selection mechanisms [206].

6.2.3 Resource Allocation

Efficient resource allocation is essential for the success of federated learning over resource-constrained wireless networks. Federated learning requires all participating devices to complete their computations and upload local updates to the central server before aggregation and starting the next round. Thus, the overall training speed is constrained by the node with the most limited communication and computational resources. This challenge is further compounded by heterogeneity in data, computation, and network resources across different nodes, including devices and servers. In response, resource allocation design in wireless federated learning targets judicious allocation of these resources to improve overall model performance and efficiency of model training. In particular, we focus on the following two types of resources in wireless federated learning environments:

- **Computational resources:** Efficient federated learning requires the effective distribution of computational capabilities among participating devices. This involves strategically assigning computational tasks to devices during the training process, to minimize the overall training time and enhance the efficiency of model updates.
- **Communication resources:** The success of federated learning also depends on the effective management of communication resources, including subchannels, bandwidth, power, and other network resources. Proper allocation of these resources reduces latency, minimizes transmission distortion, and ensures that updates are both timely and accurate across the network.

Unlike traditional machine learning settings, federated learning over wireless networks must consider communication factors like packet errors, transmission delays, and energy consumption, which directly affect the performance and convergence speed of the learning process. An effective strategy would be to explicitly consider wireless packet errors and optimize resource allocation, such as transmit power and resource block (RB) allocation, to reduce error rates and improve federated learning performance [207]. The optimization problem is formulated to minimize federated learning training loss while adhering to constraints on delay and energy consumption. By deriving the expected convergence rate of the federated learning algorithm, we can establish a relationship between packet error rates and federated learning performance, which helps in simplifying the problem into a mixed-integer non-linear programming problem. The Hungarian algorithm is subsequently used to solve this problem, determining the optimal user selection and RB allocation strategy that minimize training loss. The approach optimizes transmit power allocation to manage packet errors effectively, thereby enhancing the overall convergence rate and reducing the training time for federated learning in wireless networks.

Building on the need for efficient communication and computation management, a resource-allocation strategy for federated learning is proposed in [208], which jointly optimizes device scheduling and bandwidth allocation to maximize model accuracy within a constrained training-time deadline. A convergence-bound model is introduced to guide resource allocation by analyzing the relationship between training loss, the number of training rounds, and the number of clients scheduled per round. This problem is decomposed into two interrelated sub-problems: bandwidth allocation and device scheduling. For bandwidth allocation, the strategy involves assigning more bandwidth to clients with poor channel conditions or lower computational power, minimizing upload delays and preventing these clients from becoming bottlenecks. In terms of device scheduling, a greedy algorithm is employed to prioritize clients with the shortest update times, incrementally adding them until the inclusion of more clients no longer benefits the convergence speed. This joint approach significantly improves training efficiency and model performance across various data distributions and network conditions by effectively managing both communication and computational resources.

Fairness is a critical concern in resource allocation for federated learning. Owing to significant differences in computing power, communication capabilities, and data quality across clients, simply allowing all clients to participate in every training round is impractical, as it can lead to resource waste and a decrease in overall training efficiency. To address this issue, the q-fair federated learning (q-FFL) method is proposed in [209], which achieves different fairness strategies by adjusting the parameter q. Specifically, when $q = 0$, the optimization objective is to maximize the overall average performance; whereas, when $q > 0$ and the value is relatively large, the strategy tends to achieve max-min fairness, reducing performance disparities between clients and significantly lowering the variance in accuracy across clients. This method assigns higher importance to clients with a larger loss during optimization, thereby indirectly influencing the allocation of computational and communication resources in favor of these clients. This strategy effectively reduces the impact of device-performance differences on overall training performance, ensuring that weaker clients are not marginalized during long-term training. Particularly in scenarios with high data heterogeneity, the q-FFL method ensures that the local data from each device contributes its share to the global model training, thereby enhancing the overall performance and robustness of the federated learning system.

Traditional approaches to resource allocation in federated learning often optimize each training round independently, overlooking the long-term effects of resource usage and the varying significance of different training stages. However, experiments in [210] revealed a "later-is-better" phenomenon in typical machine learning tasks, where the later stages of training have a more substantial impact on model performance. Specifically, training in these later stages tends to result in higher accuracy, lower training loss, and improved model robustness compared to earlier stages. This is because, as the model converges closer to its optimal state, each update becomes more valuable in fine-tuning the model's parameters. Building on this observation, a heuristic guideline for long-term resource allocation under total energy constraints is proposed, called the online client selection and bandwidth allocation (OCEAN) algorithm. OCEAN leverages current wireless-channel information to optimize joint client selection and bandwidth allocation over a long time scale. In particular, the OCEAN algorithm strategically reserves more resources for later stages of training and incrementally adds new clients in each training round. It explores the correlation between client sampling, bandwidth allocation, and final model quality during long-term training, providing an effective strategy for wireless bandwidth allocation under client energy constraints. This approach has been shown to significantly enhance model accuracy, particularly in non-i.i.d. settings.

6.3 Decentralized Federated Learning

This section focuses on decentralized training through direct device-to-device communications, enabling collaborative learning without central-server assistance. However, real-world device-to-device communications are prone to packet loss and transmission errors. To address this, a soft decentralized SGD (Soft-DSGD) is proposed in [211], which is compatible with lightweight and unreliable communication protocols, such as the user datagram protocol (UDP). Soft-DSGD updates model parameters using partially received messages and optimizes mixing weights for consensus update based on communication link reliability. Despite the unreliable communication, Soft-DSGD achieves the same convergence rate as traditional decentralized SGD with perfectly reliable connections realized through, e.g., the transmission control protocol (TCP) and accelerates convergence by leveraging available communication links.

6.3.1 Basic Concept

For simplicity, we assume a uniform weight w_i across all N devices. Thus, the global objective of decentralized federated learning is formulated as

$$f^* \triangleq \min_{\mathbf{x} \in \mathbb{R}^d} \left[f(\mathbf{x}) = \frac{1}{N} \sum_{i=1}^{N} f_i(\mathbf{x}) \right], \tag{6.15}$$

where $f_i \colon \mathbb{R}^d \to \mathbb{R}$ defines the local objective, which is only known by the ith device. Moreover, the network topology is represented by an undirected connected graph $\mathcal{G}$, where the devices can only communicate along the edges.

In the decentralized federated framework [212], each device maintains its own local parameters, $\mathbf{x}_i \in \mathbb{R}^d$, and computes the local gradient, $\mathbf{g}_i = \nabla f_i(\mathbf{x}_i, \xi_i)$, based on ξ_i sampled from the local dataset $\mathcal{D}_i$. After that, parameters are exchanged with the neighboring devices via device-to-device communications. Formally, the two steps in each training iteration are

1. **Local training:** $\mathbf{x}_i^{t+\frac{1}{2}} = \mathbf{x}_i^t - \eta \mathbf{g}_i^t$, where η denotes the learning rate.
2. **Consensus update:** $\mathbf{x}_i^{t+1} = \sum_{j=1}^{N} w_{i,j} \mathbf{x}_j^{t+\frac{1}{2}}$, where $w_{i,j}$ is the (i,j)th item of the mixing weights matrix $\mathbf{W} \in \mathbb{R}^{N \times N}$. If two devices i and j are neither neighbors nor identical, then $w_{i,j} = 0$.

To ensure convergence, $\mathbf{W}$ is often set to be symmetric and doubly stochastic, i.e., $\mathbf{W}^T = \mathbf{W}$ and $\mathbf{W}\mathbf{1} = \mathbf{1}$, where $\mathbf{1}$ indicates an N-dimensional vector of all ones. The spectrum gap of $\mathbf{W}$ is also required to be strictly positive, i.e., $\max\{\|\lambda_2(\mathbf{W})\|, \|\lambda_N(\mathbf{W})\|\} < 1$, where $\lambda_k(\mathbf{W})$ denotes the kth largest eigenvalue of $\mathbf{W}$.

6.3.2 Decentralized Federated Learning Design

Existing decentralized optimization methods often rely on reliable communication networks, yet real-world systems are challenged by packet loss and transmission errors. Wireless federated learning exacerbates this issue due to adverse wireless conditions such as noise, fading, and interference. Typically, reliable communication protocols like TCP address these issues through acknowledgment (ACK), retransmission, and time-out mechanisms, but they incur significant overhead and slow down training due to frequent retransmissions. To overcome these limitations, a lightweight transmission protocol, i.e., the UDP protocol, is adopted to provide a connectionless but unreliable packet delivery service instead of using these reliable but heavyweight communication protocols. Each device communicates with the rest of the network via soft and unreliable communication links, where packet loss and transmission errors occur randomly. A robust decentralized training algorithm, called Soft-DSGD, is developed to deal with the unreliable communications. The devices update their model parameters with partially received messages, and the mixing weights in the consensus updates are optimized according to the reliability matrix of the different communication links.

System Model

Instead of modeling the communication network as a graph $\mathcal{G}$ where each device can only communicate with its neighbors along an edge, as in TCP, we assume a communication network where each device can send and receive messages from the rest of the network. However, delivery is not guaranteed in UDP as the transmitted packets are exposed to unreliable communication conditions, and no mechanism for reliability enhancement like packet retransmission is implemented. We use a matrix $\mathbf{P} = [p_{i,j}] \in \mathbb{R}^{N \times N}$ to describe the level of link reliability in the communication network, where $p_{i,j}$ represents the probability of successful transmission from the ith device to the jth device and $p_{i,i} = 0$, for all $i \in \{1, 2, \ldots, N\}$. Parameter exchange among the devices consists of the following three steps.

1. **Dividing parameters into packets:** The number of parameters in machine learning models, especially deep neural networks, is usually too huge to be transmitted in a single packet. We assume that the parameters are randomly grouped into multiple packets and transmitted independently.
2. **Broadcasting:** The packets are broadcast to the rest of the network without targeted recipients. With UDP, it is unknown whether the packet has successfully reached its destination.
3. **Stochastic receiving:** Due to the unreliability of the communication links, packets may be dropped or contaminated randomly when received at any device. With the help of an error detection code, such as a checksum, the receiver can detect errors in the packet, in which case the packet is declared lost and is discarded. Otherwise, the packet is successfully received.

In summary, after the ith device broadcasts packets of its parameters $\mathbf{x}_i$ to the network, they are randomly received by devices in the rest of network. For instance, at the jth device, the received data, $\mathbf{z}_{i \to j}$, may only include parts of $\mathbf{x}_i$. Therefore, it can be expressed as $\mathbf{z}_{i \to j} = \mathbf{m}_{i \to j} \odot \mathbf{x}_i$, where $\odot$ denotes element-wise multiplication; $\mathbf{m}_{i \to j}(k) = 1$ if the kth parameter broadcast by the ith device is successfully received at the jth device, and $\mathbf{m}_{i \to j}(k) = 0$ otherwise.

Algorithm 6.4 Soft-DSGD: SGD-Based Decentralized Federated Learning with Soft Communication Links

---

1: Initialize $\mathbf{x}_i$, for all $i \in \{1, 2, \ldots, N\}$, a learning rate $\eta \geq 0$, and mixing weight matrix $\mathbf{W}$.

2: **for** each round $t = 1, 2, \ldots$ **do**

3:     **for** each client $i = 1, 2, \ldots, N$ **in parallel do**

4:         Local training: Update its parameters locally by

$$\mathbf{x}_i^{t+\frac{1}{2}} = \mathbf{x}_i^t - \eta \mathbf{g}_i^t. \tag{6.16}$$

5:         Broadcasting: Broadcast its parameters $\mathbf{x}_i^{t+\frac{1}{2}}$ to the rest of network.

6:         Stochastic receiving: Receive messages from other devices by

$$\mathbf{z}_{j \to i}^{t+\frac{1}{2}} = \mathbf{m}_{j \to i}^t \odot \mathbf{x}_j^{t+\frac{1}{2}}, j \neq i. \tag{6.17}$$

7:         Handle transmission failures: Replace the missing values with stale data via

$$\hat{\mathbf{z}}_{j \to i}^{t+\frac{1}{2}} = \mathbf{z}_{j \to i}^{t+\frac{1}{2}} + (1 - \mathbf{m}_{j \to i}^t) \odot \mathbf{x}_i^{t+1}. \tag{6.18}$$

8:         Consensus update: Update its parameters with the partially received messages from other devices by

$$\mathbf{x}_i^{t+1} = w_{i,i}\mathbf{x}_i^{t+\frac{1}{2}} + \sum_{j=1, j \neq i}^{N} w_{i,j}\hat{\mathbf{z}}_{j \to i}^{t+\frac{1}{2}}. \tag{6.19}$$

Algorithm

Soft-DSGD, illustrated in Algorithm 6.4, is designed to address unreliability in UDP transmissions. We adopt the vanilla decentralized training framework [212], where each device maintains its own local parameters and conducts both a local SGD update and a consensus update in each training iteration. The key challenges are how to conduct a consensus update with only partially received messages from other devices and how to optimize the mixing weight matrix $\mathbf{W}$ according to the link reliability matrix $\mathbf{P}$.

Filling lost packets with local parameters. To deal with transmission failures in UDP, we replace the lost packets with local parameters at each device. In particular, the ith device will fill the missing parameters in $\mathbf{z}_{j \to i}$ with those in $\mathbf{x}_i$, and the filled message $\hat{\mathbf{z}}_{j \to i}$ can expressed as

$$\hat{\mathbf{z}}_{j \to i}^{t+\frac{1}{2}} = \mathbf{z}_{j \to i}^{t+\frac{1}{2}} + (1 - \mathbf{m}_{j \to i}^t) \odot \mathbf{x}_i^{t+\frac{1}{2}}. \tag{6.20}$$

With the filled parameters, the consensus update can be expressed as

$$\begin{aligned}
\mathbf{x}_i^{t+1} &= w_{i,i}\mathbf{x}_i^{t+\frac{1}{2}} + \sum_{j=1, j \neq i}^{N} w_{i,j}\hat{\mathbf{z}}_{j \to i}^{t+\frac{1}{2}} \\
&= \mathbf{x}_i^{t+\frac{1}{2}} + \sum_{j=1, j \neq i}^{N} w_{i,j}\mathbf{m}_{j \to i}^t \odot (\mathbf{x}_j^{t+\frac{1}{2}} - \mathbf{x}_i^{t+\frac{1}{2}}).
\end{aligned} \tag{6.21}$$

Note that since we use local parameters to replace the lost values, there is no additional memory required to store the historically received data.

Due to the randomness of the communication network, the consensus update step becomes stochastic. The following lemma introduces two important matrices $\overline{\mathbf{W}}$ and $\overline{\mathbf{W}^2}$, which characterize the first and second moments of the update parameters and also play an important role in the analysis of the convergence rate.

Lemma 6.1 *With the updating rule, the expectations of* $\mathbf{X}_{t+1}$ *and* $(\mathbf{X}_l^{t+1})^T(\mathbf{X}_l^{t+1})$ *can be expressed as*

$$\mathbb{E}\left[\mathbf{X}_{t+1}\right] = \overline{\mathbf{W}}(\mathbf{X}_t - \eta\mathbf{G}_t) \tag{6.22}$$

and

$$\mathbb{E}\left[(\mathbf{X}_l^{t+1})^T(\mathbf{X}_l^{t+1})\right] = (\mathbf{X}_l^t - \eta\mathbf{G}_l^t)^T\overline{\mathbf{W}^2}(\mathbf{X}_l^t - \eta\mathbf{G}_l^t), \tag{6.23}$$

where $\overline{\mathbf{W}}$ *and* $\overline{\mathbf{W}^2}$ *are defined as*

$$\overline{\mathbf{W}}[i,j] \triangleq \begin{cases} 1 - \sum_{l=1,l\neq i}^N w_{i,l}p_{i,l}, & \text{if } i = j, \\ w_{i,j}p_{i,j}, & \text{otherwise,} \end{cases} \tag{6.24}$$

and

$$\overline{\mathbf{W}^2}[i,j] \triangleq \begin{cases} 1 - 2\sum_{l=1}^N p_{i,l}\left(w_{i,l} - w_{i,l}^2\right) \\ \quad + \sum_{l=1}^N \sum_{m=1,m\neq l}^N p_{i,l}w_{i,l}p_{i,m}w_{i,m}, & \text{if } i = j, \\ \sum_{l=1}^N w_{i,l}p_{i,l}w_{j,l}p_{j,l} + 2p_{i,j}w_{i,j} \\ \quad -p_{i,j}w_{i,j}\sum_{l=1}^N \left(p_{i,l}w_{i,l} + p_{j,l}w_{j,l}\right), & \text{otherwise.} \end{cases} \tag{6.25}$$

Lemma 6.1 illustrates the first- and second-order of statistics of the Soft-DSGD update. In expectation, the consensus updates with an unreliable communication network are equivalent to reliable consensus updates with $\overline{\mathbf{W}}$ as the weight matrix, which is also a doubly stochastic matrix.

Optimizing the mixing matrix. In vanilla decentralized SGD, the convergence rate of training largely depends on the mixing matrix $\mathbf{W}$. From Lemma 6.1, the average mixing weight $\overline{\mathbf{W}}$ depends not only on $\mathbf{W}$ but also on the link reliability matrix $\mathbf{P}$. We employ two methods for selecting the mixing matrix $\mathbf{W}$ based on the availability of $\mathbf{P}$.

1. If the link reliability matrix $\mathbf{P}$ is unavailable, each link will be treated equally, and uniform mixing weights will be adopted, i.e., $\mathbf{W} = \mathbf{J} = \frac{1}{N}\mathbf{1}\mathbf{1}^T$.
2. If the link reliability matrix $\mathbf{P}$ is available (e.g., maintained at a coordinator), we can optimize $\mathbf{W}$ for faster convergence. As will be shown in the following, the convergence of Soft-DSGD depends on the largest eigenvalue of matrix $\overline{\mathbf{W}^2} - \frac{1}{N}\mathbf{1}\mathbf{1}^T$. Therefore, we optimize $\mathbf{W}$ to minimize the largest eigenvalue of this matrix, formulated as

$$\begin{aligned} \min_{\mathbf{W}} \quad & \lambda_{\max}(\overline{\mathbf{W}^2} - \tfrac{1}{N}\mathbf{1}\mathbf{1}^T) \\ \text{s.t.} \quad & 0 \leq w_{i,j} \leq 1, \ \mathbf{W}^T = \mathbf{W}, \ \mathbf{W}\mathbf{1} = \mathbf{1}, \end{aligned} \tag{6.26}$$

which is a convex optimization problem and can be solved efficiently.

Convergence Analysis

We analyze the convergence of Soft-DSGD and prove that even in unreliable communication networks, Soft-DSGD achieves the same asymptotic convergence rate as vanilla decentralized SGD with perfect communications. To establish this result, we first make assumptions on the functions f and f_i, which are assumed to possess the following properties.

Assumption 6.1 (L-smoothness) *Each local objective $f_i(\cdot)$ is smooth and with L-Lipschitzian gradients, i.e., there exists a constant $L > 0$, such that for any $\mathbf{x}$ and $\mathbf{y} \in \mathbb{R}^d$, we have*

$$\|\nabla f_i(\mathbf{x}) - \nabla f_i(\mathbf{y})\| \leq L\|\mathbf{x} - \mathbf{y}\|. \tag{6.27}$$

Assumption 6.2 (Bounded variance) *We assume that there exist constants $\sigma > 0$ and $\zeta > 0$, such that for any $\mathbf{x} \in \mathbb{R}^d$, we have*

$$\mathbb{E}\|\nabla f_i(\mathbf{x}, \xi) - \nabla f_i(\mathbf{x})\| \leq \sigma^2 \tag{6.28}$$

and

$$\frac{1}{N}\sum_{i=1}^{N}\|\nabla f_i(\mathbf{x}) - \nabla f(\mathbf{x})\| \leq \zeta^2. \tag{6.29}$$

Hence, σ^2 bounds the variance of stochastic gradients at each device, and ζ^2 bounds the discrepancy of data distributions at different devices.

Assumption 6.3 (Unbiased stochastic gradients) *Stochastic gradients obtained at each device are unbiased estimates of the real gradients of the local objectives, such that*

$$\mathbb{E}\left[\mathbf{g}_i\right] = \nabla f_i(\mathbf{x}_i). \tag{6.30}$$

These assumptions on the objectives are widely used in the non-convex decentralized optimization literature [212, 213, 214] and are valid in most applications. Besides the above assumptions on the functions, we make additional assumptions on the unreliable communication network.

Assumption 6.4 (Symmetric matrix) *The probability for successful transmission from the ith device to the jth device is the same as the probability from the jth device to the ith device, i.e., $\mathbf{P}^T = \mathbf{P}$.*

Assumption 6.5 (Independent and stable links) *The packet transmissions on different links are independent, and the link reliability matrix $\mathbf{P}$ remains fixed during training.*

These assumptions on communication networks are reasonable and easy to satisfy in practice. Due to channel reciprocity, the link reliability is the same for transmissions in either direction on the same communication link. In addition, the assumption on independence of the links is valid as long as the distance between devices is much larger than the wavelength of the signal, and the link reliability remains stable if the devices are static during the training. Note that Soft-DSGD can be directly applied in a dynamic environment, where the link reliability matrix changes with time. We concentrate on the static reliability matrix to make the analysis easy to understand.

Based on the above assumptions and lemmas, the convergence rate for the proposed decentralized training algorithm with an unreliable communication network can be demonstrated in the following theorem, proved in [211].

Theorem 6.1 (Convergence of Soft-DSGD) *Suppose that all local models are initialized with $\mathbf{x}_0 \in \mathbb{R}^d$. Under Assumptions 6.1–6.5, if the learning rate satisfies $\eta L \leq \min\{1, \sqrt{\rho^{-1}} - 1\}$, then after T iterations, we have*

$$\frac{1}{T}\sum_{t=1}^{T}\|\nabla f(\bar{\mathbf{x}}_t)\|^2 \leq \left(\frac{\mathbb{E}[f(\bar{\mathbf{x}}_T)] - \mathbb{E}[f(\bar{\mathbf{x}}_0)]}{\eta T}\right.$$

$$+ \frac{\eta L}{N}\sigma^2 + \frac{2\eta L\kappa}{N}\sigma^2 + \frac{6L\kappa\eta\zeta^2}{N}\left.\right)\frac{1-D}{1-2D}$$

$$+ \left(L^2 + \frac{2L\kappa}{\eta N} + \frac{2(3N+1)L^3\eta\kappa}{N}\right)$$

$$\left(\frac{2\eta^2\sigma^2\rho}{1-\rho} + \frac{6\eta^2\zeta^2\rho}{(1-\sqrt{\rho})^2}\right)\frac{1}{1-2D}, \tag{6.31}$$

where $D = \frac{6\eta^2 L^2\rho}{(1-\sqrt{\rho})^2}$, $\kappa = 2\max_i \sum_{j=1}^{N} p_{i,j}\left(1 - p_{i,j}\right) w_{i,j}$, and ρ is the largest eigenvalue of the matrix $\overline{\mathbf{W}^2} - \mathbf{J}$.

The unreliability of the unreliable communication network is reflected in the terms containing κ. If all the communication links are deterministic with $p_{i,j} = 0$ or 1, then $\kappa = 0$, and the results will be consistent with the convergence bound for vanilla decentralized SGD. In addition, the convergence bound depends on ρ to a large degree, which justifies our mixing weight optimization method.

Furthermore, if the learning rate is configured properly, it can achieve a linear speedup in terms of the number of devices, matching the same rate as vanilla decentralized SGD, as indicated in the following corollary.

Corollary 6.1 *Under the same conditions as Theorem 6.1, if the learning rate η is set as $\eta = \sqrt{\frac{N}{T}}$, after total T iterations, we have*

$$\frac{1}{T}\sum_{t=1}^{T}\|\nabla f(\bar{\mathbf{x}}_t)\|^2 = \mathcal{O}\left(\frac{1}{\sqrt{NT}}\right) + \mathcal{O}\left(\frac{N}{T}\right), \tag{6.32}$$

where all other constants are subsumed in $\mathcal{O}$.

Consistency with Vanilla Decentralized SGD

Recall that vanilla decentralized SGD converges at the asymptotic rate of $\mathcal{O}\left(\frac{1}{\sqrt{NT}}\right) + \mathcal{O}\left(\frac{N}{T}\right)$ [212]. Hence, decentralized SGD with unreliable communications can achieve the same asymptotic convergence rate as vanilla decentralized SGD, which assumes a reliable communication network. Therefore, the asymptotic convergence is not negatively affected by unreliability in the communication network.

Experiments

We conduct experiments on image classification tasks for evaluation. We train residual network-20 (ResNet-20) models [100] on the CIFAR-10 dataset [215]. Soft-DSGD is compared with vanilla decentralized SGD using TCP, where a communication graph $\mathcal{G}(V, E)$ is first constructed, and the devices only exchange information with their neighbors.

With TCP as the communication protocol, the receiver will send the ACK to the transmitter once it successfully receives the packet. Otherwise, the transmitter will resend the last packet. If there are multiple neighbors, the transmitter needs to collect the ACK messages from all its neighbours to ensure reliability.

We first compare Soft-DSGD with vanilla decentralized SGD using TCP. To obtain reliable results, we run the algorithms on five randomly generated communication graphs. Each graph consists of 16 devices, which are randomly located in a unit square. Figure 6.10 shows the convergence with respect to the number of communication rounds. For Soft-DSGD, the number of communication rounds is the same as the number of iterations, while the packets need to be retransmitted in the event of packet loss or transmission errors with TCP. Therefore, the required number of communication rounds is larger compared with Soft-DSGD. From the figure, the communication rounds required for TCP to reach 90% training accuracy (or 85% test accuracy) are twice as many as the communication rounds required for Soft-DSGD.

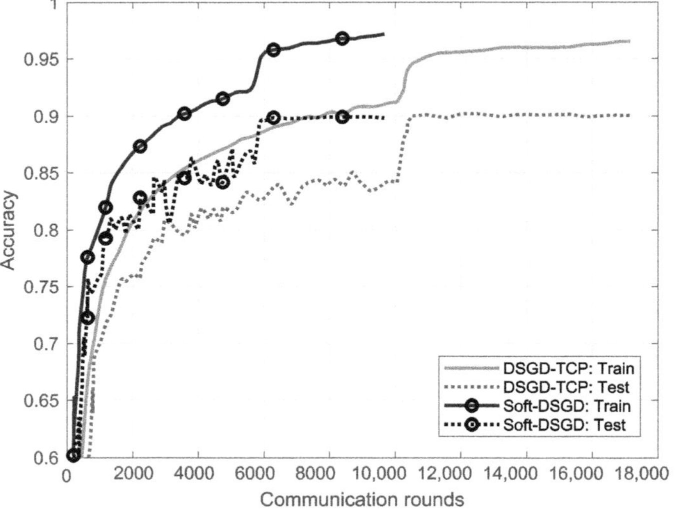

Figure 6.10 Performance of DSGD-TCP and Soft-DSGD algorithms in training and testing accuracy over communication round.

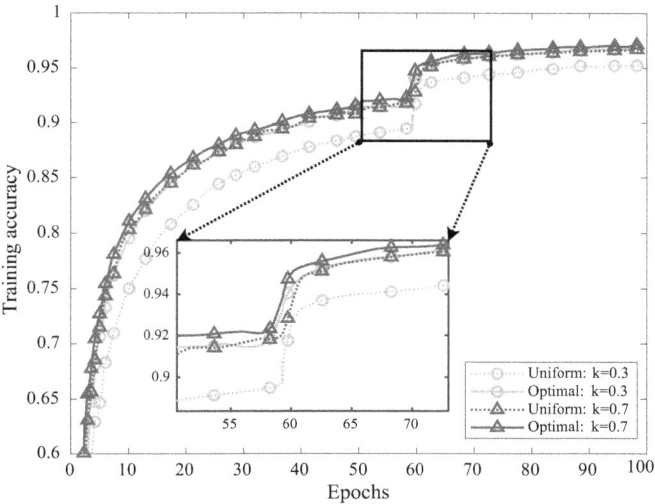

Figure 6.11 Training accuracy with uniform and optimal mixing weights under different link-reliability matrices.

We further investigate the effects of optimizing the mixing weight matrix with respect to the link reliability matrix. We evaluate Soft-DSGD using optimal and uniform weights with different link reliability matrices. The probability of successful transmission for each link is defined in a way that decays with the distance between the devices, i.e., $p_{i,j} = p_{j,i} = k^{\left(\frac{d_{i,j}}{0.4}\right)^2}$, where $d_{i,j}$ represents the distance between the ith device and the jth device. Increasing k enhances the transmission success rate for a given distance, while decreasing k reduces the probability. In our experiments, we continue to use the random communication graphs in a unit square. We keep the positions of the devices fixed and change k to generate different link reliability matrices.

To compare two weight-setting approaches, Figure 6.11 illustrates the training accuracy with uniform and optimal mixing weights under different link reliability matrices. From the figures, the training procedures with both types of weights slow down with the degradation of the communication links. In addition, when $k = 0.7$ and the average reliability of the links is high, the performance gap between the uniform and optimal weights is quite small. But as the link quality degrades with $k = 0.5$ and $k = 0.3$, the performance gap between the two approaches increases. This is because when there are many links with little probability of transmission success, the uniform weights model still assigns equal weights to these links, which impedes convergence.

6.4 Over-the-Air Computation for Federated Learning

In this section, we focus on the integration of over-the-air computation into federated learning. We begin by introducing the basic concepts of over-the-air computation and its application to federated learning. Subsequently, we delve into the system optimization for over-the-air federated learning (AirFL), addressing key aspects such as transceiver design, learning rate optimization, and the user selection method in scenarios where only a subset of users can participate.

6.4.1 Basic Concept

Over-the-Air Computation

The rapid expansion of internet of things (IoT) devices has significantly increased the scale of federated learning within the IoT ecosystem, placing unprecedented demands on communication resources and often exceeding the capabilities of existing systems. Traditional federated learning approaches with orthogonal multiple access (OMA) struggle to meet the demands of widespread access under stringent latency and resource constraints, as aggregating models from numerous clients presents substantial challenges for communication design and optimization.

To address these issues, over-the-air computation has been proposed as an innovative solution that leverages resource reuse by allowing large-scale devices to access the medium non-orthogonally [216]. By exploiting the signal superposition properties of multiple-access channels, over-the-air computation converts communication channels into computational resources [217], enabling synchronized computing and communication for distributed data. Unlike conventional OMA schemes, this approach permits simultaneous transmission from all devices on the same radio resource, thereby enhancing low-latency data aggregation even in bandwidth-limited networks. Figure 6.12 presents a comparative analysis of traditional sequential communication and computation using OMA versus over-the-air computation.

Specifically, the signal superposition property of an ideal multiple-access channel involving N transmitters is given by

$$r = \sum_{i=1}^{N} u_i, \tag{6.33}$$

where u_i is the transmitted signal of transmitter i, and r is the received signal at the receiver. Then we consider performing a computing task over a signal vector $\mathbf{s} = [s_1, s_2, \ldots, s_N]^T$, denoted by $\phi : \mathbb{R}^N \to \mathbb{R}$, where each signal component $s_i \in \mathbb{R}$ is available at the transmitter i. It has been demonstrated in [218] that any real-valued multivariate function $\phi(\cdot)$ can be expressed in a finite summation of nomographic functions, given by

$$\phi(\mathbf{s}) = \psi\left(\sum_{i=1}^{N} \varphi_i(s_i)\right), \tag{6.34}$$

where $\varphi_i : \mathbb{R} \to \mathbb{R}$ represents pre-processing functions, and $\psi : \mathbb{R} \to \mathbb{R}$ designates a post-processing function.

Therefore, the computing task $\phi(s_1, s_2, \ldots, s_N)$ can be performed via over-the-air computation by going through the following procedure:

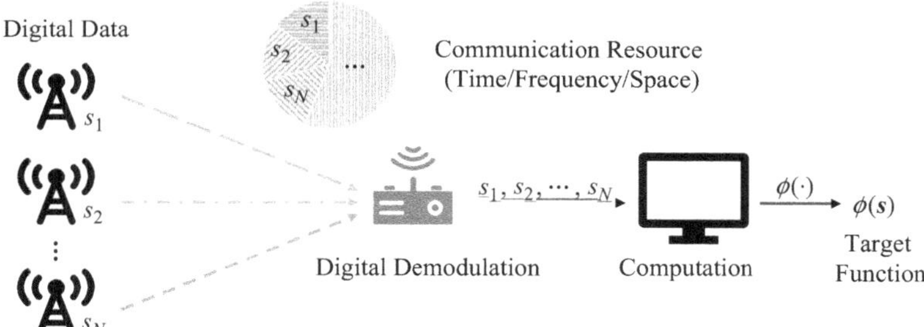

(a) Sequential communication and computation with OMA.

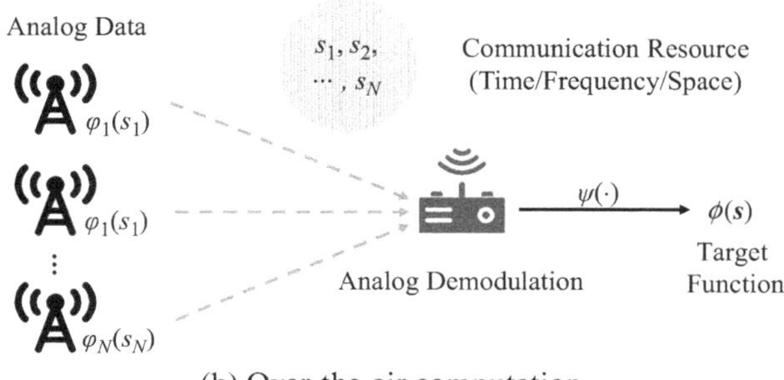

(b) Over-the-air computation.

Figure 6.12 Two paradigms for wireless-data aggregation over a multiple-access channel.

1. **Pre-processing at the transmitter:** Each transmitter individually pre-processes its signal and simultaneously transmits $u_i = \varphi_i(s_i)$ to the receiver.
2. **Post-processing at the receiver:** The receiver receives the sum of signals $r = \sum_{i=1}^{N} u_i$ and then processes it using the function $\psi(r)$.

In doing so, we integrate the transmission and computation within a single communication round, i.e., at a symbol level. Additionally, the initial task $\phi(\cdot)$ for the receiver is decomposed into $(N + 1)$ smaller lightweight tasks represented by $\{\varphi_1(\cdot), \ldots, \varphi_N(\cdot), \psi(\cdot)\}$, which are then accomplished by each transmitter or receiver. Consequently, this streamlined approach notably enhances communication efficiency and computation efficiency.

Example 6.2 Table 6.2 lists several representative real-valued multivariate functions. Decompose each function into its structured form involving the pre-processing function $\varphi_i(\cdot)$ and the post-processing function $\psi(\cdot)$.

Solution

1. Arithmetic mean:

$$\varphi_i(s_i) = s_i \text{ and } \psi(r) = \frac{r}{N},$$

where the pre-processing function is the identity function, and the post-processing function scales the sum by $1/N$.

Table 6.2 Real-valued multivariate functions for over-the-air computation.

Name	Expression
Arithmetic mean	$\phi = \frac{1}{N} \sum_{i=1}^{N} s_i$
Weighted sum	$\phi = \sum_{i=1}^{N} w_i s_i$
Geometric mean	$\phi = \left(\prod_{i=1}^{N} s_i \right)^{1/N}$
Polynomial	$\phi = \sum_{i=1}^{N} w_i s_i^{\beta_i}$
ℓ_p norm	$\phi = \left(\sum_{i=1}^{N} |s_i|^p \right)^{1/p}$

2. Weighted sum:

$$\varphi_i(s_i) = w_i s_i \text{ and } \psi(r) = r,$$

where the pre-processing function scales each s_i by a weight w_i, and the postprocessing function is the identity function.
3. Geometric mean:

$$\varphi_i(s_i) = \ln(s_i) \text{ and } \psi(r) = e^{r/N},$$

where the pre-processing function takes the natural logarithm of s_i, and the post-processing function exponentiates the result after dividing by N.
4. Polynomial:

$$\varphi_i(s_i) = w_i s_i^{\beta_i} \text{ and } \psi(r) = r,$$

where the pre-processing function involves raising s_i to a power of β_i and scaling it by w_i, while the post-processing function is the identity function.
5. ℓ_p norm:

$$\varphi_i(s_i) = |s_i|^p \text{ and } \psi(r) = r^{1/p},$$

where the pre-processing function computes the pth power of the absolute value of s_i, and the post-processing function takes the pth root.

In practical applications, this decomposition method enables the majority of computations to be carried out during the transmission phase, thereby substantially reducing the bandwidth requirements and transmission latency.

Over-the-Air Federated Learning

In wireless federated learning, over-the-air computation can be employed during the aggregation phase, wherein distributed clients transmit their model updates to a central server via a multiple-access channel [219]. This process iterates over multiple communication rounds until the global model converges. To be specific, the local model updates can be represented as either model parameters or gradient vectors, aligning with the model-averaging and gradient-averaging approaches, respectively. The target aggregated global update in wireless federated learning is given by

$$\nabla f(\mathbf{x}) = \sum_{i=1}^{N} w_i \nabla f_i(\mathbf{x}) \tag{6.35}$$

or

$$\mathbf{x} = \sum_{i=1}^{N} w_i \mathbf{x}_i, \tag{6.36}$$

where w_i denotes the weight, $\mathbf{x}_i$ and $\mathbf{x}$ denote the vector of parameters for the local and global models, respectively, while $\nabla f_i(\mathbf{x})$ and $\nabla f(\mathbf{x})$ denote the stochastic gradients of the local and global models, respectively.

AirFL leverages the signal superposition property of multiple-access channels to compute weighted averages of locally computed updates at selected devices, as formulated in (6.34). This approach improves communication efficiency by integrating transmission and computation into a unified process, thereby reusing limited communication resources more effectively. Besides improving communication efficiency, AirFL also reduces privacy leakage. Since individual updates are aggregated over the air, they remain inaccessible to the server.

Compared to systems with separate sequential communication and computation designs, AirFL tightly integrates these processes, making it crucial to effectively manage computation errors that arise during transmission. Consequently, optimizing transmission with a focus on computational accuracy becomes essential for enhancing system performance. Next we will explore key strategies to optimize AirFL systems, including transceiver design, learning rate design, and joint client selection and beamforming design.

6.4.2 Over-the-Air Federated Learning System Design

We investigate a basic wireless federated learning system consisting of one server and N distributed clients. For ease of presentation, we denote the global gradient vector in the tth communication round as $\mathbf{g}^t = \nabla f(\mathbf{x}^t) \in \mathbb{R}^B$ and local gradient vector $\mathbf{g}_i^t = \nabla f_i(\mathbf{x}^t) \in \mathbb{R}^B$, where B is the dimension of these vectors. They can be decomposed by $\mathbf{g}^t = [g_1^t, g_2^t, \ldots, g_B^t]^T$ and $\mathbf{g}_i^t = [g_{i,1}^t, g_{i,2}^t, \ldots, g_{i,B}^t]^T$, respectively. Thus, we have $\mathbf{g}^t = \sum_{i=1}^{N} w_i \mathbf{g}_i^t$, where w_i is equal to n_i/n in (6.2).

If all local gradients $\mathbf{g}_i^t$ are transmitted error-free, the standard gradient descent algorithm updates the global model as follows:

$$\mathbf{x}^{t+1} = \mathbf{x}^t - \eta_t \mathbf{g}^t, \tag{6.37}$$

where η_t is the learning rate at the server.

Assuming that the central server has M antennas, each client has a single antenna, and all clients participate in gradient aggregation, we define the pre-processing and post-processing functions of over-the-air computation as

$$\varphi_i(s_i) = w_i \times s_i \tag{6.38}$$

and

$$\psi(r) = r. \tag{6.39}$$

Consequently, the received signal at the server in the tth communication round is given by

$$\mathbf{y}_b^t = \sum_{i=1}^{N} \mathbf{h}_i^t a_i^t s_{i,b}^t + \mathbf{z}_b^t, \tag{6.40}$$

where $b \in \{1, 2, \ldots, B\}$ indexes the vector dimension, $\mathbf{h}_i^t \in \mathbb{C}^{M \times 1}$ is the channel fading coefficient from client i to the central server in the tth round, $\mathbf{z}_b^t \sim \mathcal{CN}(0, \sigma_0^2)$ represents the additive white Gaussian noise with variance σ_0^2, a_i^t is the transmit power of client i, and $s_{i,b}^t = w_i g_{i,b}^t$ is the information-bearing symbol after pre-processing at the transmitter.

Defining $\mathbf{s}_i^t = [s_{i,1}^t, s_{i,2}^t, \ldots, s_{i,b}^t]^T$, we have $\mathbf{s}_i^t = w_i \mathbf{g}_i^t$. As a result, $\mathbf{g}^t = \sum_{i=1}^{N} \mathbf{s}_i^t$. With linear post-processing at the receiver, the estimated global gradient $\hat{g}_b^t$ is calculated as

$$\begin{aligned}
\hat{g}_b^t &= \frac{1}{\sqrt{\alpha_t}} \mathrm{Re} \left\{ \{\mathbf{v}^t\}^H \mathbf{y}_b^t \right\} \\
&= \frac{1}{\sqrt{\alpha_t}} \mathrm{Re} \left\{ \{\mathbf{v}^t\}^H \sum_{i=1}^{N} \mathbf{h}_i^t a_i^t s_{i,b}^t \right\} + \frac{1}{\sqrt{\alpha_t}} u_b^t,
\end{aligned} \tag{6.41}$$

where $\mathbf{v}^t \in \mathbb{C}^{M \times 1}$ is the receive beamforming vector with unit power $\|\mathbf{v}^t\|_2 = 1$, α_t is the receiver scaling factor, and $u_b^t = \mathrm{Re}\left\{ \{\mathbf{v}^t\}^H \mathbf{z}_b^t \right\} \sim \mathcal{N}(0, \sigma_0^2/2)$.

To obtain an unbiased estimate of $\mathbf{g}^t$ from $\hat{g}_b^t$, the design of the beamforming vector $\mathbf{v}^t$ and the transmit power a_i^t in (6.41) require careful consideration. To achieve this, we employ the typical channel inversion scheme at the transmitter, wherein the transmit power a_i^t is given by

$$a_i^t = \sqrt{\alpha_t} \frac{\left(\{\mathbf{v}^t\}^H \mathbf{h}_i^t \right)^*}{\left| \{\mathbf{v}^t\}^H \mathbf{h}_i^t \right|^2}. \tag{6.42}$$

Substituting a_i^t into (6.41), we can obtain an estimate of the global gradient, given by

$$\hat{g}_b^t = \sum_{i=1}^{N} s_{i,b}^t + \hat{u}_b^t, \tag{6.43}$$

where $\hat{u}_b^t = [1/(\sqrt{\alpha_t})] u_b^t \sim \mathcal{N}(0, [1/(2\alpha_t)]\sigma_0^2)$. By defining $\hat{\mathbf{u}}^t = [\hat{u}_1^t, \hat{u}_2^t, \ldots, \hat{u}_B^t]^T$, we obtain this equation in vectorized form, expressed as

$$\hat{\mathbf{g}}^t = \sum_{i=1}^{N} \mathbf{s}_i^t + \hat{\mathbf{u}}^t = \mathbf{g}^t + \hat{\mathbf{u}}^t. \tag{6.44}$$

Therefore, $\mathbb{E}_{\hat{\mathbf{u}}^t}[\hat{\mathbf{g}}^t] = \mathbf{g}^t$ holds true, given that the noise $\hat{\mathbf{u}}^t$ has zero mean and is independent of $\hat{\mathbf{g}}^t$. Consequently, $\hat{\mathbf{g}}^t$ serves as an unbiased estimate of the full gradient $\mathbf{g}^t$.

Considering (6.37), we utilize the subsequent SGD algorithm at the central server to iteratively update the global model, and finally we have

$$\mathbf{x}^{t+1} = \mathbf{x}^t - \eta_t \hat{\mathbf{g}}^t. \tag{6.45}$$

Transceiver Design

The additive noise present in $\hat{\mathbf{g}}^t$ adversely affects both the convergence rate and the learning accuracy of AirFL. To mitigate this issue and enhance performance, it is essential to minimize the noise variance in $\hat{\mathbf{g}}^t$. According to (6.43), the noise variance is given by $(1/2\alpha_t)\sigma_0^2$, where α_t is the receiver scaling factor. Therefore, minimizing the noise variance is equivalent to maximizing α_t. Additionally, the receive beamforming vector $\mathbf{v}_t$ plays a significant role in system performance. By jointly optimizing α_t and $\mathbf{v}_t$, we can further improve the effectiveness of AirFL.

Assuming that the transmit power of each client is constrained in every communication round by

$$\left\| a_i^t \mathbf{s}_i^t \right\|_2^2 \leq P, \quad \text{for all } i \in \{1, 2, \ldots, N\}, \tag{6.46}$$

the optimization problem for the receiver in communication round t can be formulated as

$$\begin{aligned} \max_{\mathbf{v}^t, \alpha_t} \ & \alpha_t \\ \text{s.t.} \ & \left\| a_i^t \mathbf{s}_i^t \right\|_2^2 \leq P, \quad \text{for all } i \in \{1, 2, \ldots, N\}, \\ & \left\| \mathbf{v}^t \right\|_2^2 = 1. \end{aligned} \tag{6.47}$$

We consider a basic scenario in which the channel state information (CSI) is fully accessible to the server. The source symbols $\{s_{i,1}^t, s_{i,2}^t, \ldots, s_{i,b}^t\}$ at client i are assumed to remain static across every communication round, and the average signal power is expressed by $\mathbb{E}_{i,t,b}[\|s_{i,b}^t\|^2] = \chi_i$. Given the typically large dimension of s_i^t, denoted by B, an approximation can be made as $\|s_i^t\|_2^2 = B\chi_i$. In this scenario, the optimization of the receiver in (6.47) can be reformulated as

$$\begin{aligned} \max_{\mathbf{v}^t, \alpha_t} \ & \alpha_t \\ \text{s.t.} \ & \alpha_t \leq \frac{P}{B\chi_i} |\{\mathbf{v}^t\}^H \mathbf{h}_i^t|^2, \quad \text{for all } i \in \{1, 2, \ldots, N\}, \\ & \left\| \mathbf{v}^t \right\|_2^2 = 1. \end{aligned} \tag{6.48}$$

However, the assumption of stationary source symbols is impractical in federated learning. The stochastic gradient of the transmitted symbols in AirFL diminishes gradually as the model training converges. Consequently, optimizing the transceiver based solely on CSI is suboptimal.

To account for the non-stationarity of the local gradient $\mathbf{g}_i^t$, we decompose it into two components: the direction vector $\bar{\mathbf{g}}_i^t$ and the magnitude E_i^t, defined as

$$E_i^t = \left\| \mathbf{g}_i^t \right\|_2 \tag{6.49}$$

and

$$\bar{\mathbf{g}}_i^t = \frac{\mathbf{g}_i^t}{E_i^t}. \tag{6.50}$$

It is evident that $\|\bar{\mathbf{g}}_i^t\|_2^2 = 1$ holds consistently for all i and t, thereby attributing the non-stationarity observed in $\mathbf{g}_i^t$ to E_i^t. Hence, we denote E_i^t as the data state information (DSI).

To address the challenges arising from non-stationarity, a DSI-and-CSI-aware design can be considered. In this approach, the central server is required to gather both CSI and DSI from client i. Analogous to the transformation discussed in (6.48), the optimization of transceiver parameters in (6.47) can be reformulated as

$$\begin{aligned}
\max_{\mathbf{v}^t, \alpha_t} \quad & \alpha_t \\
\text{s.t.} \quad & \alpha_t \leq \frac{P}{w_i^2} |\{\mathbf{v}^t\}^H \bar{\mathbf{h}}_i^t|^2, \quad \text{for all } i \in \{1, 2, \ldots, N\}, \\
& \|\mathbf{v}^t\|_2^2 = 1,
\end{aligned} \tag{6.51}$$

where $\bar{\mathbf{h}}_i^t = \mathbf{h}_i^t / E_i^t$ is the effective channel coefficient, and $w_i = n_i/n$ can be known as a prior at the server, as it remains independent of t. Details of solving the optimization problem in (6.51) can be found in [220].

Learning-Rate Design

Let us turn to the design of the learning rate η_t in (6.45). We begin by presenting one additional assumption commonly used in the convergence analysis of federated learning algorithms, in addition to Assumption 6.1 in Section 6.3.

Assumption 6.6 (μ-strongly convexity)   *Each local objective $f_i(\cdot)$ is μ-strongly convex, i.e., there exists a constant $\mu \geq 0$, such that for any $\mathbf{x}$ and $\mathbf{y} \in \mathbb{R}^d$, we have*

$$\|\nabla f_i(\mathbf{x}) - \nabla f_i(\mathbf{y})\| \geq \mu \|\mathbf{x} - \mathbf{y}\|. \tag{6.52}$$

Under Assumptions 6.1 and 6.6, it follows that $\mu \leq L$. Moreover, it can be derived that the global objective $f(\cdot)$ is also L-smooth and μ-strongly convex.

In traditional machine learning, the SGD algorithm computes gradients using mini-batches, resulting in a constant noise variance determined by the batch size. To mitigate this noise, a diminishing learning rate is typically employed. However, the noise variance in $\hat{\mathbf{g}}^t$, as defined in (6.44), varies across communication rounds due to non-stationary local gradients and time-varying wireless channels. We demonstrate that a diminishing learning rate can still guarantee the convergence of AirFL and propose a method to accelerate convergence.

Theorem 6.2 (Diminishing learning rate)   *Under Assumptions 6.1 and 6.6, assume that the learning rate η_t for all t follows*

$$\eta_t = \frac{\beta}{\tau + t}, \tag{6.53}$$

where $\tau > 0$ denotes the time delay parameter, $\beta > 1/\mu$ denotes the learning rate scaling factor, and the initial learning rate $\eta_1 \leq 1/L$. Then, the upper bound of the optimality gap $\Delta_{t+1} = \mathbb{E}[f(\mathbf{x}_{t+1})] - f^*$ for the AirFL algorithm is given by

$$\Delta_{t+1} \leq \max \left\{ \frac{\beta^2 LB\sigma_0^2}{4\bar{\alpha}_t(\beta\mu - 1)(\tau + t + 1)}, \frac{\Delta_t(\tau + t)}{\tau + t + 1} \right\}, \tag{6.54}$$

where $f^* = f(\mathbf{x}^*)$, with $\mathbf{x}^*$ representing the optimal parameter for the federated learning problem. Additionally, $\bar{\alpha}_t = \mathbb{E}_{\hat{\mathbf{u}}}[\alpha_t]$, where the expectation is computed over all received noise $\hat{\mathbf{u}}'_t$ for all $t' < t$.

For simplicity, we designate the two terms in (6.54) as $J_1(t, \bar{\alpha}_t)$ and $J_2(t, \Delta_t)$. As t approaches infinity, if the transmit power is bounded away from zero, Δ_{t+1} in (6.54) approaches zero with a sub-linear convergence rate of $\mathcal{O}(1/t)$; a detailed proof is provided in [220]. To enhance convergence performance, it is necessary to fine-tune the parameters τ and β in (6.53).

First, we fine-tune τ, under the assumption that β remains fixed. Considering that $J_2(t, \Delta_t)$ does not involve any variables related to wireless transmission, when the values of $\bar{\alpha}_t$ are significantly large for all t, $J_1(t, \bar{\alpha}_t)$ becomes very small, leading to the optimality gap

$$\Delta_{t+1} \leq \frac{\Delta_1(\tau + 1)}{\tau + t + 1} = J_2(t, \Delta_1), \tag{6.55}$$

which is the optimal convergence performance achievable through the use of the diminishing learning rate η_t. In such scenarios, a smaller τ is favored to minimize the aforementioned gap.

However, for small $\bar{\alpha}_t$, the $J_1(t, \bar{\alpha}^t)$ term dominates the gap, as can be seen from

$$\Delta_{t+1} \leq \frac{\beta^2 LB\sigma_0^2}{4\bar{\alpha}_t(\beta\mu - 1)(\tau + t + 1)} = J_1(t, \bar{\alpha}_t). \tag{6.56}$$

In this scenario, in contrast, a larger τ is required to mitigate the impact of $J_1(t, \bar{\alpha}_t)$. In practice, it is advisable to slightly increase the value of τ to ensure robust convergence, particularly when facing poor channel conditions or a limited transmit power budget.

Second, we address the fine-tuning of β with a fixed τ. It is worth mentioning that β appears only in $J_1(t, \bar{\alpha}_t)$. Therefore, fine-tuning β is equivalent to minimizing $[\beta^2/(\beta\mu - 1)]$ for all t, with the solution given by

$$\beta = \frac{2}{\mu}. \tag{6.57}$$

Finally, in fine-tuning both τ and β, the constraint $\eta_1 \leq 1/L$ must be satisfied. A practical approach is to start with $\beta = 2/\mu$ and then fine-tune τ within the range $[2L/\mu, +\infty)$, using a slightly larger initial value.

Joint Client Selection and Beamforming Design

In the aforementioned scenario, all clients upload their gradients simultaneously, aiming to minimize aggregation errors. However, this approach often leads to significant delays because the overall latency is determined by the computation time of the slowest client. Furthermore, under the channel inversion scheme, poor channel conditions can result in inefficient power usage, causing an increase in the aggregated mean squared error (MSE) and potentially degrading model performance.

Key observations in [46, 221] highlight that aggregation errors can significantly impact learning performance, while increasing the number of participating devices can accelerate training convergence. Building on these observations, the aggregation performance of AirFL can be enhanced on two fronts:

1. Maximize the number of devices selected in each round to enhance the convergence rate of the distributed training process.
2. Minimize model aggregation errors to enhance the learning performance.

Considering this, we assume that the server selects only a subset of clients for gradient aggregation and denote by S_t the index set of selected clients in the tth communication round. Accordingly, we modify the pre-processing and post-processing functions in (6.38) and (6.39) of the wireless computation to

$$\varphi_i(x) = n_i \times x \tag{6.58}$$

and

$$\psi(x) = \frac{1}{\sum_{i \in S_t} n_i} \times x. \tag{6.59}$$

The symbol vector for each local gradient before pre-processing, $\mathbf{g}_i^t = \nabla f_i(\mathbf{x}^t) \in \mathbb{C}^{B \times 1}$, is assumed to be normalized with unit variance, i.e., $\mathbb{E}[\mathbf{g}_i^t \mathbf{g}_i^{t,H}] = \mathbf{I}_B$. For each dimension $b = 1, 2, \ldots, B$, the chosen devices transmit the pre-processed signal $s_{i,b}^{',t} = n_i g_{i,b}^t \in \mathbb{C}$ to the central server, rather than $w_i g_{i,b}^t$. The remaining parameter settings are consistent with the aforementioned content. Then, similar to (6.40), the signal received by the server is expressed as

$$\mathbf{y}_b^t = \sum_{i \in S_t} \mathbf{h}_i^t a_i^t s_{i,b}^{',t} + \mathbf{z}_b^t. \tag{6.60}$$

With linear combination at the receiver using the beamforming vector $\mathbf{v}^t$ and the power scaling factor α_t, the computation of the partially aggregated gradient estimation $\hat{g}_b^t$ is given by

$$
\begin{aligned}
\hat{g}_b^t &= \frac{1}{\sqrt{\alpha_t}} \mathrm{Re}\left\{ \{\mathbf{v}^t\}^H \mathbf{y}_b^t \right\} \\
&= \frac{1}{\sqrt{\alpha_t}} \mathrm{Re}\left\{ \{\mathbf{v}^t\}^H \sum_{i \in S_t} \mathbf{h}_i^t a_i^t x_{i,b}^{',t} \right\} + \frac{1}{\sqrt{\alpha_t}} \{\mathbf{v}^t\}^H \mathbf{z}_b^t.
\end{aligned}
\tag{6.61}
$$

Let $g_b^t = \sum_{i \in S_t} s_{i,b}^{',t}$ represent the target function to be calculated through over-the-air computation. The distortion of $\hat{g}_b^t$ determines the performance of over-the-air computation for global model aggregation in the FedAvg algorithm and is quantified by the MSE, expressed as

$$\mathrm{MSE}(\hat{g}_b^t, g_b^t) = \mathbb{E}\left[|\hat{g}_b^t - g_b^t|^2\right]$$

$$= \sum_{i \in S_t} \left| \frac{\{\mathbf{v}^t\}^H \mathbf{h}_i^t a_i^t n_i}{\sqrt{\alpha_t}} - n_i \right|^2 + \frac{\sigma_0^2}{\alpha_t} \|\mathbf{v}^t\|^2. \tag{6.62}$$

Meanwhile, the transmit power constraint for device t is expressed as

$$\mathbb{E}[|a_i^t s_{i,b}^{',t}|^2] = |a_i^t n_i|^2 \le P', \tag{6.63}$$

where $P' > 0$ is the maximum transmit power within one time slot. Additionally, we still employ a channel inversion scheme, maintaining the transmit power a_i^t as per (6.42). By incorporating the transmit power constraint from (6.63) into the transmit power a_i^t, we can derive a closed-form expression for the receiver scaling factor, given by

$$\alpha_t = \min_{i \in S_t} \frac{P' \left|\{\mathbf{v}^t\}^H \mathbf{h}_i^t\right|^2}{n_i^2}. \tag{6.64}$$

Hence the MSE in (6.62) follows

$$\mathrm{MSE}(\hat{g}_b^t, g_b^t; S_t, \mathbf{v}^t) = \frac{\|\{\mathbf{v}^t\}^H\|^2 \sigma_0^2}{\alpha_t} = \frac{\sigma_0^2}{P'} \max_{i \in S_t} n_i^2 \frac{\|\mathbf{v}^t\|^2}{\left|\{\mathbf{v}^t\}^H \mathbf{h}_i^t\right|^2}. \tag{6.65}$$

Building on the pivotal insights outlined in [46], we aim to maximize the number of selected devices while ensuring the MSE requirements for over-the-air computation. The combinatorial optimization problem is formulated as

$$\begin{aligned} \max_{S_t, \mathbf{v}^t} \quad & |S_t| \\ \text{s.t.} \quad & \frac{\sigma_0^2}{P'} \max_{i \in S_t} n_i^2 \frac{\|\mathbf{v}^t\|^2}{\left|\{\mathbf{v}^t\}^H \mathbf{h}_i^t\right|^2} \le \gamma, \end{aligned} \tag{6.66}$$

where $\gamma > 0$ is the MSE requirement for partial gradient aggregation.

It is evident that the optimization problem in (6.66) is NP-hard. However, this optimization problem can be transformed into a non-convex optimization problem with a sparse objective function and low-rank constraints, which can be readily solved via a difference-of-convex-functions representation framework in [222].

6.5 Federated Learning with the ADMM Framework

In this section, we explore federated learning using the ADMM framework. We begin by introducing the fundamental concepts of the ADMM framework and its application in federated learning. To improve communication and computational efficiency, we then discuss optimization strategies for ADMM-based federated learning. This includes communication-efficient ADMM-based federated learning, inexact ADMM-based federated learning, and a combination of both approaches, referred to as inexact communication-efficient ADMM-based federated learning.

6.5.1 Basic Concept

ADMM Framework

ADMM is a widely used algorithm for convex optimization problems, introduced to tackle issues with separable objectives or constraints, as summarized in [223]. The core idea of ADMM is to decompose complex optimization problems into simpler sub-problems and solve them iteratively while ensuring consistency with augmented Lagrange multipliers. ADMM alternates between updating primal, dual, and scaled Lagrange multipliers and is particularly amenable to distributed implementation.

We consider an optimization problem of the form

$$\min_{\mathbf{x}\in\mathbb{R}^d, \mathbf{z}\in\mathbb{R}^q} \quad f(\mathbf{x}) + g(\mathbf{z}),$$
$$\text{s.t.} \quad \mathbf{Ax} + \mathbf{Bz} = \mathbf{c}, \tag{6.67}$$

where $\mathbf{A} \in \mathbb{R}^{p\times d}$, $\mathbf{B} \in \mathbb{R}^{p\times q}$, and $\mathbf{c} \in \mathbb{R}^p$. The essential component for implementing ADMM is the augmented Lagrangian function of the problem in (6.67), defined as

$$\mathcal{L}(\mathbf{x}, \mathbf{z}, \boldsymbol{\pi}) \triangleq f(\mathbf{x}) + g(\mathbf{z}) + \langle \mathbf{Ax} + \mathbf{Bz} - \mathbf{c}, \boldsymbol{\pi} \rangle + \frac{\rho}{2} \|\mathbf{Ax} + \mathbf{Bz} - \mathbf{c}\|^2, \tag{6.68}$$

where $\boldsymbol{\pi}$ denotes the Lagrange multiplier, $\rho > 0$ denotes the penalty parameter, and $\langle \mathbf{Ax} + \mathbf{Bz} - \mathbf{c}, \boldsymbol{\pi} \rangle$ denotes the inner product of vectors $(\mathbf{Ax} + \mathbf{Bz} - \mathbf{c})$ and $\boldsymbol{\pi}$. Utilizing the augmented Lagrange function, ADMM iterates through the following steps for a given initial point $(\mathbf{x}^0, \mathbf{z}^0, \boldsymbol{\pi}^0)$ and any $t \geq 0$:

$$\mathbf{x}^{t+1} = \arg\min_{\mathbf{x}\in\mathbb{R}^d} \mathcal{L}(\mathbf{x}, \mathbf{z}^t, \boldsymbol{\pi}^t),$$
$$\mathbf{z}^{t+1} = \arg\min_{\mathbf{z}\in\mathbb{R}^q} \mathcal{L}(\mathbf{x}^{t+1}, \mathbf{z}, \boldsymbol{\pi}^t), \tag{6.69}$$
$$\boldsymbol{\pi}^{t+1} = \boldsymbol{\pi}^t + \rho(\mathbf{Ax}^{t+1} + \mathbf{Bz}^{t+1} - \mathbf{c}).$$

Example 6.3 The least absolute shrinkage and selection operator (LASSO) optimization problem is commonly utilized for extracting essential features from high-dimensional data, formulated as

$$\min_{\mathbf{w}} \frac{1}{2}\|\mathbf{Xw} - \mathbf{y}\|_2^2 + \lambda\|\mathbf{w}\|_1,$$

where $\mathbf{w}$ is the vector of regression coefficients, $\mathbf{X}$ is the input feature matrix, $\mathbf{y}$ is the target vector, and $\lambda > 0$ is the regularization parameter.

a) Reformulate the problem into a form suitable for solving with ADMM.
b) Provide the corresponding ADMM update steps.

Solution

a) We begin by reformulating the LASSO problem by introducing an auxiliary
variable $\mathbf{z}$ and expressing it as a constrained optimization problem:

$$\min_{\mathbf{w},\mathbf{z}} \frac{1}{2}\|\mathbf{X}\mathbf{w} - \mathbf{y}\|_2^2 + \lambda\|\mathbf{z}\|_1, \quad \text{s.t.} \quad \mathbf{w} = \mathbf{z}.$$

Next, we define the augmented Lagrangian function for the ADMM method.
Let $f(\mathbf{w}) = \frac{1}{2}\|\mathbf{X}\mathbf{w} - \mathbf{y}\|_2^2$ and $g(\mathbf{z}) = \lambda\|\mathbf{z}\|_1$. The constraint can be expressed as
$\mathbf{A}\mathbf{w} + \mathbf{B}\mathbf{z} = \mathbf{c}$, where $\mathbf{A} = \mathbf{I}$, $\mathbf{B} = -\mathbf{I}$, and $\mathbf{c} = \mathbf{0}$. Thus, the augmented
Lagrangian function is expressed as

$$\mathcal{L}(\mathbf{w},\mathbf{z},\boldsymbol{\pi}) = \frac{1}{2}\|\mathbf{X}\mathbf{w} - \mathbf{y}\|_2^2 + \lambda\|\mathbf{z}\|_1 + \langle \boldsymbol{\pi}, \mathbf{w} - \mathbf{z}\rangle + \frac{\rho}{2}\|\mathbf{w} - \mathbf{z}\|_2^2,$$

where $\rho > 0$ is a penalty parameter.

b) The ADMM update steps proceed as follows. First, we update $\mathbf{w}$ by minimizing
the augmented Lagrangian function while keeping $\mathbf{z}$ and $\boldsymbol{\pi}$ fixed. The update
rule for $\mathbf{w}$ is represented by

$$\mathbf{w}^{t+1} = \arg\min_{\mathbf{w}} \frac{1}{2}\|\mathbf{X}\mathbf{w} - \mathbf{y}\|_2^2 + \frac{\rho}{2}\|\mathbf{w} - \mathbf{z}^t + \frac{\boldsymbol{\pi}^t}{\rho}\|_2^2,$$

which is a standard regularized linear regression problem.

Next, we update $\mathbf{z}$ by minimizing the augmented Lagrangian function while
keeping $\mathbf{w}$ and $\boldsymbol{\pi}$ fixed. The update rule for $\mathbf{z}$ is given by

$$\mathbf{z}^{t+1} = \arg\min_{\mathbf{z}} \lambda\|\mathbf{z}\|_1 + \frac{\rho}{2}\|\mathbf{w}^{t+1} - \mathbf{z} + \frac{\boldsymbol{\pi}^t}{\rho}\|_2^2,$$

which can be solved using the iterative shrinkage thresholding algorithm
discussed in Section 2.4 of Chapter 2.

Finally, we update the Lagrange multiplier $\boldsymbol{\pi}$ as

$$\boldsymbol{\pi}^{t+1} = \boldsymbol{\pi}^t + \rho(\mathbf{w}^{t+1} - \mathbf{z}^{t+1}),$$

which adjusts the Lagrange multiplier based on the difference between $\mathbf{w}^{t+1}$ and
$\mathbf{z}^{t+1}$ to enforce the constraint $\mathbf{w} = \mathbf{z}$.

ADMM-Based Federated Learning

Recalling the global objective defined in (6.2), the objective of federated learning is to
learn an optimal parameter $\mathbf{x}^*$ on the central server to minimize the global objective
$f: \mathbb{R}^d \to \mathbb{R}$, expressed as

$$\mathbf{x}^* \triangleq \arg\min_{\mathbf{x}\in\mathbb{R}^d} \left[f(\mathbf{x}) = \sum_{i=1}^{N} w_i f_i(\mathbf{x}) \right]. \tag{6.70}$$

We observe that federated learning is implemented in a distributed manner, and its
optimization problem can be expressed as a series of sub-optimization problems on
the client side. This structure is particularly well suited for the ADMM framework.

By introducing auxiliary variables $\mathbf{x}_i = \mathbf{x}$, the federated learning problem in (6.70) can be reformulated into

$$\min_{\mathbf{x}, \mathbf{x}_1, \ldots, \mathbf{x}_N \in \mathbb{R}^d} \sum_{i=1}^{N} w_i f_i(\mathbf{x}_i),$$

$$\text{s.t.} \quad \mathbf{x}_i = \mathbf{x}, \text{ for all } i \in \{1, 2, \ldots, N\}. \tag{6.71}$$

Obviously, the problem in (6.71) conforms to the ADMM problem format, and its augmented Lagrangian function written as

$$\mathcal{L}(\mathbf{x}, \mathbf{X}, \Pi) = \sum_{i=1}^{N} L(\mathbf{x}, \mathbf{x}_i, \boldsymbol{\pi}_i), \tag{6.72}$$

where $\mathbf{X} = (\mathbf{x}_1, \mathbf{x}_2, \ldots, \mathbf{x}_N)$, $\Pi = (\boldsymbol{\pi}_1, \boldsymbol{\pi}_2, \ldots, \boldsymbol{\pi}_N)$, and $L(\mathbf{x}, \mathbf{x}_i, \boldsymbol{\pi}_i)$ is given by

$$L(\mathbf{x}, \mathbf{x}_i, \boldsymbol{\pi}_i) = w_i f_i(\mathbf{x}_i) + \langle \mathbf{x}_i - \mathbf{x}, \boldsymbol{\pi}_i \rangle + \frac{\rho_i}{2} \|\mathbf{x}_i - \mathbf{x}\|^2, \tag{6.73}$$

where $\boldsymbol{\pi}_i \in \mathbb{R}^p$, for all $i \in \{1, 2, \ldots, N\}$ are the Lagrange multipliers, and $\rho_i > 0$, for all $i \in \{1, 2, \ldots, N\}$ are the penalty parameters.

By applying the ADMM framework to the reformulated federated learning problem in (6.71), we perform the following iterative updates with the initialization $(\mathbf{x}^0, \mathbf{X}^0, \Pi^0)$:

$$\mathbf{x}^{t+1} = \arg\min_{\mathbf{x} \in \mathbb{R}^d} \mathcal{L}(\mathbf{x}, \mathbf{X}^t, \Pi^t),$$

$$\mathbf{x}_i^{t+1} = \arg\min_{\mathbf{x}_i \in \mathbb{R}^d} L(\mathbf{x}^{t+1}, \mathbf{x}_i, \boldsymbol{\pi}_i^t), \quad \text{for all } i \in \{1, 2, \ldots, N\}, \tag{6.74}$$

$$\boldsymbol{\pi}_i^{t+1} = \boldsymbol{\pi}_i^t + \rho_i(\mathbf{x}_i^{t+1} - \mathbf{x}^{t+1}), \quad \text{for all } i \in \{1, 2, \ldots, N\},$$

which is termed FedADMM and summarized in Algorithm 6.5.

6.5.2 Enhanced ADMM-Based Federated Learning

Communication-Efficient ADMM-Based Federated Learning

In SGD-based federated learning illustrated in Algorithm 6.1, FedSGD, communication occurs between the local clients and the central server at each step. Since FedADMM naively applies the ADMM framework to federated learning, it also alternates between global model update and local training at each step. Specifically, the central server broadcasts the weight $\mathbf{x}^{t+1}$ and $\boldsymbol{\pi}_i^t$ to all local clients and each client, in turn, uploads its weight $\mathbf{x}_i^t$ to the central server in each communication round.

In federated learning, the efficiency of the learning process is significantly influenced by the number of communication rounds; excessive communication leads to prolonged learning time and increased communication resource consumption. To address this issue, in FedAvg, local clients are permitted to update their parameters multiple times before transmitting their weights to a central server [46]. Similarly, [224] introduced a communication-efficient ADMM-based federated learning

---

Algorithm 6.5 FedADMM: ADMM-Based Federated Learning

1: Initialize $\mathbf{x}_i^0, \boldsymbol{\pi}_i^0, \rho_i > 0$, for all $i \in \{1, 2, \ldots, N\}$. Set $t \Leftarrow 0$.
2: **for** each round $t = 0, 1, 2, \ldots$ **do**
3:     Local uploading: Each client sends the parameters $\mathbf{x}_i^t$ and $\boldsymbol{\pi}_i^t$ to the central server.
4:     Global model update: The central server calculates the average parameter $\mathbf{x}^{t+1}$ by

$$\mathbf{x}^{t+1} = \arg\min_{\mathbf{x}} \mathcal{L}(\mathbf{x}, \mathbf{X}^t, \Pi^t). \tag{6.75}$$

5:     Global model broadcasting: The central server broadcasts the parameter $\mathbf{x}^{t+1}$ to every client.
6:     **for** each client $i = 1, 2, \ldots, N$ **in parallel do**
7:         Local updates: Update its parameters by

$$\mathbf{x}_i^{t+1} = \arg\min_{\mathbf{x}_i} L(\mathbf{x}^{t+1}, \mathbf{x}_i, \boldsymbol{\pi}_i^t), \tag{6.76}$$

$$\boldsymbol{\pi}_i^{t+1} = \boldsymbol{\pi}_i^t + \rho_i(\mathbf{x}_i^{t+1} - \mathbf{x}^{t+1}). \tag{6.77}$$

(CEADMM) algorithm, summarized in Algorithm 6.6. In CEADMM, aggregation at the central server occurs only after multiple client updates, thus reducing the number of communication rounds. The CEADMM algorithm specifies that communication occurs only when $t \in \mathcal{T} = \{0, t_0, 2t_0, \ldots\}$, where t_0 is a predefined positive constant. Consequently, the number of communication rounds can be notably reduced, thus accelerating convergence substantially.

For the local updates in CEADMM, we introduce an auxiliary point $\mathbf{y}^{t+1} = \mathbf{x}^{\tau_t+1}$, where $\tau_t = \lfloor t/t_0 \rfloor$ represents the largest integer not exceeding t/t_0. Moreover, the global model update possesses a closed-form solution given by

$$\mathbf{x}^{t+1} = \arg\min_{\mathbf{x}} \mathcal{L}(\mathbf{x}, \mathbf{X}^t, \Pi^t) = \sum_{i=1}^{N} \frac{\rho_i \mathbf{x}_i^t}{\rho} + \sum_{i=1}^{N} \frac{\boldsymbol{\pi}_i^t}{\rho}, \tag{6.78}$$

where $\rho = \sum_{i=1}^{N} \rho_i$, and is thus directly written out in (6.79).

To analyze the convergence property of CEADMM, we consider the sequence $\{(\mathbf{y}^t, \mathbf{X}^t, \Pi^t)\}$ generated by CEADMM with $\rho_i > 2w_i L_i$, for all $i \in \{1, 2, \ldots, N\}$. The following results are valid under mild assumptions, and the detailed proof can be found in [224].

1. The sequence $\{(\mathbf{y}^t, \mathbf{X}^t, \Pi^t)\}$ is bounded, i.e. $\|\mathbf{y}^t\| \leq +\infty$, $\|\mathbf{x}_i^t\| \leq +\infty$ and $\|\boldsymbol{\pi}_i^t\| \leq +\infty$ for all i and t. Moreover, any accumulating point $(\mathbf{y}^\infty, \mathbf{X}^\infty, \Pi^\infty)$ of this sequence is a stationary point of the reformulated problem in (6.71), where $\mathbf{y}^\infty$ is a stationary point of the problem in (6.70).
2. If we further assume that $\mathbf{y}^\infty$ is isolated, then the entire sequence $\{(\mathbf{y}^t, \mathbf{X}^t, \Pi^t)\}$ converges to $(\mathbf{y}^\infty, \mathbf{X}^\infty, \Pi^\infty)$.

Algorithm 6.6 CEADMM: Communication-Efficient ADMM-Based Federated Learning

---

1: Initialize $\mathbf{x}_i^0, \boldsymbol{\pi}_i^0, \rho_i > 0$, for all $i \in \{1, 2, \ldots, N\}$, an integer $t_0 > 0$. Set $t \Leftarrow 0$.

2: **for** each round $t = 0, 1, 2, \ldots$ **do**

3:    **if** $t \in \mathcal{T} = \{0, t_0, 2t_0, \ldots\}$ **then**

4:       Local uploading: Each client sends the parameters $\mathbf{x}_i^t$ and $\boldsymbol{\pi}_i^t$ to the central server.

5:       Global model update: The central server calculates the average parameter $\mathbf{x}^{t+1}$ by

$$\mathbf{x}^{t+1} = \sum_{i=1}^{N} \frac{\rho_i \mathbf{x}_i^t}{\rho} + \sum_{i=1}^{N} \frac{\boldsymbol{\pi}_i^t}{\rho}. \tag{6.79}$$

6:       Global model broadcasting: The central server broadcasts the parameter $\mathbf{x}^{t+1}$ to every client.

7:    **for** each client $i = 1, 2, \ldots, N$ **in parallel do**

8:       Local updates: By letting

$$\mathbf{y}^{t+1} = \mathbf{x}^{\tau_t + 1}, \quad \text{where} \quad \tau_t = \lfloor t/t_0 \rfloor t_0,$$

9:       update its parameters by

$$\mathbf{x}_i^{t+1} = \arg\min_{\mathbf{x}_i} w_i f_i(\mathbf{x}_i) + \langle \mathbf{x}_i - \mathbf{y}^{t+1}, \boldsymbol{\pi}_i^t \rangle$$

$$+ \frac{\rho_i}{2} \|\mathbf{x}_i - \mathbf{y}^{t+1}\|^2, \tag{6.80}$$

$$\boldsymbol{\pi}_i^{t+1} = \boldsymbol{\pi}_i^t + \rho_i(\mathbf{x}_i^{t+1} - \mathbf{y}^{t+1}). \tag{6.81}$$

Linearized Inexact ADMM-Based Federated Learning

In CEADMM, each client i is required to compute $\mathbf{x}_i^{t+1}$ and $\boldsymbol{\pi}_i^{t+1}$ after receiving the global parameter $\mathbf{y}^{t+1}$. While the calculation of $\boldsymbol{\pi}_i^{t+1}$ based on the closed-form solution is efficient, determining $\mathbf{x}_i^{t+1}$ involves solving the sub-problem outlined in (6.80), which generally lacks a closed-form solution and is computationally intensive. To expedite local computation, we choose to solve this sub-problem approximately.

To obtain an approximate solution of the sub-problem, we perform a second-order Taylor expansion of f_i at a point $\mathbf{z}_i^t$ close to $\mathbf{x}_i$, yielding

$$h_i(\mathbf{x}_i; \mathbf{z}_i^t, \mathbf{H}_i) = f_i(\mathbf{z}_i^t) + \langle \nabla f_i(\mathbf{z}_i^t), \mathbf{x}_i - \mathbf{z}_i^t \rangle + \frac{1}{2}(\mathbf{x}_i - \mathbf{z}_i^t)^T \mathbf{H}_i(\mathbf{x}_i - \mathbf{z}_i^t). \tag{6.82}$$

Then, the sub-problem in (6.80) can be approximated by

$$\mathbf{x}_i^{t+1} = \arg\min_{\mathbf{x}_i} w_i h_i(\mathbf{x}_i; \mathbf{z}_i^t, \mathbf{H}_i) + \langle \mathbf{x}_i - \mathbf{y}^{t+1}, \boldsymbol{\pi}_i^t \rangle + \frac{\rho_i}{2} \|\mathbf{x}_i - \mathbf{x}^{t+1}\|^2$$

$$= \mathbf{z}_i^t - (w_i \mathbf{H}_i + \rho_i \mathbf{I})^{-1} \left[\rho_i(\mathbf{z}_i^t - \mathbf{y}^{t+1}) + w_i \nabla f_i(\mathbf{z}_i^t) + \boldsymbol{\pi}_i^t \right], \tag{6.83}$$

where $\mathbf{H}_i \succeq \mathbf{0}$ (i.e., $\mathbf{H}_i$ is positive semidefinite) can be chosen to satisfy $\mathbf{H}_i \approx \nabla^2 f_i$. If f_i is L_i-smooth, then $\mathbf{H}_i$ can be chosen as $\mathbf{H}_i \approx L_i \mathbf{I}$.

For choosing the local point $\mathbf{z}_i^t$, there are two potential candidates: the previous local parameter $\mathbf{x}_i^t$ or the updated parameter $\mathbf{x}_i^{t+1}$ from the central server.

Choice 1: If $\mathbf{z}_i^t = \mathbf{x}_i^t$, denoting $g_i^t = w_i \nabla f_i(\mathbf{x}_i^t)$, then (6.83) turns into

$$\mathbf{x}_i^{t+1} = \mathbf{x}_i^t - (w_i \mathbf{H}_i + \rho_i \mathbf{I})^{-1}\left[\rho_i(\mathbf{x}_i^t - \mathbf{y}^{t+1}) + w_i \nabla f_i(\mathbf{x}_i^t) + \boldsymbol{\pi}_i^t\right]. \qquad (6.84)$$

Choice 2: If $\mathbf{z}_i^t = \mathbf{y}^{t+1}$, then (6.83) turns into

$$\mathbf{x}_i^{t+1} = \mathbf{y}^{t+1} - (w_i \mathbf{H}_i + \rho_i \mathbf{I})^{-1}\left[w_i \nabla f_i(\mathbf{y}^{t+1}) + \boldsymbol{\pi}_i^t\right]. \qquad (6.85)$$

In addition, the following settings are assumed:

1. $t_0 = 1$; Hence, aggregation and broadcasting are conducted at each step, signifying that $\mathbf{y}^{t+1} = \mathbf{x}^{t+1}$.
2. $\rho_i = w_i/\eta$, for all $i \in \{1, 2, \ldots, N\}$; therefore, given learning rate η, (6.78) turns into

$$\mathbf{x}^{t+1} = \sum_{i=1}^{N} w_i \mathbf{x}_i^t + \eta \sum_{i=1}^{N} \boldsymbol{\pi}_i^t. \qquad (6.86)$$

3. $\mathbf{z}_i^t = \mathbf{y}^{t+1}$ and $\mathbf{H}_i = \mathbf{0}$; This implies that we utilize Choice 2 while linearizing f_i. Taking into account the previously mentioned configurations $\mathbf{y}^{t+1} = \mathbf{x}^{t+1}$ and $\rho_i = w_i/\eta$, (6.85) becomes

$$\mathbf{x}_i^{t+1} = \mathbf{x}^{t+1} - \frac{w_i}{\rho_i} \nabla f_i(\mathbf{y}^{t+1}) - \frac{\boldsymbol{\pi}_i^t}{\rho_i} = \mathbf{x}^{t+1} - \eta \nabla f_i(\mathbf{x}^{t+1}) - \frac{\eta}{w_i} \boldsymbol{\pi}_i^t. \qquad (6.87)$$

Based on these settings, we derive the linearized inexact ADMM-based federated learning (LIADMM) algorithm in Algorithm 6.7.

In contrast to (6.4) and (6.3) in FedSGD, both (6.88) and (6.89) in LIADMM are characterized by an additional term related to the dual parameters $\boldsymbol{\pi}_i^t$. Therefore, the simplest FedSGD can be regarded as a special case of linearized inexact ADMM. A comparison of these two frameworks is illustrated in Figure 6.13.

Inexact Communication-Efficient ADMM-Based Federated Learning

LIADMM represents the simplest form of inexact ADMM, allowing clients to solve the subproblem in (6.80) imprecisely, thereby reducing computational complexity. However, this algorithm sets $t_0 = 1$, which is not communication-efficient. In order to reduce computational complexity and communication overhead simultaneously, we build upon CEADMM and set local epochs $t_0 > 1$. Moreover, with $t_0 > 1$, the approximate function $h_i(\mathbf{x}_i^{t+1}; \mathbf{x}_i^t, \mathbf{H}_i)$ tends to be more accurate than $h_i(\mathbf{x}^{t+1}; \mathbf{x}_i^t, \mathbf{H}_i)$, approaching $h_i(\mathbf{x}_i^t; \mathbf{x}_i^t, \mathbf{H}_i) = f_i(\mathbf{x}_i^t)$. Therefore, we adopt Choice 1, setting $\mathbf{z}_i^t = \mathbf{x}_i^t$. Then the inexact communication-efficient ADMM-based federated learning (ICEADMM) algorithm can be summarized in Algorithm 6.8.

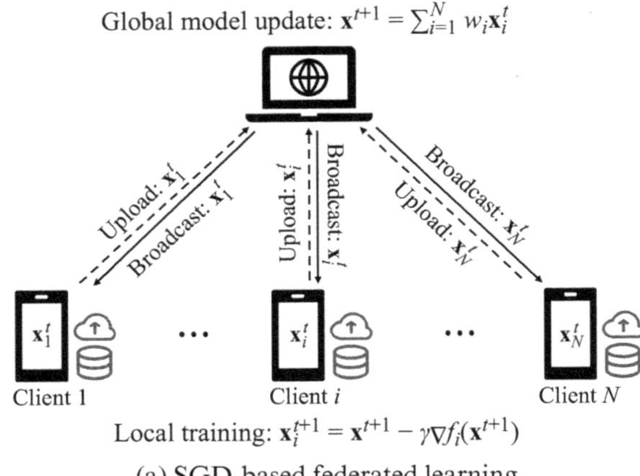

(a) SGD-based federated learning.

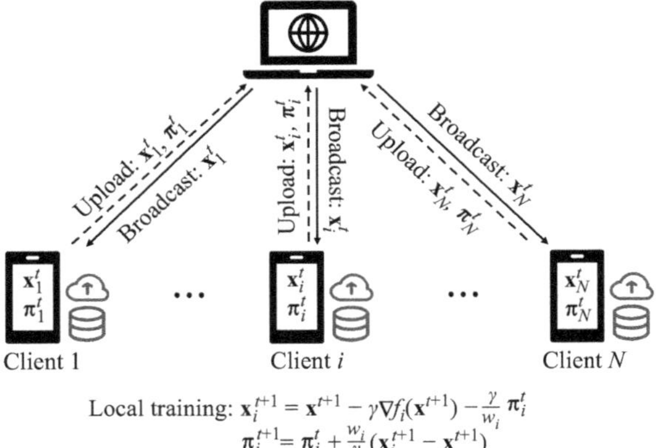

(b) Linearized inexact ADMM-based federated learning.

Figure 6.13 Illustrative comparison of FedSGD and LIADMM.

Supposing that every f_i, for all $i \in \{1, 2, \ldots, N\}$ is L_i-smooth where $i > 0$, then there exists a Θ_i such that $L_i \mathbf{I} \succeq \Theta_i \succeq \mathbf{0}$, i.e. $L_i \mathbf{I} - \Theta_i$ and Θ_i are both positive semidefinite, and for any $\mathbf{x}_i$ and $\mathbf{z}_i \in \mathbb{R}^d$, we have

$$f_i(\mathbf{x}_i) \leq f_i(\mathbf{z}_i) + \langle \nabla f_i(\mathbf{z}_i), \mathbf{x}_i - \mathbf{z}_i \rangle + \frac{1}{2}(\mathbf{x}_i - \mathbf{z}_i)^T \Theta_i (\mathbf{x}_i - \mathbf{z}_i), \tag{6.93}$$

where the existence is obvious since we at least can choose $\Theta_i = L_i \mathbf{I}$. To analyze the convergence property of ICEADMM, we consider the sequence $\{(\mathbf{y}^t, \mathbf{X}^t, \Pi^t)\}$ generated by ICEADMM with $\mathbf{H}_i = \Theta_i$ and $\rho_i > 3\sqrt{2} w_i L_i$, for all $i \in \{1, 2, \ldots, N\}$. The following results are valid under mild assumptions, and the detailed proof can be founded in [224].

1. The sequence $\{(\mathbf{y}^t, \mathbf{X}^t, \Pi^t)\}$ is bounded, and any accumulation point $(\mathbf{y}^\infty, \mathbf{X}^\infty, \Pi^\infty)$ of this sequence is a stationary point of the reformulated problem in (6.71), where $\mathbf{y}^\infty$ is a stationary point of the problem in (6.70).

Algorithm 6.7 LIADMM: Linearized Inexact ADMM-Based Federated Learning

---

1: Initialize $\mathbf{x}_i^0, \boldsymbol{\pi}_i^0$, a learning rate $\eta > 0$. Set $t \Leftarrow 0$.
2: **for** each round $t = 0, 1, 2, \ldots$ **do**
3:      Local uploading: Each client sends the parameters $\mathbf{x}_i^t$ and $\boldsymbol{\pi}_i^t$ to the central server.
4:      Global model update: The central server calculates the average parameter $\mathbf{x}^{t+1}$ by

$$\mathbf{x}^{t+1} = \sum_{i=1}^{N} w_i \mathbf{x}_i^t + \eta \sum_{i=1}^{N} \boldsymbol{\pi}_i^t. \tag{6.88}$$

5:      Global model broadcasting: The central server broadcasts the parameter $\mathbf{x}^{t+1}$ to every client.
6:      **for** each client $i = 1, 2, \ldots, N$ **in parallel do**
7:          Local updates: Update its parameters by

$$\mathbf{x}_i^{t+1} = \mathbf{x}^{t+1} - \eta \nabla f_i(\mathbf{x}^{t+1}) - \frac{\eta}{w_i} \boldsymbol{\pi}_i^t, \tag{6.89}$$

$$\boldsymbol{\pi}_i^{t+1} = \boldsymbol{\pi}_i^t + \frac{w_i}{\eta}(\mathbf{x}_i^{t+1} - \mathbf{x}^{t+1}). \tag{6.90}$$

---

2. If we further assume that $\mathbf{y}^\infty$ is isolated, then the entire sequence $\{(\mathbf{y}^t, \mathbf{X}^t, \Pi^t)\}$ converges to $(\mathbf{y}^\infty, \mathbf{X}^\infty, \Pi^\infty)$.

In summary, we have introduced three ADMM-based federated learning algorithms: CEADMM, LIADMM, and ICEADMM. These approaches effectively address challenges such as heavy communication loads and high computational complexity. Additionally, inexact ADMM-based federated learning algorithm, proposed by [225], mitigates the straggler effect by incorporating a client selection step, while FedGiA, introduced by [226], integrates the SGD framework with ADMM to handle scenarios involving clients under poor communication conditions.

6.6   Exercises

Exercise 6.1   Describe how federated learning enables machine learning models to be trained across many distributed devices with local data samples without sharing data. Discuss the privacy and bandwidth benefits of this approach compared to traditional centralized machine learning.

Exercise 6.2   The SGD algorithm is a fundamental optimization method for the efficient training of machine learning models. The update rule is defined as

$$\mathbf{x}^{t+1} = \mathbf{x}^t - \eta_t \nabla f(\mathbf{x}^t, \xi^t),$$

where the objective $f(\cdot)$ is assume to be L-smooth and μ-strongly convex.

Algorithm 6.8 ICEADMM: Inexact Communication-Efficient ADMM-Based Federated Learning

---

1: Initialize $\mathbf{x}_i^0, \boldsymbol{\pi}_i^0, \rho_i > 0, \mathbf{H}_i \succeq \mathbf{0}$, for all $i \in \{1, 2, \ldots, N\}$, an integer $t_0 > 0$. Set $t \Leftarrow 0$.

2: **for** each round $t = 0, 1, 2, \ldots$ **do**

3:      **if** $t \in \mathcal{T} = \{0, t_0, 2t_0, \ldots\}$ **then**

4:          Local uploading: Each client sends the parameters $\mathbf{x}_i^t$ and $\boldsymbol{\pi}_i^t$ to the central server.

5:          Global model update: The central server calculates the average parameter $\mathbf{x}^{t+1}$ by

$$\mathbf{x}^{t+1} = \sum_{i=1}^{N} \frac{\rho_i \mathbf{x}_i^t}{\rho} + \sum_{i=1}^{N} \frac{\boldsymbol{\pi}_i^t}{\rho}. \tag{6.91}$$

6:          Global model broadcasting: The central server broadcasts the parameter $\mathbf{x}^{t+1}$ to every client.

7:      **for** each client $i = 1, 2, \ldots, N$ **in parallel do**

8:          Local updates: By letting

$$\mathbf{y}^{t+1} = \mathbf{x}^{\tau_t + 1}, \quad \text{where} \quad \tau_t = \lfloor t/t_0 \rfloor t_0,$$

9:          update its parameters by

$$\begin{aligned}
\mathbf{x}_i^{t+1} &= \mathbf{x}_i^t - (w_i \mathbf{H}_i + \rho_i \mathbf{I})^{-1} \left[\rho_i(\mathbf{x}_i^t - \mathbf{y}^{t+1}) + w_i \nabla f_i(\mathbf{x}_i^t) + \boldsymbol{\pi}_i^t \right], \\
\boldsymbol{\pi}_i^{t+1} &= \boldsymbol{\pi}_i^t + \rho_i(\mathbf{x}_i^{t+1} - \mathbf{y}^{t+1}).
\end{aligned} \tag{6.92}$$

(a) Under ideal conditions, the stochastic gradient $\nabla f(\mathbf{x}^t, \xi^t) = \nabla f(\mathbf{x}^t)$. Prove that if the constant learning rate $\eta_t = \mu/L^2$, then it holds that

$$\|\mathbf{x}^t - \mathbf{x}^*\|^2 \leq \left(1 - \frac{\mu^2}{L^2}\right)^t \|\mathbf{x}^0 - \mathbf{x}^*\|^2,$$

thereby demonstrating the linear convergence rate.

(b) More practically, we assume that the stochastic gradient is unbiased in that $\mathbb{E}[\nabla f(\mathbf{x}^t, \xi^t)] = \nabla f(\mathbf{x}^t)$, and that the variance is bounded as

$$\mathbb{E}[\|\nabla f(\mathbf{x}^t, \xi^t) - \nabla f(\mathbf{x}^t)\|^2] \leq \sigma^2.$$

Prove that if the constant learning rate $\eta_t = \eta \leq \mu/L^2$, then it holds that

$$\mathbb{E}[\|\mathbf{x}^t - \mathbf{x}^*\|^2] \leq (1 - \eta\mu)^t \|\mathbf{x}^0 - \mathbf{x}^*\|^2 + \frac{\eta\sigma^2}{\mu} \left[1 - (1 - \eta\mu)^t\right].$$

Exercise 6.3 The FedSGD algorithm is a foundational approach in federated learning, facilitating decentralized model training by aggregating gradients computed locally across multiple clients. The local training and global model aggregation are formally expressed as

$$\mathbf{g}_i^t = \nabla f_i(\mathbf{x}^t, \xi_i^t)$$

and

$$\mathbf{x}^{t+1} = \mathbf{x}^t - \eta_t \mathbf{g}^t = \mathbf{x}^t - \eta_t \sum_{i=1}^{N} w_i \mathbf{g}_i^t.$$

Similar to Exercise 6.2, we assume that each local stochastic gradient is unbiased in that $\mathbb{E}[f_i(\mathbf{x}^t, \xi_i^t)] = \nabla f_i(\mathbf{x}^t)$ and that the variance is bounded as

$$\mathbb{E}[\|f_i(\mathbf{x}^t, \xi_i^t) - \nabla f_i(\mathbf{x}^t)\|^2] \le \sigma_i^2.$$

Prove that under the Assumptions 6.1 and 6.6, if the constant learning rate $\eta_t = \eta \le \mu/L^2$, then it holds that

$$\mathbb{E}[\|\mathbf{x}^t - \mathbf{x}^*\|^2] \le (1 - \eta\mu)^t \|\mathbf{x}^0 - \mathbf{x}^*\|^2 + \frac{\eta\bar{\sigma}^2}{\mu}\left[1 - (1 - \eta\mu)^t\right],$$

where $\bar{\sigma}^2 = \sum_{i=1}^{N} w_i^2 \sigma_i^2$ denotes the weighted sum of local variance.

Exercise 6.4 Develop a Python program to implement the FedProx algorithm, simulating a scenario where data is distributed in a non-i.i.d. manner across multiple clients. The program should simulate the training and communication process between clients and the central server, and output the test accuracy of the global model after each round of communication. Experiment with different values of the proximal term coefficient μ to observe its effect on the training performance.

The implementation should include the following:

(a) Non-i.i.d. data: Use a Dirichlet distribution to partition the data unevenly across clients.
(b) Proximal term: Add a proximal term to the loss function to constrain local updates from drifting too far from the global model.
(c) Plotting results: After multiple communication rounds, generate a plot showing test accuracy versus communication rounds to track the training performance over time.

In this setup, the number of clients is $N = 20$, and there are $T = 100$ communication rounds, with each client training locally for $E = 5$ epochs per round. Data can be split using a Dirichlet distribution with the parameter $\alpha = 0.5$ to simulate non-i.i.d. conditions, and a proximal term with coefficient $\mu = 0, 0.01, 0.1$ can be added. The MNIST dataset can be automatically downloaded via PyTorch. The code snippet has been provided in https://github.com/le-liang/wcmlbook/blob/main/ch6/Exercise_6.4/.

Exercise 6.5 Discuss how wireless communication constraints affect the design and efficiency of federated learning systems. What are some strategies that can be implemented to mitigate these effects?

Exercise 6.6 Develop a Python program to simulate the impact of packet loss on federated learning using the FedAvg algorithm. Implement a CNN model for

MNIST classification, and distribute the data among 100 clients using a Dirichlet distribution to create a non-i.i.d. scenario. Simulate packet loss at rates of 1%, 5%, and 10%, and observe the effects on model accuracy and convergence over multiple communication rounds. Track the global model's test accuracy and loss, and visualize the impact of different packet loss rates. The code snippet has been provided in https://github.com/le-liang/wcmlbook/blob/main/ch6/Exercise_6.6/.

Exercise 6.7   Demonstrate that the objective function of the optimization problem presented in (6.26) is convex. Formally, a function $f: \mathbb{R}^n \to \mathbb{R}$ is convex if and only if for any $\mathbf{x}, \mathbf{y} \in \mathbb{R}^n$ and $\lambda \in [0, 1]$, the following inequality holds:

$$f(\lambda \mathbf{x} + (1 - \lambda)\mathbf{y}) \le \lambda f(\mathbf{x}) + (1 - \lambda)f(\mathbf{y}).$$

Exercise 6.8   Calculate the expectation of $\mathbf{X}_{t+1}$ and $(\mathbf{X}_l^{t+1})^T(\mathbf{X}_l^{t+1})$ under the Soft-DSGD updating scheme in (6.21).

Exercise 6.9   Given the convergence results of Soft-DSGD as stated in Theorem 6.1, prove Corollary 6.1.

Exercise 6.10   Conduct a comparative analysis of the performance of federated learning under sequential communication and computation using OMA versus over-the-air computation. The analysis should address the following aspects:

(a) Compare the differences between the two federated learning methods in modulation type and communication resource consumption.
(b) Compare the risks of privacy leakage of the two federated learning methods.
(c) Compare the performance of the two federated learning methods in terms of aggregation accuracy.

Exercise 6.11   DSI-and-CSI-aware transceiver design can be formulated as an optimization problem in (6.51).

(a) Show that the problem in (6.51) can be equivalently transformed into the following problem:

$$\max_{\mathbf{v}_t} \min_{i} \quad \frac{P}{w_i^2}\left|\mathbf{v}_t^H \bar{\mathbf{h}}_{i,t}\right|^2$$
$$\text{s.t.} \qquad \|\mathbf{v}_t\|_2^2 = 1.$$

(b) Introducing $\tilde{\mathbf{v}}_t = \tilde{\mathbf{v}}_t / \|\tilde{\mathbf{v}}_t\|_2$ and slack variable $\boldsymbol{\gamma} = [\gamma_1, \gamma_2, \ldots, \gamma_N]^T \in \mathbb{R}^N$, demonstrate that the above optimization problem can be solved via successive convex approximation. Particularly, the jth convex problem to be solved is given by

$$\min_{\tilde{\mathbf{v}}_t, \boldsymbol{\gamma}} \quad \|\tilde{\mathbf{v}}_t\|_2^2 + \lambda \|\boldsymbol{\gamma}\|_2^2$$
$$\text{s.t.} \quad -2\mathrm{Re}\left\{\mathbf{z}^H \bar{\mathbf{h}}_{i,t} \bar{\mathbf{h}}_{i,t}^H \tilde{\mathbf{v}}_t\right\}$$
$$\qquad + \left|\mathbf{z}_t^H \bar{\mathbf{h}}_{i,t}\right|^2 \le \gamma_i - w_i^2, \quad \text{for all } i \in \{1, 2, \ldots, N\},$$
$$\qquad \gamma_i \ge 0, \qquad\qquad\quad \text{for all } i \in \{1, 2, \ldots, N\},$$

where $\mathbf{z}$ is the solution of $\tilde{\mathbf{v}}_t$ in the $(j - 1)$th iteration.

Exercise 6.12 The algorithmic evolution of SGD-based and ADMM-based federated learning present notable differences and connections. Provide a comparative analysis of the FedSGD algorithm and LIADMM algorithm, focusing on their convergence efficiency and privacy.

References

[1] W. S. McCulloch and W. Pitts, "A logical calculus of the ideas immanent in nervous activity," *Bulletin of Mathematical Biophysics*, vol. 5, pp. 115–133, Dec. 1943.

[2] A. Krizhevsky, I. Sutskever, and G. E. Hinton, "Imagenet classification with deep convolutional neural networks," in *Proceedings of the Advances in Neural Information Processing Systems*, 2012, pp. 1097–1105.

[3] C. Weng, D. Yu, S. Watanabe, and B.-H. F. Juang, "Recurrent deep neural networks for robust speech recognition," in *Proceedings of the IEEE International Conference on Acoustics, Speech and Signal Processing*, 2014, pp. 5532–5536.

[4] J. Bruck and M. Blaum, "Neural networks, error-correcting codes, and polynomials over the binary n-cube," *IEEE Transactions on Information Theory*, vol. 35, no. 5, pp. 976–987, Aug. 1989.

[5] F. Jondral, "Automatic classification of high frequency signals," *Signal Processing*, vol. 9, no. 3, pp. 177–190, Oct. 1985.

[6] M. J. Demongeot, M. J. Mazoyer, M. P. Peretto, and M. D. Whitley, "Neural network synthesis using cellular encoding and the genetic algorithm," Ph.D. dissertation, Laboratoire de l'Informatique du Parallilisme, Ecole Normale Supirieure de Lyon, Lyon, France, 1994.

[7] D. Chen, B. Sheu, and T. Berger, "A compact neural network based CDMA receiver for multimedia wireless communication," in *Proceedings of the International Conference on Computer Design. VLSI in Computers and Processors*, 1996, pp. 99–103.

[8] M. Bkassiny, Y. Li, and S. K. Jayaweera, "A survey on machine-learning techniques in cognitive radios," *IEEE Communications Surveys And Tutorials*, vol. 15, no. 3, pp. 1136–1159, Oct. 2013.

[9] E. Nachmani, Y. Be'ery, and D. Burshtein, "Learning to decode linear codes using deep learning," in *Proceedings of the Annual Allerton Conference on Communication, Control, and Computing*, 2016, pp. 341–346.

[10] T. O'shea and J. Hoydis, "An introduction to deep learning for the physical layer," *IEEE Transactions on Cognitive Communications and Networking*, vol. 3, no. 4, pp. 563–575, Oct. 2017.

[11] F. A. Aoudia and J. Hoydis, "Model-free training of end-to-end communication systems," *IEEE Journal on Selected Areas in Communications*, vol. 37, no. 11, pp. 2503–2516, Nov. 2019.

[12] T. O'Shea, "Learning from data in radio algorithm design," Ph.D. dissertation, Virginia Polytechnic Institute and State University, Arlington, VA, 2017.

[13] H. He, C.-K. Wen, S. Jin, and G. Y. Li, "Deep learning-based channel estimation for beamspace mmWave massive MIMO systems," *IEEE Wireless Communications Letters*, vol. 7, no. 5, pp. 852–855, Oct. 2018.

[14] C.-K. Wen, W.-T. Shih, and S. Jin, "Deep learning for massive MIMO CSI feedback," *IEEE Wireless Communications Letters*, vol. 7, no. 5, pp. 748–751, Oct. 2018.

[15] H. Ye, G. Y. Li, and B.-H. Juang, "Power of deep learning for channel estimation and signal detection in OFDM systems," *IEEE Wireless Communications Letters*, vol. 7, no. 1, pp. 114–117, Feb. 2018.

[16] T. Gruber, S. Cammerer, J. Hoydis, and S. ten Brink, "On deep learning-based channel decoding," in *Proceedings of the*

Annual Conference on Information Sciences and Systems, 2017, pp. 1–6.

[17] S. Dorner, S. Cammerer, J. Hoydis, and S. t. Brink, "Deep learning based communication over the air," *IEEE Journal of Selected Topics in Signal Processing*, vol. 12, no. 1, pp. 132–143, Feb. 2018.

[18] H. Sun, X. Chen, Q. Shi, M. Hong, X. Fu, and N. D. Sidiropoulos, "Learning to optimize: Training deep neural networks for interference management," *IEEE Transactions on Signal Processing*, vol. 66, no. 20, pp. 5438–5453, Oct. 2018.

[19] B. Matthiesen, A. Zappone, K.-L. Besser, E. A. Jorswieck, and M. Debbah, "A globally optimal energy-efficient power control framework and its efficient implementation in wireless interference networks," *IEEE Transactions on Signal Processing*, vol. 68, pp. 3887–3902, Jun. 2020.

[20] M. Lee, Y. Xiong, G. Yu, and G. Y. Li, "Deep neural networks for linear sum assignment problems," *IEEE Wireless Communications Letters*, vol. 7, no. 6, pp. 962–965, Dec. 2018.

[21] W. Cui, K. Shen, and W. Yu, "Spatial deep learning for wireless scheduling," *IEEE Journal on Selected Areas in Communications*, vol. 37, no. 6, pp. 1248–1261, Jun. 2019.

[22] M. Eisen, C. Zhang, L. F. Chamon, D. D. Lee, and A. Ribeiro, "Learning optimal resource allocations in wireless systems," *IEEE Transactions on Signal Processing*, vol. 67, no. 10, pp. 2775–2790, Apr. 2019.

[23] R. Li, Z. Zhao, Q. Sun, I. Chih-Lin, C. Yang, X. Chen, M. Zhao, and H. Zhang, "Deep reinforcement learning for resource management in network slicing," *IEEE Access*, vol. 6, pp. 74 429–74 441, Nov. 2018.

[24] Y. He, N. Zhao, and H. Yin, "Integrated networking, caching, and computing for connected vehicles: A deep reinforcement learning approach," *IEEE Transactions on Vehicular Technology*, vol. 67, no. 1, pp. 44–55, Oct. 2017.

[25] X. Chen, Z. Zhao, C. Wu, M. Bennis, H. Liu, Y. Ji, and H. Zhang, "Multi-tenant cross-slice resource orchestration: A deep reinforcement learning approach," *IEEE Journal on Selected Areas in Communications*, vol. 37, no. 10, pp. 2377–2392, Aug. 2019.

[26] C. M. Bishop, *Pattern Recognition and Machine Learning*. Springer, 2006.

[27] K. P. Murphy, *Machine Learning: A Probabilistic Perspective*. MIT Press, 2012.

[28] I. Goodfellow, Y. Bengio, and A. Courville, *Deep Learning*. MIT Press, 2016.

[29] R. S. Sutton and A. G. Barto, *Reinforcement Learning: An Introduction*. MIT Press, 2018.

[30] S. Hu, Y.-d. Yao, and Z. Yang, "MAC protocol identification using support vector machines for cognitive radio networks," *IEEE Wireless Communications*, vol. 21, no. 1, pp. 52–60, Feb. 2014.

[31] Y. LeCun, L. Bottou, Y. Bengio, and P. Haffner, "Gradient-based learning applied to document recognition," *Proceedings of the IEEE*, vol. 86, no. 11, pp. 2278–2324, Nov. 1998.

[32] J. Schmidhuber, S. Hochreiter *et al.*, "Long short-term memory," *Neural Computation*, vol. 9, no. 8, pp. 1735–1780, 1997.

[33] A. Vaswani, "Attention is all you need," in *Proceedings of the Advances in Neural Information Processing Systems*, pp. 6000–6010, 2017.

[34] Y. Song, J. Sohl-Dickstein, D. P. Kingma, A. Kumar, S. Ermon, and B. Poole, "Score-based generative modeling through stochastic differential equations," in *Proceedings of the International Conference on Learning Representations*, 2020.

[35] J. Sohl-Dickstein, E. Weiss, N. Maheswaranathan, and S. Ganguli, "Deep unsupervised learning using nonequilibrium thermodynamics," in *Proceedings of the International Conference on Machine Learning*, 2015, pp. 2256–2265.

[36] J. Ho, A. Jain, and P. Abbeel, "Denoising diffusion probabilistic models," in *Proceedings of the Advances in Neural*

Information Processing Systems, 2020, pp. 6840–6851.

[37] P. Vincent, "A connection between score matching and denoising autoencoders," *Neural Computation*, vol. 23, no. 7, pp. 1661–1674, Jul. 2011.

[38] Y. Song and S. Ermon, "Generative modeling by estimating gradients of the data distribution," in *Proceedings of the Advances in Neural Information Processing Systems*, 2019, pp. 11 918–11 930.

[39] G. O. Roberts and R. L. Tweedie, "Exponential convergence of Langevin distributions and their discrete approximations," *Bernoulli*, vol. 2, no. 4, pp. 341–363, Dec. 1996.

[40] V. Mnih, K. Kavukcuoglu, D. Silver, A. A. Rusu, J. Veness, M. G. Bellemare, A. Graves, M. Riedmiller, A. K. Fidjeland, G. Ostrovski *et al.*, "Human-level control through deep reinforcement learning," *Nature*, vol. 518, no. 7540, pp. 529–533, Feb 2015.

[41] J. Schulman, S. Levine, P. Abbeel, M. Jordan, and P. Moritz, "Trust region policy optimization," in *Proceedings of the International Conference on Machine Learning*, 2015, pp. 1889–1897.

[42] V. Mnih, A. P. Badia, M. Mirza, A. Graves, T. Lillicrap, T. Harley, D. Silver, and K. Kavukcuoglu, "Asynchronous methods for deep reinforcement learning," in *Proceedings of the International Conference on Machine Learning*, 2016, pp. 1928–1937.

[43] V. François-Lavet, P. Henderson, R. Islam, M. G. Bellemare, J. Pineau *et al.*, "An introduction to deep reinforcement learning," *Foundations and Trends® in Machine Learning*, vol. 11, no. 3–4, pp. 219–354, Dec. 2018.

[44] J. Schulman, F. Wolski, P. Dhariwal, A. Radford, and O. Klimov, "Proximal policy optimization algorithms," *arXiv preprint arXiv:1707.06347*, 2017.

[45] P. Kairouz, H. B. McMahan, B. Avent, A. Bellet, M. Bennis, A. N. Bhagoji, K. Bonawitz, Z. Charles, G. Cormode, R. Cummings *et al.*, "Advances and open problems in federated learning,"

Foundations and Trends® in Machine Learning, vol. 14, no. 1–2, pp. 1–210, Jun. 2021.

[46] B. McMahan, E. Moore, D. Ramage, S. Hampson, and B. A. y Arcas, "Communication-efficient learning of deep networks from decentralized data," in *Proceedings of the Artificial Intelligence and Statistics*, 2017, pp. 1273–1282.

[47] C. B. Finn, "Learning to learn with gradients," Ph.D. dissertation, University of California, Berkeley, CA, 2018.

[48] T. P. Lillicrap, J. J. Hunt, A. Pritzel, N. Heess, T. Erez, Y. Tassa, D. Silver, and D. Wierstra, "Continuous control with deep reinforcement learning," *arXiv preprint arXiv:1509.02971*, 2015.

[49] T. Haarnoja, A. Zhou, P. Abbeel, and S. Levine, "Soft actor-critic: Off-policy maximum entropy deep reinforcement learning with a stochastic actor," in *Proceedings of the International Conference on Machine Learning*, 2018, pp. 1861–1870.

[50] S. Dankwa and W. Zheng, "Twin-delayed DDPG: A deep reinforcement learning technique to model a continuous movement of an intelligent robot agent," in *Proceedings of the International Conference on Vision, Image and Signal Processing*, 2019, pp. 1–5.

[51] A. Goldsmith, *Wireless Communications*. Cambridge University Press, 2005.

[52] G. L. Stüber, *Principles of Mobile Communications*, 4th ed. Springer, 2017.

[53] D. Tse and P. Viswanath, *Fundamentals of Wireless Communication*. Cambridge University Press, 2005.

[54] P. Bello, "Characterization of random time-variant linear channels," *IEEE Transactions on Communications*, vol. 11, no. 4, pp. 360–393, Dec. 1963.

[55] C. Huang *et al.*, "Artificial intelligence enabled radio propagation for communications–part II: Scenario identification and channel modeling," *IEEE Transactions on Antennas and Propagation*, vol. 70, no. 6, pp. 3955–3969, Jun. 2022.

[56] J. Huang *et al.*, "A big data enabled channel model for 5G wireless communication systems," *IEEE Transactions on Big Data*, vol. 6, no. 2, pp. 211–222, Jun. 2020.

[57] Y. Yang, Y. Li, W. Zhang, F. Qin, P. Zhu, and C.-X. Wang, "Generative-adversarial-network-based wireless channel modeling: Challenges and opportunities," *IEEE Communications Magzine*, vol. 57, no. 3, pp. 22–27, Mar. 2019.

[58] T. J. O'Shea, T. Roy, and N. West, "Approximating the void: Learning stochastic channel models from observation with variational generative adversarial networks," in *Proceedings of the IEEE International Conference on Computing, Networking and Communication*, 2019, pp. 681–686.

[59] S. Seyedsalehi, V. Pourahmadi, H. Sheikhzadeh, and A. Foumani, "Propagation channel modeling by deep learning techniques," *arXiv preprint arXiv: 1908.06767*, 2019.

[60] M. Arjovsky, S. Chintala, and L. Bottou, "Wasserstein GAN," *arXiv preprint arXiv: 1701.07875*, 2017.

[61] H. Xiao, W. Tian, W. Liu, and J. Shen, "ChannelGAN: Deep learning-based channel modeling and generating," *IEEE Wireless Communication Letters*, vol. 11, no. 3, pp. 650–654, Mar. 2022.

[62] W. Xia *et al.*, "Generative neural network channel modeling for millimeter-wave UAV communication," *IEEE Transactions on Wireless Communication*, vol. 21, no. 11, pp. 9417–9431, Nov. 2022.

[63] C. Doersch, "Tutorial on variational autoencoders," *arXiv preprint arXiv: 1606.05908*, 2016.

[64] J.-J. Van De Beek, O. Edfors, M. Sandell, S. K. Wilson, and P. O. Borjesson, "On channel estimation in OFDM systems," in *Proceedings of the IEEE Vehicular Technology Conference*, vol. 2, 1995, pp. 815–819.

[65] O. Edfors, M. Sandell, J.-J. Van de Beek, S. K. Wilson, and P. O. Borjesson, "OFDM channel estimation by singular value decomposition," *IEEE Transactions on Communications*, vol. 46, no. 7, pp. 931–939, Jul. 1998.

[66] Y. Li, L. J. Cimini, and N. R. Sollenberger, "Robust channel estimation for OFDM systems with rapid dispersive fading channels," *IEEE Transactions on Communications*, vol. 46, no. 7, pp. 902–915, Jul. 1998.

[67] Y. Liu, Z. Tan, H. Hu, L. J. Cimini, and G. Y. Li, "Channel estimation for OFDM," *IEEE Communications Surveys & Tutorials*, vol. 16, no. 4, pp. 1891–1908, 2014.

[68] Y. Li, "Pilot-symbol-aided channel estimation for OFDM in wireless systems," *IEEE Transactions on Vehicular Technology*, vol. 49, no. 4, pp. 1207–1215, Jul. 2000.

[69] Y. Li, N. Seshadri, and S. Ariyavisitakul, "Channel estimation for OFDM systems with transmitter diversity in mobile wireless channels," *IEEE Journal on Selected Areas in Communications*, vol. 17, no. 3, pp. 461–471, Mar. 1999.

[70] Y. Li, "Simplified channel estimation for OFDM systems with multiple transmit antennas," *IEEE Transactions on Wireless Communications*, vol. 1, no. 1, pp. 67–75, Jan. 2002.

[71] X. Gao, S. Jin, C.-K. Wen, and G. Y. Li, "ComNet: Combination of deep learning and expert knowledge in OFDM receivers," *IEEE Communications Letters*, vol. 22, no. 12, pp. 2627–2630, Dec. 2018.

[72] P. Dong, H. Zhang, G. Y. Li, I. S. Gaspar, and N. NaderiAlizadeh, "Deep CNN-based channel estimation for mmWave massive MIMO systems," *IEEE Journal of Selected Topics in Signal Processing*, vol. 13, no. 5, pp. 989–1000, Sep. 2019.

[73] E. Balevi, A. Doshi, A. Jalal, A. Dimakis, and J. G. Andrews, "High dimensional channel estimation using deep generative networks," *IEEE Journal on Selected Areas in Communications*, vol. 39, no. 1, pp. 18–30, Jan. 2021.

[74] A. Bora, A. Jalal, E. Price, and A. G. Dimakis, "Compressed sensing using generative models," in *Proceedings of the International Conference on Machine Learning*, 2017, pp. 537–546.

[75] M. Arvinte and J. I. Tamir, "MIMO channel estimation using score-based generative models," *IEEE Transactions on Wireless Communications*, vol. 22, no. 6, pp. 3698–3713, Jun. 2023.

[76] G. Lin, A. Milan, C. Shen, and I. Reid, "Refinenet: Multi-path refinement networks for high-resolution semantic segmentation," in *Proceedings of the IEEE Conference on Computer Vision and Pattern Recognition*, 2017, pp. 1925–1934.

[77] A. Jalal, M. Arvinte, G. Daras, E. Price, A. G. Dimakis, and J. Tamir, "Robust compressed sensing MRI with deep generative priors," in *Proceedings of the Advances in Neural Information Processing Systems*, vol. 34, 2021, pp. 14938–14954.

[78] 3rd Generation Partnership Project, "Technical Specification Group Radio Access Network; Study on channel model for frequencies from 0.5 to 100 GHz (Release 16)," *3GPP TR 38.901 V16.1.0*, Jan. 2020.

[79] W. Jin, H. He, C.-K. Wen, S. Jin, and G. Y. Li, "Adaptive channel estimation based on model-driven deep learning for wideband mmwave systems," in *Proceedings of the IEEE Global Communications Conference*, 2021, pp. 1–6.

[80] E. J. Candès, J. Romberg, and T. Tao, "Robust uncertainty principles: Exact signal recognition from highly incomplete frequency information," *IEEE Transactions on Information Theory*, vol. 52, no. 2, pp. 489–509, Feb. 2006.

[81] D. Donoho, "Compressed sensing," *IEEE Transactions on Information Theory*, vol. 52, no. 4, pp. 1289–1306, Apr. 2006.

[82] E. J. Candès and T. Tao, "Near optimal signal recovery from random projections: Universal encoding strategies?" *IEEE Transactions on Information Theory*, vol. 52, no. 12, pp. 5406–5425, Dec. 2006.

[83] E. J. Candès and M. B. Wakin, "An introduction to compressive sampling," *IEEE Signal Processing Magazine*, vol. 25, no. 2, pp. 21–30, Mar. 2008.

[84] R. G. Baraniuk, "Compressive sensing," *IEEE Signal Processing Magazine*, vol. 24, no. 4, pp. 118–121, Jul. 2007.

[85] J. Wright and Y. Ma, *High-Dimensional Data Analysis with Low-Dimensional Models: Principles, Computation, and Applications*. Cambridge University Press, 2022.

[86] E. J. Candès and T. Tao, "Decoding by linear programming," *IEEE Transactions on Information Theory*, vol. 51, no. 12, pp. 4203–4215, Dec. 2005.

[87] S. S. Chen, D. L. Donoho, and M. A. Saunders, "Atomic decomposition by basis pursuit," *SIAM Review*, vol. 43, no. 1, pp. 129–159, 2001.

[88] R. Tibshirani, "Regression shrinkage and selection via the lasso," *Journal of the Royal Statistical Society Series B: Statistical Methodology*, vol. 58, no. 1, pp. 267–288, Dec. 1996.

[89] I. Daubechies, M. Defrise, and C. De Mol, "An iterative thresholding algorithm for linear inverse problems with a sparsity constraint," *Communications on Pure and Applied Mathematics*, vol. 57, no. 11, pp. 1413–1457, Aug. 2004.

[90] A. Beck and M. Teboulle, "A fast iterative shrinkage-thresholding algorithm for linear inverse problems," *SIAM Journal on Imaging Sciences*, vol. 2, no. 1, pp. 183–202, Jan. 2009.

[91] D. L. Donoho, A. Maleki, and A. Montanari, "Message-passing algorithms for compressed sensing," in *Proceedings of the National Academy of Sciences of the United States of America*, 2009, pp. 18914–18919.

[92] Y. C. Pati, R. Rezaiifar, and P. S. Krishnaprasad, "Orthogonal matching pursuit: Recursive function approximation with applications to wavelet decomposition," in *Proceedings of the 27th Asilomar Conference on Signals, Systems and Computers*, 1993, pp. 40–44.

[93] J. A. Tropp and A. C. Gilbert, "Signal recovery from random measurements via orthogonal matching pursuit," *IEEE Transaction on Information Theory*, vol. 53, no. 12, pp. 4655–4666, Dec. 2007.

[94] Y. Kabashima, "A CDMA multiuser detection algorithm on the basis of belief propagation," *Journal of Physics A: Mathematical and General*, vol. 36, no. 43, pp. 11 111–11 121, Oct. 2003.

[95] D. J. Thouless, P. W. Anderson, and R. G. Palmer, "Solution of 'solvable model of a spin glass'," *Philosophical Magazine*, vol. 35, no. 3, pp. 593–601, Mar. 1977.

[96] D. L. Donoho, A. Maleki, and A. Montanari, "Message passing algorithms for compressed sensing: I. motivation and construction," in *Proceedings of the IEEE Information Theory Workshop on Information Theory*, 2010, pp. 1–5.

[97] M. Bayati and A. Montanari, "The dynamics of message passing on dense graphs, with applications to compressed sensing," *IEEE Transactions on Information Theory*, vol. 57, no. 2, pp. 764–785, Feb. 2011.

[98] K. Zhang, W. Zuo, Y. Chen, D. Meng, and L. Zhang, "Beyond a Gaussian denoiser: Residual learning of deep CNN for image denoising," *IEEE Transactions on Image Processing*, vol. 26, no. 7, pp. 3142–3155, Jul. 2017.

[99] S. Ramani, T. Blu, and M. Unser, "Monte-Carlo SURE: A black-box optimization of regularization parameters for general denoising algorithms," *IEEE Transactions On Image Processing*, vol. 17, no. 9, pp. 1540–1554, Sep. 2008.

[100] K. He, X. Zhang, S. Ren, and J. Sun, "Deep residual learning for image recognition," in *Proceedings of the IEEE Conference on Computer Vision and Pattern Recognition*, 2016, pp. 770–778.

[101] X. Gao, L. Dai, S. Han, C.-L. I, and X. Wang, "Reliable beamspace channel estimation for millimeter-wave massive MIMO systems with lens antenna array," *IEEE Transactions on Wireless Communications*, vol. 16, no. 9, pp. 6010–6021, Sep. 2017.

[102] J. Yang, C.-K. Wen, S. Jin, and F. Gao, "Beamspace channel estimation in mmWave systems via cosparse image reconstruction technique," *IEEE Transactions on Communications*, vol. 66, no. 10, pp. 4767–4782, Oct. 2018.

[103] C. A. Metzler, A. Maleki, and R. G. Baraniuk, "From denoising to compressed sensing," *IEEE Transactions on Information Theory*, vol. 62, no. 9, pp. 5117–5144, Sep. 2016.

[104] Y. Wei, M.-M. Zhao, M. Zhao, M. Lei, and Q. Yu, "An AMP-based network with deep residual learning for mmwave beamspace channel estimation," *IEEE Wireless Communications Letters*, vol. 8, no. 4, pp. 1289–1292, Aug. 2019.

[105] X. Wei, C. Hu, and L. Dai, "Deep learning for beamspace channel estimation in millimeter-wave massive MIMO systems," *IEEE Transactions on Communications*, vol. 69, no. 1, pp. 182–193, Jan. 2021.

[106] H. He, R. Wang, W. Jin, S. Jin, C.-K. Wen, and G. Y. Li, "Beamspace channel estimation for wideband millimeter-wave MIMO: A model-driven unsupervised learning approach," *IEEE Transactions on Wireless Communications*, vol. 22, no. 3, pp. 1808–1822, Mar. 2023.

[107] D. J. Love, R. W. Heath, V. K. N. Lau, D. Gesbert, B. D. Rao, and M. Andrews, "An overview of limited feedback in wireless communication systems," *IEEE Journal on Selected Areas in Communications*, vol. 26, no. 8, pp. 1341–1365, Oct. 2008.

[108] P.-H. Kuo, H. T. Kung, and P.-A. Ting, "Compressive sensing based channel feedback protocols for spatially-correlated massive antenna arrays," in *Proceedings of the IEEE Conference on Wireless Communications and Networking*, 2012, pp. 492–497.

[109] X. Rao and V. K. N. Lau, "Distributed compressive CSIT estimation and feedback for FDD multi-user massive MIMO systems," *IEEE Transactions on*

Signal Processing, vol. 62, no. 12, pp. 3261–3271, Jun. 2014.

[110] L. Liu, C. Oestges, J. Poutanen, K. Haneda, P. Vainikainen, F. Quitin, F. Tufvesson, and P. D. Doncker, "The COST 2100 MIMO channel model," *IEEE Wireless Communications*, vol. 19, no. 6, pp. 92–99, Dec. 2012.

[111] C. Li, W. Yin, and Y. Zhang, "User's guide for TVAL3: TV minimization by augmented Lagrangian and alternating direction algorithms," *China Association of Automobile Manufacturers Reports*, vol. 20, no. 4, pp. 46–47, 2009.

[112] T. Wang, C.-K. Wen, S. Jin, and G. Y. Li, "Deep learning-based CSI feedback approach for time-varying massive MIMO channels," *IEEE Wireless Communications Letters*, vol. 8, no. 2, pp. 416–419, Apr. 2019.

[113] K. Xu and F. Ren, "CSVideoNet: A real-time end-to-end learning framework for high-frame-rate video compressive sensing," in *Proceedings of the IEEE Workshop on Applications of Computer Vision*, 2018, pp. 1680–1688.

[114] H. Ye, F. Gao, J. Qian, H. Wang, and G. Y. Li, "Deep learning-based denoise network for CSI feedback in FDD massive MIMO systems," *IEEE Communications Letters*, vol. 24, no. 8, pp. 1742–1746, Aug. 2020.

[115] C. Lu, W. Xu, S. Jin, and K. Wang, "Bit-level optimized neural network for multi-antenna channel quantization," *IEEE Wireless Communications Letters*, vol. 9, no. 1, pp. 87–90, Jan. 2020.

[116] J. Guo, J. Wang, C.-K. Wen, S. Jin, and G. Y. Li, "Compression and acceleration of neural networks for communications," *IEEE Wireless Communications*, vol. 27, no. 4, pp. 110–117, Jul. 2020.

[117] J. Guo, C.-K. Wen, and S. Jin, "Deep learning-based CSI feedback for beamforming in single- and multi-cell massive MIMO systems," *IEEE Journal on Selected Areas in Communications*, vol. 39, no. 7, pp. 1872–1884, Jul. 2021.

[118] 3rd Generation Partnership Project, "Technical Specification Group Radio Access Network; Physical layer procedures for data (Release 16)," *3GPP TR 38.214 V16.2.0*, Jul. 2020.

[119] J. Céspedes, P. M. Olmos, M. Sánchez-Fernández, and F. Perez-Cruz, "Expectation propagation detection for high-order high-dimensional MIMO systems," *IEEE Transactions on Communications*, vol. 62, no. 8, pp. 2840–2849, Aug. 2014.

[120] C. Jeon, R. Ghods, A. Maleki, and C. Studer, "Optimality of large MIMO detection via approximate message passing," in *Proceedings of the IEEE International Symposium on Information Theory*, 2015, pp. 1227–1231.

[121] J. Ma and L. Ping, "Orthogonal AMP," *IEEE Access*, vol. 5, pp. 2020–2033, Jan. 2017.

[122] M. Hansen, B. Hassibi, A. G. Dimakis, and W. Xu, "Near-optimal detection in MIMO systems using Gibbs sampling," in *Proceedings of the IEEE Global Telecommunications Conference*, 2009, pp. 1–6.

[123] T. Datta, N. A. Kumar, A. Chockalingam, and B. S. Rajan, "A novel Monte-Carlo-sampling-based receiver for large-scale uplink multiuser MIMO systems," *IEEE Transactions on Vehicular Technology*, vol. 62, no. 7, pp. 3019–3038, Sep. 2013.

[124] M. Bayati, M. Lelarge, and A. Montanari, "Universality in polytope phase transitions and message passing algorithms," *The Annals of Applied Probability*, vol. 25, no. 2, pp. 753–822, Apr. 2015.

[125] T. P. Minka, "Expectation propagation for approximate Bayesian inference," in *Proceedings of the Conference on Uncertainty in Artificial Intelligence*, 2001, pp. 362–369.

[126] M. J. Wainwright and M. I. Jordan, "Graphical models, exponential families, and variational inference," *Foundations and Trends®in Machine Learning*, vol. 1, no. 1–2, pp. 1–305, Nov. 2008.

[127] C. E. Rasmussen and C. K. I. Williams, *Gaussian Processes for Machine Learning*. MIT Press, 2006.

[128] B. Farhang-Boroujeny, H. Zhu, and Z. Shi, "Markov chain Monte Carlo algorithms for CDMA and MIMO communication systems," *IEEE Transactions on Signal Processing*, vol. 54, no. 5, pp. 1896–1909, May. 2006.

[129] N. M. Gowda, S. Krishnamurthy, and A. Belogolovy, "Metropolis–Hastings random walk along the gradient descent direction for MIMO detection," in *Proceedings of the IEEE International Conference on Communications*, 2021, pp. 1–7.

[130] M. Welling and Y. W. Teh, "Bayesian learning via stochastic gradient Langevin dynamics," in *Proceedings of the International Conference on Machine Learning*, 2011, pp. 681–688.

[131] X. Zhou, L. Liang, J. Zhang, C.-K. Wen, and S. Jin, "Near-optimal MIMO detection using gradient-based MCMC in discrete spaces," *arXiv preprint arXiv:2407.06042*, 2024.

[132] R. Zhang, X. Liu, and Q. Liu, "A Langevin-like sampler for discrete distributions," in *Proceedings of the International Conference on Machine Learning*, 2022, pp. 26 375–26 396.

[133] G. O. Roberts and O. Stramer, "Langevin diffusions and Metropolis–Hastings algorithms," *Methodology and Computing in Applied Probability*, vol. 4, pp. 337–357, Dec. 2002.

[134] D. A. Levin and Y. Peres, *Markov Chains and Mixing Times*. American Mathematical Society, 2017, vol. 107.

[135] J. Ma, L. Liu, X. Yuan, and L. Ping, "On orthogonal AMP in coded linear vector systems," *IEEE Transactions on Wireless Communications*, vol. 18, no. 12, pp. 5658–5672, Dec. 2019.

[136] M. Senst and G. Ascheid, "How the framework of expectation propagation yields an iterative IC-LMMSE MIMO receiver," in *Proceedings of the IEEE Global Communications Conference*, 2011, pp. 1–6.

[137] N. Samuel, T. Diskin, and A. Wiesel, "Learning to detect," *IEEE Transactions on Signal Processing*, vol. 67, no. 10, pp. 2554–2564, 2019.

[138] J. Jaldén and B. Ottersten, "The diversity order of the semidefinite relaxation detector," *IEEE Transactions on Information Theory*, vol. 54, no. 4, pp. 1406–1422, Apr. 2008.

[139] H. He, C.-K. Wen, S. Jin, and G. Y. Li, "Model-driven deep learning for MIMO detection," *IEEE Transactions on Signal Processing*, vol. 68, pp. 1702–1715, Mar. 2020.

[140] M. Khani, M. Alizadeh, J. Hoydis, and P. Fleming, "Adaptive neural signal detection for massive MIMO," *IEEE Transactions on Wireless Communications*, vol. 19, no. 8, pp. 5635–5648, Aug. 2020.

[141] M. Goutay, F. A. Aoudia, and J. Hoydis, "Deep hypernetwork-based MIMO detection," in *Proceedings of the IEEE International Workshop on Signal Processing Advances in Wireless Communications*, 2020, pp. 1–5.

[142] J. Zhang, C.-K. Wen, and S. Jin, "Adaptive MIMO detector based on hypernetwork: Design, simulation, and experimental test," *IEEE Journal on Selected Areas in Communications*, vol. 40, no. 1, pp. 65–81, Jan. 2022.

[143] C. E. Shannon, "A mathematical theory of communication," *The Bell System Technical Journal*, vol. 27, no. 3, pp. 379–423, Jul. 1948.

[144] P. Elias, "Coding for noisy channels," in *Proceedings of the IRE International Convention Record*, 1955, pp. 37–46.

[145] L. Bahl, J. Cocke, F. Jelinek, and J. Raviv, "Optimal decoding of linear codes for minimizing symbol error rate (corresp.)," *IEEE Transactions on Information Theory*, vol. 20, no. 2, pp. 284–287, Mar. 1974.

[146] E. Arikan, "Channel polarization: A method for constructing capacity-achieving codes for symmetric binary-input memoryless channels," *IEEE*

Transactions on Information Theory, vol. 55, no. 7, pp. 3051–3073, Jun. 2009.

[147] S. Cammerer, T. Gruber, J. Hoydis, and S. Ten Brink, "Scaling deep learning-based decoding of polar codes via partitioning," in *Proceedings of the IEEE Global Communications Conference*, 2017, pp. 1–6.

[148] H. Kim, Y. Jiang, R. Rana, S. Kannan, S. Oh, and P. Viswanath, "Communication algorithms via deep learning," in *Proceedings of the International Conference on Learning Representations*, 2018, pp. 1–17.

[149] Y. He, J. Zhang, S. Jin, C.-K. Wen, and G. Y. Li, "Model-driven DNN decoder for turbo codes: Design, simulation, and experimental results," *IEEE Transactions on Communications*, vol. 68, no. 10, pp. 6127–6140, Oct. 2020.

[150] E. Nachmani, E. Marciano, L. Lugosch, W. J. Gross, D. Burshtein, and Y. Be'ery, "Deep learning methods for improved decoding of linear codes," *IEEE Journal of Selected Topics in Signal Processing*, vol. 12, no. 1, pp. 119–131, Feb. 2018.

[151] V. G. Satorras and M. Welling, "Neural enhanced belief propagation on factor graphs," in *Proceedings of the International Conference on Artificial Intelligence and Statistics*, 2021, pp. 685–693.

[152] K. Ghavami and M. Naraghi-Pour, "MIMO detection with imperfect channel state information using expectation propagation," *IEEE Transactions on Vehicular Technology*, vol. 66, no. 9, pp. 8129–8138, Sep. 2017.

[153] H. Ye, L. Liang, and G. Y. Li, "Circular convolutional auto-encoder for channel coding," in *Proceedings of the IEEE International Conference on Signal Processing Advances in Wireless Communications*, 2019, pp. 1–5.

[154] A. Felix, S. Cammerer, S. Dörner, J. Hoydis, and S. ten Brink, "OFDM-autoencoder for end-to-end learning of communications systems," in *Proceedings of the IEEE International Conference on Signal Processing Advances in Wireless Communications*, 2018, pp. 1–5.

[155] T. J. O'Shea, L. Pemula, D. Batra, and T. C. Clancy, "Radio transformer networks: Attention models for learning to synchronize in wireless systems," in *Proceedings of the Asilomar Conference on Signals, Systems and Computers*, 2016, pp. 662–666.

[156] H. Ye, G. Y. Li, and B.-H. Juang, "Deep learning based end-to-end wireless communication systems without pilots," *IEEE Transactions on Cognitive Communications and Networking*, vol. 7, no. 3, pp. 702–714, Sep. 2021.

[157] T.-Y. Lin, A. RoyChowdhury, and S. Maji, "Bilinear CNN models for fine-grained visual recognition," in *Proceedings of the IEEE International Conference on Computer Vision*, 2015, pp. 1449–1457.

[158] H. Ye, L. Liang, G. Y. Li, and B.-H. Juang, "Deep learning-based end-to-end wireless communication systems with conditional GANs as unknown channels," *IEEE Transactions on Wireless Communications*, vol. 19, no. 5, pp. 3133–3143, Feb. 2020.

[159] M. Mirza and S. Osindero, "Conditional generative adversarial nets," *arXiv preprint arXiv:1411.1784*, 2014.

[160] C. Berrou, A. Glavieux, and P. Thitimajshima, "Near Shannon limit error-correcting coding and decoding: Turbo-codes," in *Proceedings of the IEEE International Conference on Communications*, 1993, pp. 1064–1070.

[161] D. J. MacKay and R. M. Neal, "Near Shannon limit performance of low density parity check codes," *Electronics letters*, vol. 33, no. 6, pp. 457–458, Jul. 1996.

[162] T. Richardson and R. Urbanke, *Modern Coding Theory*. Cambridge University Press, 2008.

[163] Y. Jiang, H. Kim, H. Asnani, S. Kannan, S. Oh, and P. Viswanath, "Learn codes: Inventing low-latency codes via recurrent neural networks," *IEEE Journal on Selected Areas in Information Theory*, vol. 1, no. 1, pp. 207–216, May 2020.

[164] ——, "Turbo autoencoder: Deep learning based channel codes for point-to-point communication channels," in *Proceedings*

of the *Advances in Neural Information Processing Systems*, 2019, pp. 1–11.

[165] Z. Qin, L. Liang, Z. Wang, S. Jin, X. Tao, W. Tong, and G. Y. Li, "AI empowered wireless communications: From bits to semantics," *Proceedings of the IEEE*, pp. 1–32, 2024.

[166] R. Carnap, Y. Bar-Hillel *et al.*, *An Outline of a Theory of Semantic Information*. Research Laboratory of Electronics, Massachusetts Institute of Technology, 1952.

[167] I. D. Melamed, "Measuring semantic entropy," in *Tagging Text with Lexical Semantics: Why, What, and How?*, 1997.

[168] X. Liu, W. Jia, W. Liu, and W. Pedrycz, "AFSSE: An interpretable classifier with axiomatic fuzzy set and semantic entropy," *IEEE Transactions on Fuzzy Systems*, vol. 28, no. 11, pp. 2825–2840, Oct. 2019.

[169] H. Xie, Z. Qin, G. Y. Li, and B.-H. Juang, "Deep learning enabled semantic communication systems," *IEEE Transactions on Signal Processing*, vol. 69, pp. 2663–2675, Apr. 2021.

[170] K. Papineni, S. Roukos, T. Ward, and W.-J. Zhu, "BLEU: A method for automatic evaluation of machine translation," in *Proceedings of the 40th annual meeting of the Association for Computational Linguistics*, 2002, pp. 311–318.

[171] J. Devlin, "BERT: Pre-training of deep bidirectional transformers for language understanding," *arXiv preprint arXiv:1810.04805*, 2018.

[172] Z. Weng and Z. Qin, "Semantic communication systems for speech transmission," *IEEE Journal on Selected Areas in Communications*, vol. 39, no. 8, pp. 2434–2444, Jun. 2021.

[173] E. Bourtsoulatze, D. B. Kurka, and D. Gündüz, "Deep joint source-channel coding for wireless image transmission," *IEEE Transactions on Cognitive Communications and Networking*, vol. 5, no. 3, pp. 567–579, Sep. 2019.

[174] J. Johnson, A. Alahi, and L. Fei-Fei, "Perceptual losses for real-time style transfer and super-resolution," in

Proceedings of the European Conference on Computer Vision, 2016, pp. 694–711.

[175] Y. Sheng, H. Ye, L. Liang, S. Jin, and G. Y. Li, "Semantic communication for cooperative perception based on importance map," *Journal of the Franklin Institute*, vol. 361, no. 6, p. 106739, Feb. 2024.

[176] A. H. Lang, S. Vora, H. Caesar, L. Zhou, J. Yang, and O. Beijbom, "Pointpillars: Fast encoders for object detection from point clouds," in *Proceedings of IEEE Conference on Computer Vision and Pattern Recognition*, 2019, pp. 12 697–12 705.

[177] D. Erhan, C. Szegedy, A. Toshev, and D. Anguelov, "Scalable object detection using deep neural networks," in *Proceedings of the IEEE conference on computer vision and pattern recognition*, 2014, pp. 2147–2154.

[178] Y. Hu, S. Fang, Z. Lei, Y. Zhong, and S. Chen, "Where2comm: Communication-efficient collaborative perception via spatial confidence maps," in *Proceedings of the Advances in neural information processing systems*, 2022, pp. 4874–4886.

[179] W. Yu and R. Lui, "Dual methods for nonconvex spectrum optimization of multicarrier systems," *IEEE Transactions on Communication*, vol. 54, no. 7, pp. 1310–1322, Jul. 2006.

[180] Q. Shi, M. Razaviyayn, Z.-Q. Luo, and C. He, "An iteratively weighted MMSE approach to distributed sum-utility maximization for a MIMO interfering broadcast channel," *IEEE Transactions on Signal Processing*, vol. 59, no. 9, pp. 4331–4340, Sep. 2011.

[181] D. Bertsekas and J. Tsitsiklis, *Parallel and Distributed Computation: Numerical Methods*. Athena Scientific, 2015.

[182] K. Shen and W. Yu, "FPLinQ: A cooperative spectrum sharing strategy for device-to-device communications," in *Proceedings of the IEEE International Symposium on Information Theory*, 2017, pp. 2323–2327.

[183] F. Liang, C. Shen, W. Yu, and F. Wu, "Towards optimal power control via ensembling deep neural networks," *IEEE Transactions on Communications*, vol. 68, no. 3, pp. 1760–1776, Mar. 2020.

[184] J. Gilmer, S. S. Schoenholz, P. F. Riley, O. Vinyals, and G. E. Dahl, "Neural message passing for quantum chemistry," in *Proceedings of the International Conference on Machine Learning*, 2017, pp. 1–14.

[185] W. L. Hamilton, *Graph Representation Learning*. Morgan & Claypool Publishers, 2020.

[186] T. N. Kipf and M. Welling, "Semi-supervised classification with graph convolutional networks," in *Proceedings of the International Conference on Learning Representations*, 2017, pp. 1–14.

[187] P. Veličković, G. Cucurull, A. Casanova, A. Romero, P. Liò, and Y. Bengio, "Graph attention networks," in *Proceedings of the International Conference on Learning Representations*, 2018, pp. 1–12.

[188] K. Xu, W. Hu, J. Leskovec, and S. Jegelka, "How powerful are graph neural networks?" in *Proceedings of the International Conference on Learning Representations*, 2019, pp. 1–17.

[189] H. Ye, G. Y. Li, and B.-H. F. Juang, "Deep reinforcement learning based resource allocation for V2V communications," *IEEE Transactions on Vehicular Technology*, vol. 68, no. 4, pp. 3163–3173, Apr. 2019.

[190] L. Liang, H. Ye, and G. Y. Li, "Spectrum sharing in vehicular networks based on multi-agent reinforcement learning," *IEEE Journal on Selected Area in Communications*, vol. 37, no. 10, pp. 2282–2292, Oct. 2019.

[191] 3rd Generation Partnership Project, "Technical Specification Group Radio Access Network; Study enhancement 3GPP Support for 5G V2X Services; (Release 15)," *3GPP TR 22.886 V15.1.0*, Mar. 2017.

[192] ——, "Technical Specification Group Radio Access Network; Study LTE-Based V2X Services; (Release 14)," *3GPP TR 36.885 V14.0.0*, Jun. 2016.

[193] ——, "Technical Specification Group Radio Access Network; NR; Study on NR Vehicle-to-Everything(V2X); (Release 16)," *3GPP TR 38.885 V16.0.0*, Mar. 2019.

[194] C. Finn, P. Abbeel, and S. Levine, "Model-agnostic meta-learning for fast adaptation of deep networks," in *Proceedings of International Conference on Machine Learning*, 2017, pp. 1126–1135.

[195] A. Nichol, J. Achiam, and J. Schulman, "On first-order meta-learning algorithms," *arXiv preprint arXiv:1803.02999*, 2018.

[196] L. Zintgraf, K. Shiarlis, V. Kurin, K. Hofmann, and S. Whiteson, "Fast context adaptation via meta-learning," in *Proceedings of the International Conference on Machine Learning*, 2019, pp. 7693–7702.

[197] F. Alet, T. Lozano-Perez, and L. P. Kaelbling, "Modular meta-learning," in *Proceedings of the International Conference on Machine Learning*, 2018, pp. 856–868.

[198] T. Li, A. K. Sahu, M. Zaheer, M. Sanjabi, A. Talwalkar, and V. Smith, "Federated optimization in heterogeneous networks," in *Proceedings of Machine learning and systems*, 2020, pp. 429–450.

[199] Z. Zhao, Y. Mao, Y. Liu, L. Song, Y. Ouyang, X. Chen, and W. Ding, "Towards efficient communications in federated learning: A contemporary survey," *Journal of the Franklin Institute*, vol. 360, no. 12, pp. 8669–8703, Aug. 2023.

[200] S. Zheng, C. Shen, and X. Chen, "Design and analysis of uplink and downlink communications for federated learning," *IEEE Journal on Selected Areas in Communications*, vol. 39, no. 7, pp. 2150–2167, Dec. 2020.

[201] P. Han, S. Wang, and K. K. Leung, "Adaptive gradient sparsification for efficient federated learning: An online learning approach," in *Proceedings of the IEEE International Conference on Distributed Computing Systems*, 2020, pp. 300–310.

[202] Z. Wang, Z. Zhang, Y. Tian, Q. Yang, H. Shan, W. Wang, and T. Q. S. Quek, "Asynchronous federated learning over wireless communication networks," *IEEE Transactions on Wireless Communications*, vol. 21, no. 9, pp. 6961–6978, Mar. 2022.

[203] C. Briggs, Z. Fan, and P. Andras, "Federated learning with hierarchical clustering of local updates to improve training on non-iid data," in *Proceedings of the International Joint Conference on Neural Networks*, 2020, pp. 1–9.

[204] B. Soltani, V. Haghighi, A. Mahmood, Q. Z. Sheng, and L. Yao, "A survey on participant selection for federated learning in mobile networks," in *Proceedings of the ACM Workshop on Mobility in the Evolving Internet Architecture*, 2022, pp. 19–24.

[205] T. Nishio and R. Yonetani, "Client selection for federated learning with heterogeneous resources in mobile edge," in *Proceedings of the IEEE International Conference on Communications*, 2019, pp. 1–7.

[206] Y. Deng, F. Lyu, J. Ren, H. Wu, Y. Zhou, Y. Zhang, and X. Shen, "AUCTION: Automated and quality-aware client selection framework for efficient federated learning," *IEEE Transactions on Parallel and Distributed Systems*, vol. 33, no. 8, pp. 1996–2009, Dec. 2022.

[207] M. Chen, Z. Yang, W. Saad, C. Yin, H. V. Poor, and S. Cui, "A joint learning and communications framework for federated learning over wireless networks," *IEEE Transactions on Wireless Communications*, vol. 20, no. 1, pp. 269–283, Oct. 2021.

[208] W. Shi, S. Zhou, Z. Niu, M. Jiang, and L. Geng, "Joint device scheduling and resource allocation for latency constrained wireless federated learning," *IEEE Transactions on Wireless Communications*, vol. 20, no. 1, pp. 453–467, Jan. 2021.

[209] T. Li, M. Sanjabi, A. Beirami, and V. Smith, "Fair resource allocation in federated learning," *arXiv preprint arXiv:1905.10497*, 2019.

[210] J. Xu and H. Wang, "Client selection and bandwidth allocation in wireless federated learning networks: A long-term perspective," *IEEE Transactions on Wireless Communications*, vol. 20, no. 2, pp. 1188–1200, Oct. 2021.

[211] H. Ye, L. Liang, and G. Y. Li, "Decentralized federated learning with unreliable communications," *IEEE Journal of Selected Topics in Signal Processing*, vol. 16, no. 3, pp. 487–500, Feb. 2022.

[212] X. Lian, C. Zhang, H. Zhang, C.-J. Hsieh, W. Zhang, and J. Liu, "Can decentralized algorithms outperform centralized algorithms? A case study for decentralized parallel stochastic gradient descent," in *Proceedings of the Advances in Neural Information Processing Systems*, 2017, pp. 5330–5340.

[213] H. Tang, S. Gan, C. Zhang, T. Zhang, and J. Liu, "Communication compression for decentralized training," in *Proceedings of the Advances in Neural Information Processing Systems*, 2018, pp. 7652–7662.

[214] A. Koloskova, S. Stich, and M. Jaggi, "Decentralized stochastic optimization and gossip algorithms with compressed communication," in *Proceedings of the International Conference on Machine Learning*, 2019, pp. 3478–3487.

[215] A. Krizhevsky, G. Hinton *et al.*, *Learning Multiple Layers of Features from Tiny Images*. Master's thesis. University of Toronto, 2009.

[216] M. Goldenbaum, H. Boche, and S. Stańczak, "Harnessing interference for analog function computation in wireless sensor networks," *IEEE Transactions on Signal Processing*, vol. 61, no. 20, pp. 4893–4906, Oct. 2013.

[217] G. Zhu, J. Xu, K. Huang, and S. Cui, "Over-the-air computing for wireless data aggregation in massive IoT," *IEEE Wireless Communications*, vol. 28, no. 4, pp. 57–65, Aug. 2021.

[218] R. C. Buck, *Approximate Complexity and Functional Representation*. University of Wisconsin-Madison. Mathematics Research Center, 1976.

[219] G. Zhu, Y. Wang, and K. Huang, "Broadband analog aggregation for

low-latency federated edge learning," *IEEE Transactions on Wireless Communications*, vol. 19, no. 1, pp. 491–506, Oct. 2019.

[220] H. Guo, A. Liu, and V. K. Lau, "Analog gradient aggregation for federated learning over wireless networks: Customized design and convergence analysis," *IEEE Internet of Things Journal*, vol. 8, no. 1, pp. 197–210, Jun. 2020.

[221] J. Wang and G. Joshi, "Cooperative SGD: A unified framework for the design and analysis of communication-efficient SGD algorithms," *arXiv preprint arXiv:1808.07576*, 2018.

[222] K. Yang, T. Jiang, Y. Shi, and Z. Ding, "Federated learning via over-the-air computation," *IEEE Transactions on Wireless Communications*, vol. 19, no. 3, pp. 2022–2035, Jan. 2020.

[223] S. Boyd, N. Parikh, and E. Chu, *Distributed Optimization and Statistical Learning via the Alternating Direction Method of Multipliers. Foundations and Trend® in Machine Learning*, vol. 3, no. 1, pp. 1–122, Jul. 2011.

[224] S. Zhou and G. Y. Li, "Communication-efficient ADMM-based federated learning," *arXiv preprint arXiv:2110.15318*, Oct. 2021.

[225] ——, "Federated learning via inexact ADMM," *IEEE Transactions on Pattern Analysis and Machine Intelligence*, vol. 45, no. 8, pp. 9699–9708, Feb. 2023.

[226] ——, "FedGiA: An efficient hybrid algorithm for federated learning," *IEEE Transactions on Signal Processing*, vol. 71, pp. 1493–1508, Apr. 2023.

Index

For EU product safety concerns, contact us at Calle de José Abascal, 56–1°,
28003 Madrid, Spain or eugpsr@cambridge.org.

www.ingramcontent.com/pod-product-compliance
Ingram Content Group UK Ltd.
Pitfield, Milton Keynes, MK11 3LW, UK
UKHW061850181225
466196UK00009B/186